The Economics of Organic Farming
An International Perspective

The Economics of Organic Farming
An International Perspective

Edited by

N.H. Lampkin and S. Padel
*Department of Agricultural Sciences,
University of Wales, Aberystwyth, UK*

CAB INTERNATIONAL

CABI is a trading name of CAB International

CABI Head Office
Nosworthy Way
Wallingford
Oxfordshire OX10 8DE
UK

Tel: +44 (0)1491 832111
Fax: +44 (0)1491 833508
Email: cabi@cabi.org
Web site: www.cabi.org

CABI North American Office
875 Massachusetts Avenue
7th Floor
Cambridge, MA 02139
USA

Tel: +1 617 395 4056
Fax: +1 617 354 6875
Email: cabi-nao@cabi.org

A catalogue record for this book is available from the British Library, London, UK.

ISBN-13: 978-0-85198-911-2
ISBN-10: 0-85198-911-X

First published 1994
Transferred to print on demand 2007

Printed and bound in the UK by CPI Antony Rowe, Eastbourne.

Contents

Contributors

MOLLY D. ANDERSON: *Research Assistant Professor, School of Nutrition, Tufts University, Medford, MA 02155, USA*

DAVID I. BATEMAN: *Professor of Agricultural Economics, Department of Agricultural Sciences, University of Wales, Aberystwyth, Dyfed SY23 3DD, UK*

HARTMUT BOSSEL: *Professor of Environmental Systems Analysis, Department of Mathematics, University Gesamthochschule Kassel, Mönchebergstrasse 11, D-34125 Kassel, Germany*

JÜRGEN BRAUN: *Research Assistant, Department of Agricultural Economics (Institut 410A), University of Hohenheim, D-70593 Stuttgart, Germany*

STEPHAN DABBERT: *Professor of Agricultural Economics, Department of Agricultural Economics (Institut 410A), University of Hohenheim, D-70593 Stuttgart, Germany*

ALEX DUBGAARD: *Associate Professor, Department of Economics and Natural Resources, The Royal Veterinary and Agricultural University, Roligshedsvej 26, DK-1958 Frederiksberg C (Copenhagen), Denmark*

BERNHARD FREYER: *Head of Department for Landscape Economics and Ecology, Research Institute for Organic Agriculture (FIBL), Bernhardsberg, CH-4104 Oberwil/BL, Switzerland*

JOHN HENNING: *Associate Professor of Agricultural Economics, Department of Agricultural Economics, McGill University, Macdonald Campus, Ste Anne de Bellevue, Quebec H9X 3V9, Canada*

HANNE HOLST: *Research Assistant, Department of Economics and Natural Resources, The Royal Veterinary and Agricultural University, Roligshedsvej 26, DK-1958 Frederiksberg C (Copenhagen), Denmark*

NICOLAS H. LAMPKIN: *Lecturer in Agricultural Economics, Department of*

Agricultural Sciences, University of Wales, Aberystwyth, Dyfed SY23 3DD, UK

PETER MIDMORE: *Lecturer in Agricultural Economics, Department of Agricultural Sciences, University of Wales, Aberystwyth, Dyfed SY23 3DD, UK*

IRENE MÜHLEBACH: *Farm Adviser, Landwirtschaftliche Schule Muri, CH-5630 Muri/Aargau, Switzerland*

JÖRG MÜHLEBACH: *Farm Adviser, Landwirtschaftliche Schule Muri, CH-5630 Muri/Aargau, Switzerland*

SUSANNE PADEL: *Research Associate, Department of Agricultural Sciences, University of Wales, Aberystwyth, Dyfed SY23 3DD, UK*

STEVEN E. PETERS: *Research Agronomist, Rodale Institute Research Center, RD 1, Box 323, Kutztown, PA 19530, USA*

RUDOLF RANTZAU: *Section Leader for Organic Farming, Extensification and Soil Conservation, Lower Saxony Ministry of Food, Agriculture and Forestry, Calenbergerstrasse 2, D-30169 Hanover, Germany*

OTTO SCHMID: *Senior Organic Farm Adviser and Lecturer, Research Institute for Organic Agriculture (FIBL), Bernhardsberg, CH-4104 Oberwil/BL, Switzerland*

LUDGER SCHULZE PALS: *Civil Servant, Ministry for Environment, Regional Policy and Agriculture in North-Rhine Westphalia, Referat II95, Schwanstrasse 3, D-40476 Düsseldorf, Germany*

WILLIAM B. TATE: *Freelance Analyst, Juniper Cottage, Weekes Lane, West Brabourne, Ashford, Kent TN25 5LZ, UK*

HARTMUT VOGTMANN: *President of the Hessian Office for Regional Development and Agriculture, Kölnische Strasse 48–50, D-34117 Kassel, Germany*

ELS WYNEN: *Researcher/Consultant, Eco Landuse Systems, 3 Ramage Place, Flynn, ACT 2615, Australia*

ULI ZERGER: *Research Assistant, Faculty of Agriculture, International Rural Development and Environmental Protection, University Gesamthochschule Kassel, Steinstrasse 19, D-37213 Witzenhausen, Germany*

Preface

The economics of organic farming is a subject on which it is often difficult to obtain reliable and useful information. In part, this has been due to the relatively low priority given to the economic aspects of organic farming by researchers and the organic movement, particularly when compared with the traditional emphasis on improving soil fertility, animal and human health, and the environment. Where the research has been done, the problem is knowing where to look for it, as much of the information is contained in internal reports or in specialist journals and organic farming magazines which are not readily available. Yet when all the reports and studies which have been produced in various countries are taken together, a considerable amount of information is, in fact, available.

This book has emerged as a consequence of the difficulties which we experienced in finding information when we first started researching the economics of organic farming. Our intention has been to produce a book where as many as possible of the different studies which exist could be presented in a single volume, making it easy for the reader to compare methods, results and conclusions. We asked the authors of recent studies to contribute a summary of their work or a literature review as a chapter. As a result, studies from Germany, Britain, Denmark, Switzerland, Australia, Canada and the United States have been brought together in individual chapters. These are grouped into sections on established organic farming systems, the conversion process, the impacts of widespread adoption, and the policy context of organic farming. Where possible, references to studies from other countries have been included in the overview chapters. We believe that this opportunity to compare results from different countries opens a new perspective on the subject, allowing the typical characteristics of organic farming to be seen more clearly.

Susanne Padel
Nicolas H. Lampkin

Currency Conversion Table

Country	Currency	A$	AS	C$	DK	FM	DM	Fl	NK	SK	SF	£	US$	ECU[a]	ECU[b]
Australia	A$	1.00	8.49	0.91	4.67	3.95	1.21	1.35	5.20	5.65	1.02	0.46	0.69	0.62	n/a
Austria	10 AS	1.18	10.00	1.07	5.50	4.65	1.42	1.59	6.13	6.65	1.20	0.54	0.81	0.73	n/a
Belgium	100 BF	3.97	33.71	3.62	18.56	15.70	4.80	5.37	20.68	22.46	4.05	1.84	2.74	2.47	2.03
Canada	C$	1.10	9.32	1.00	5.13	4.34	1.33	1.48	5.71	6.21	1.12	0.51	0.76	0.68	n/a
Denmark	10 DK	2.14	18.17	1.95	10.00	8.46	2.58	2.89	11.14	12.10	2.18	0.99	1.48	1.33	1.07
Finland	10 FM	2.53	21.49	2.31	11.83	10.00	3.05	3.42	13.17	14.30	2.58	1.17	1.75	1.57	n/a
France	10 FF	2.44	20.71	2.22	11.40	9.65	2.95	3.30	12.70	13.79	2.49	1.13	1.68	1.52	1.25
Germany	DM	0.83	7.03	0.76	3.87	3.27	1.00	1.12	4.31	4.68	0.85	0.38	0.57	0.52	0.42
Greece	100 Dr	0.58	4.90	0.53	2.69	2.28	0.70	0.78	3.00	3.26	0.59	0.27	0.40	0.36	0.30
Ireland	I£	2.07	17.59	1.89	9.68	8.19	2.50	2.80	10.79	11.71	2.12	0.96	1.43	1.29	1.02
Italy	1000 L	0.85	7.17	0.77	3.97	3.37	1.03	1.15	4.42	4.80	0.87	0.39	0.59	0.53	0.44
Netherlands	Fl	0.74	6.28	0.67	3.46	2.92	0.89	1.00	3.85	4.18	0.76	0.34	0.51	0.46	0.38
New Zealand	NZ$	0.81	6.87	0.74	3.78	3.20	0.98	1.09	4.21	4.57	0.83	0.37	0.56	0.50	n/a
Norway	10 NK	1.92	16.30	1.75	8.98	7.60	2.32	2.60	10.00	10.86	1.96	0.89	1.33	1.20	n/a
Portugal	100 Es	0.82	6.98	0.75	3.84	3.25	0.99	1.11	4.28	4.65	0.84	0.38	0.57	0.51	0.42
Spain	100 Pt	1.00	8.52	0.92	4.70	3.97	1.21	1.36	5.23	5.68	1.03	0.46	0.69	0.63	0.52
Sweden	10 SK	1.77	15.02	1.61	8.27	6.99	2.14	2.39	9.21	10.00	1.81	0.82	1.22	1.10	n/a
Switzerland	SF	0.98	8.32	0.89	4.58	3.87	1.18	1.32	5.10	5.54	1.00	0.45	0.68	0.61	n/a
UK	£	2.16	18.36	1.97	10.11	8.55	2.61	2.92	11.26	12.23	2.21	1.00	1.49	1.35	1.09
US	US$	1.45	12.31	1.32	6.78	5.73	1.75	1.96	7.55	8.20	1.48	0.67	1.00	0.90	n/a
European Union	ECU[a]	1.61	13.64	1.46	7.51	6.36	1.94	2.17	8.37	9.09	1.64	0.74	1.11	1.00	0.82[c]
	ECU[b]	n/a	n/a	n/a	9.35	n/a	2.35	2.65	n/a	n/a	n/a	0.92	n/a	1.22[c]	1.00

[a]Market rate.
[b]Administered 'green' rate for agricultural policy transactions.
[c]Unweighted average for comparison purposes only.
Sources: *Financial Times*, 17 January 1994; *Agra-Europe*, 21 January 1994.

Part 1

Organic Farming Concepts, History and Research Issues

1 Organic Farming: Sustainable Agriculture in Practice

Nicolas H. Lampkin

Introduction

Organic farming systems have attracted increasing attention over the last two decades because they are perceived to offer some solutions to the problems currently besetting the agricultural sectors of industrialized countries. Organic farming has the potential to provide benefits in terms of environmental protection, conservation of non-renewable resources, improved food quality, reduction in output of surplus products and the reorientation of agriculture towards areas of market demand (Lampkin, 1990). Some governments in Europe have recognized and responded to these potential benefits by encouraging farmers to adopt organic farming practices, either directly through financial incentives or indirectly through support for research, extension and marketing initiatives. As a consequence, the organic sector throughout Europe is expanding rapidly (Fig. 1.1), with similar trends, if less marked, elsewhere in the world.

At the same time, the organic sector is still very small: less than 0.5% of the agricultural industry in most countries, only exceeding 1% in the countries which have seen strongest growth in the last two years (e.g. Germany 2–3%; Austria 2–3%), although much higher, localized concentrations exist in these countries (see, for example, Chapter 25).

There is considerable latent interest among farmers in conversion to organic farming. In the UK, 38% of more than 26,000 farmers who responded to a postal questionnaire stated that they would consider or are considering organic production on all or part of their land in future

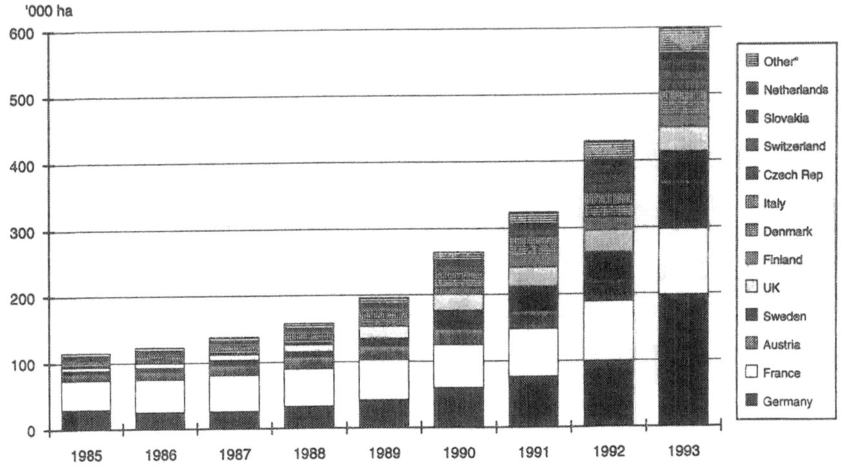

Fig. 1.1. Organically farmed land area in Europe, 1985–1993. ('Other' = countries with <10,000 ha in 1993.)

(NatWest, 1992). However, farmers are reluctant to convert because of the perceived high costs of doing so, and the risks involved, even though premium markets for organically produced food are now making it possible for organic producers to earn similar incomes to their conventional counterparts. Concern that the premium markets may not be sustainable is a major factor in this.

Despite the attention which has been paid to organic farming over the last few years, very little accessible information actually exists on what the costs and risk factors involved in farming organically and in converting to an organic farming system really are. The purpose of this book is to make the assessments which have been carried out in various countries around the world more readily available to agricultural decision-makers and researchers.

What is Organic Farming?

Organic farming can be defined as an approach to agriculture where the aim is:

> to create integrated, humane, environmentally and economically sustainable agricultural production systems, which maximize reliance on farm-derived renewable resources and the

> management of ecological and biological processes and interactions, so as to provide acceptable levels of crop, livestock and human nutrition, protection from pests and diseases, and an appropriate return to the human and other resources employed.

As such, the objective of sustainability lies at the heart of organic farming and is one of the major factors determining the acceptability or otherwise of specific production practices. The term 'sustainable' is used in a wider sense, to encompass not just conservation of non-renewable resources (soil, energy, minerals) but also issues of environmental and social sustainability. The term 'organic' is best thought of as referring not to the type of inputs used, but to the concept of the farm as an organism, in which all the component parts – the soil minerals, organic matter, microorganisms, insects, plants, animals and humans – interact to create a coherent whole. In many European countries, organic agriculture is known as ecological agriculture, reflecting this reliance on ecosystem management rather than external inputs, chemical or otherwise.

Detailed descriptions of the principles and practices of organic farming can be found in the many books on the subject published around the world in recent years (e.g. Lampkin, 1990; Neuerburg and Padel, 1992). The major factor which distinguishes organic farming from other approaches to sustainable agriculture is the existence of both legislated and voluntary standards and certification procedures to define a clear dividing line between organic and other farming systems, primarily for marketing purposes (see Chapter 2).

The key characteristics of organic farming include:

- protecting the long-term fertility of soils by maintaining organic matter levels, fostering soil biological activity and careful mechanical intervention;
- providing crop nutrients indirectly by using relatively insoluble nutrient sources which are made available to the plant by the action of soil microorganisms;
- nitrogen self-sufficiency through the use of legumes and biological nitrogen fixation, as well as effective recycling of organic materials including crop residues and livestock wastes;
- weed, disease and pest control relying primarily on crop rotations, natural predators, diversity, organic manuring, resistant varieties and limited (preferably minimal) thermal, biological and chemical intervention;
- the extensive management of livestock, paying full regard to their evolutionary adaptations, behavioural needs and animal welfare

issues with respect to nutrition, housing, health, breeding and rearing;
- careful attention to the impact of the farming system on the wider environment and the conservation of wildlife and natural habitats.

Why Organic Farming?

Modern organic farming represents a merging of a number of different streams of thought (see Chapter 2, this volume; also Boeringa, 1980; Merrill, 1983). In 1924, the Austrian philosopher Dr Rudolf Steiner presented an alternative vision of agriculture arising from his spiritual science of anthroposophy (Steiner, 1924). This formed the basis of the development of bio-dynamic agriculture which is now practised in many countries around the world. At about the same time, Dr Hans Müller founded a movement for agricultural reform in Switzerland centred around Christian concepts of land stewardship and preservation of family farms. Later, Dr Hans-Peter Rusch contributed important ideas relating to soil fertility and soil microbiology which led to the further development of organic-biological agriculture in central Europe (Rusch, 1968).

In the United Kingdom and other English-speaking parts of the world, scientists such as Sir George Stapledon (Conford, 1988) and Sir Albert Howard (Howard, 1943) were influenced by the agricultural ideas of Rudolf Steiner, although they did not adopt them directly. However, Stapledon's work with alternate husbandry systems and Howard's work on the role of organic matter in soils and composting were to provide a powerful stimulus to the founding of the organic movement by Lady Eve Balfour (Balfour, 1943). Howard and Balfour's ideas emphasizing the role of a healthy, fertile soil in the production of healthy crops and livestock and the link with human health were pursued first by J. I. Rodale and then by his son Robert Rodale in the United States.

Since the 1960s, wider concerns have influenced the development of organic agriculture:

- social concerns over the loss of jobs and rural population decline;
- environmental concerns over the loss of wildlife species and habitats, pollution, and the use of non-renewable resources;
- health concerns over pesticide residues and other food-quality issues.

It is inevitable, given the historical development of organic farming that it has taken some time for the ideas to fuse into a coherent concept which is now practised world-wide. This is reflected in the terminology where words such as 'biological', 'ecological', 'bio-dynamic', 'organic-biological', 'natural' and 'sustainable' are broadly synonymous with organic (Lockeretz, 1990).

However, the organic (biological/ecological) approach is no longer the only alternative to conventional production systems currently being advocated. Low-input sustainable agriculture (LISA) has received considerable research funding and support in the United States (see Chapter 10). Reduced use, but not elimination, of certain chemical inputs is a key factor distinguishing LISA from organic farming. Lower-input and organic systems are both covered under the headings 'sustainable' and 'alternative' (NRC, 1989; Edwards *et al.*, 1990; Francis *et al.*, 1990). Increasingly, researchers in Europe are working on the development of 'integrated' farming systems (derived from 'integrated pest management' where the use of chemical and biological controls is integrated in an 'optimal' way) (Vereijken, 1990).

Integrated agriculture allows for conventional systems, particularly individual practices, to be modified without the complete change in perspective which organic systems are believed to entail, and is therefore seen as a compromise between intensive conventional and organic approaches. However, the international working group Integrated Arable Farming Systems recognizes the potential of organic farming in its objectives for integrated farming (Vereijken and Royle, 1989):

- Shifting the emphasis from greater production to cost reduction and improvement of quality of both products and production ways, through substitution of expensive and potentially harmful inputs, especially fertilizers and pesticides, by both agricultural and ecological knowledge, labour and non-chemical husbandry techniques.
- Encouragement and conservation of flora and fauna in and around the fields, to stabilize the agro-ecosystem as a major preventive measure against the outbreak of pests, weeds and diseases.
- Some forms of biological or organic farming may be considered as the most radical approach of these integrated objectives. In the long term it may also appear the most desirable approach, provided the necessary technical and economic improvements can be made.

It remains to be seen to what extent these different approaches to sustainable agriculture can be reconciled, and in particular the extent to which they can be seen to meet society's environmental and other objectives as well as the private objectives of the individual farmer (see Chapter 4).

The Scope of this Book

This book focuses primarily on organic farming as defined above, because its specific characteristics, including the developed premium market and related legislative and policy context, mean that there is now a significant body of economic research on this topic. The more loosely defined concepts of LISA and integrated agriculture have yet to yield clearly defined cohorts of producers which can be used for comparative purposes, particularly with respect to financial and economic analyses.

The scope of this book is further restricted to consideration of organic farming in the context of developed economies. Although there are strong arguments for the application in developing countries of the ecological approaches on which organic farming is based, the economic context is very different and no attempt has been made to develop these issues here.

The book is divided into several sections, bringing together contributions and experiences from several countries in Europe, North America and Australasia:

- an introductory section looking at the development of organic agriculture world-wide, the methodological problems associated with comparisons between farming systems, and the role of economics in assessing the wider costs and benefits of organic farming;
- a section devoted to the economic performance of established organic farming in several countries, including the impacts on yields, prices, costs and incomes;
- a section examining the specific costs and other factors associated with the transition to organic farming, including investments in restructuring the farm business, extension needs and the impact of social and institutional barriers to conversion;
- a section providing some preliminary assessments of the possible impacts of widespread conversion to organic production; and
- a final section which examines the policy context of organic farming, in particular the introduction of financial aid schemes to assist farmers converting to organic farming in many European countries, and the potential role of organic farming in meeting the new environmental and output/public expenditure reduction objectives of policy-makers in developed countries.

References

Balfour, E.B. (1943) *The Living Soil.* Faber and Faber, London.
Boeringa, R. (ed.) (1980) Alternative Methods of Agriculture. *Agriculture and*

Environment Special Issue, 5. Elsevier, Amsterdam.

Conford, P. (ed.) (1988) *The Organic Tradition: An Anthology of Writings on Organic Farming 1900–1950.* Green Classics, Green Books, Bideford, Devon.

Edwards, C.A., Lal, R., Madden, P., Miller, R.H. and House, G. (eds) (1990) *Sustainable Agricultural Systems.* Soil and Water Conservation Society, Ankeny, Iowa.

Francis, C.A., Flora, C.B. and King, L.D. (eds) (1990) *Sustainable Agriculture in Temperate Zones.* John Wiley, New York.

Howard, A. (1943) *An Agricultural Testament.* Oxford University Press, London.

Lampkin, N.H. (1990) *Organic Farming.* Farming Press, Ipswich.

Lockeretz, W. (1990) Major issues confronting sustainable agriculture. In: Francis, C.A., Flora, C.B. and King, L.D. (eds) *Sustainable Agriculture in Temperate Zones.* John Wiley, New York, pp. 423–438.

Merrill, M.C. (1983) Eco-agriculture: a review of its history and philosophy. *Biological Agriculture and Horticulture* 1, 181–210.

NatWest (1992) *National Farm Survey.* National Westminster Bank, Coventry.

Neuerburg, W. and Padel, S. (1992) *Organisch-biologischer Landbau in der Praxis.* BLV-Verlag, Munich.

NRC (1989) *Alternative Agriculture.* National Research Council, Board on Agriculture, Committee on the Role of Alternative Farming Methods in Modern Production Agriculture, National Academy Press, Washington, DC.

Rusch, H.-P. (1968) *Bodenfruchtbarkeit: eine Studie biologischen Denkens.* Verlag K. F. Haug, Heidelberg.

Steiner, R. (1924) *Agriculture: A Course of Eight Lectures.* Rudolf Steiner Press/Bio Dynamic Agricultural Association, London.

Vereijken, P. (1990) Proceedings of the 1990 annual meeting of the working group 'Integrated Arable Farming Systems'. *Schweizerische Landwirtschaftliche Forschung* 29, 1–436.

Vereijken, P. and Royle, D.J. (eds) (1989) Current status of integrated farming systems research in western Europe. In: *Proceedings of the 1989 Annual Meeting of the Working Group 'Integrated Arable Farming Systems'.* WPRS Bulletin, Centre for Agrobiological Research, Wageningen.

2 The Development of the Organic Industry and Market: An International Perspective

William B. Tate

The Origins of Organic Farming and the Development of the Organic Movement

Modern organic agriculture began in Europe in the 1920s and struggled in its early years in the face of the all powerful chemical lobby. In the worst case of persecution, the bio-dynamic agricultural movement was banned by the Nazis from 1940 under pressure from influential German chemical companies (Peter and Ghesquiere, 1988). Governments were concerned with maximizing agricultural output for economic, social and political reasons; environmental concern was negligible. However, in the last quarter century recognition of persistent agricultural overproduction in the West and a belated realization of the fragility of the planet has led to organic agriculture being viewed in a better light. A favourable demand–supply relationship for organic produce in the last decade in particular has enabled a great expansion in hectarage under certified organic cultivation (see Chapter 1).

The development of the international organic movement over the last 70 years can be summarized in three main phases. The years 1924–1970 were a period of struggle and financial difficulty establishing organic agriculture in a largely hostile environment, and a period in which core works were written. Between 1970 and 1980 key organic symbol schemes were set up as consumer demand increased, and 'green' awareness grew; retail outlets multiplied. From 1980 organic agriculture gained acceptance; national and international standards were set, and governments introduced organic aid schemes for farmers. The confirmation of the newly found respectability of organic agriculture came when the Swedish, Danish and German governments introduced schemes encouraging their farmers to convert to organic agriculture in the late 1980s, followed shortly by legislation in the United States and the European Community defining organic farming in legal terms.

Key figures in the organic movement

The international development of organic cultivation and consumption began with the five giant personalities who inspired the movement: Rudolf Steiner from Austria, Hans Müller, the German-Swiss, Lady Eve Balfour in Britain; J.I. Rodale in the United States; and Masanobu Fukuoka in Japan.

The influence of *Rudolf Steiner* (1861–1925) has been profound in many areas, especially education. Steiner prompted the growth of the earliest and for some time the most popular form of organic agriculture in the world, bio-dynamic cultivation. He was trained in mathematics, physics and chemistry and did not have an agricultural background. The bio-dynamic movement with its symbol of Demeter, the Greek goddess of agriculture, arose from a course of lectures Steiner gave in 1924, a year before his death (Steiner, 1924). Steiner had accepted an invitation to speak to an audience of German farmers at Koberwitz. They were concerned about the trend towards 'scientific' farming, with its growing use of chemicals (Davy, 1980). In the lectures, Steiner pointed out the failings of the materialistic approach to agriculture. He stressed the greater significance of the natural forces that form all the constituents within living nature. Even by the early twentieth century, farmers had noticed falling soil fertility, declining seed viability and poor animal health.

Steiner advised working with natural forces rather than opposing them and suggested special herbal preparations to use for this purpose. These preparations aim to enhance the decomposition process in manures and compost, which in turn when applied to soil will improve plant, animal and human life.

Bio-dynamic farming is mixed. There is a suitable balance of animals and crops, a system of recycling, and benign methods of pest and disease control. Farming by bio-dynamic principles has been called 'organic plus' because of its commitment to Steiner's spiritual teachings ('anthroposophy'), its use of bio-dynamic preparations, and its self-sufficient farming philosophy. Strict bio-dynamic certification standards mean that bio-dynamic produce, carrying the Demeter logo, is universally accepted as genuinely organic.

Hans Müller (1891–1988) was both a politician and a farmers' leader. He wanted to end what he saw as the exploitation of farmers by middlemen, by forging a direct link between producers and consumers.

In 1946 Müller founded BIO Gemüse AVG, a Swiss cooperative of small farmers using organic-biological methods. The theory of organic-biological agriculture was outlined by H.P. Rusch in his book *Bodenfruchtbarkeit* (Soil fertility) in 1968. Rusch pointed to the key role of microbiology in soil fertility, and emphasized the need for soil fertilization from the fresh dung produced by free-ranging livestock. Cultivation had to be shallow and light so as not to destroy vital soil organisms.

German practitioners of the Müller–Rusch farming techniques founded Bioland, the organic-biological farming group, in 1971. By 1990 Bioland was the German organic group with the largest hectarage and the greatest number of certified organic producers. Bioland follows Müller's strong preference for direct dealing with organic consumers.

Lady Eve Balfour (1899–1990) was the moving force behind the foundation in 1946, and subsequent development, of the Soil Association, Britain's leading organic organization. She published her best-selling work, *The Living Soil*, in 1943. Lady Eve's central principle was that the health of the soil and the health of man are inseparable. Between 1938 and 1970 she conducted what became known as the Haughley Experiment into organic farming, on her farm in Suffolk (Balfour, 1976).

Lady Eve's work with the Soil Association lasted 44 years. Initially the Association had attracted a large international membership but then declined into a phase of financial difficulties and internal division. In the 1980s the Soil Association emerged as a vigorous campaigning organization with widely respected certification standards (SA, 1993). Lady Eve Balfour is the personality identified above all others with the growth of the organic movement in Britain.

In the United States, *J.I. Rodale* publicized the organic farming ideas of Eve Balfour and Albert Howard through the hugely successful magazine *Organic Gardening*, which grew to a circulation of over 2 million by the 1980s. The success of this magazine financed the establishment of the Rodale Research Institute, which pioneered organic farming research in the USA in the 1970s and 1980s, and also led to the publication of the *New Farm* magazine aimed specifically at farmers. Rodale's work was continued by his son Robert Rodale, who coined the term 'regenerative' agriculture, until Robert's death in Moscow in 1990 while attempting to establish interest in sustainable agriculture as part of the dramatic reforms taking place in the former Soviet Union at that time.

In Japan, *Masanobu Fukuoka* developed a very different approach to organic farming, both philosophically and in practice, and is best known to a western readership through his book *One Straw Revolution*.

The Development of the Market for Organic Food

Consumer demand

In the early years of the organic movement producers were isolated. They were usually long on principles and short of finance and had to make do with inferior land in remoter areas. There was no certification and no prospect of a price premium for organic produce. Organic farmers were often the subject of derision from the rest of the farming community and the general public alike. Nowadays the organic movement has a loud political voice and is respected by the media who are sympathetic to green issues and increasingly critical of conventional agricultural practices. Commercially the increased numbers of organic farmers makes the formation of marketing cooperatives possible, which strengthens their hand when dealing with wholesalers, and can lead to machinery sharing and joint buying of certain inputs. Price premiums for certain organic produce exist, for example organic milling wheat tends to carry a minimum 100 per cent premium over non-organic, which compensates for the lower yield. Care must, however, be taken over international generalizations; there is usually a limited market premium for instance in Germany for fresh organic meat and milk, but none at all in France where organic certification standards for animal products do not exist.

In the late 1980s and early 1990s, various market research reports pointed to strong and rapid growth in consumer demand for organic produce, both in North America and in Europe (Table 2.1). These pointed to organic farming taking a significant share of the total retail market, with estimates ranging from 2% to 5% by the year 2000. The increase in consumer demand was attributed to two factors: the food scares, particularly in the UK (*Salmonella*, *Listeria* and BSE) and in the USA (alar); and the growth of the environmental movement and 'green' consumerism (Mintel, 1989). However, a number of factors have combined to slow down growth in the last two years, including the recession in Europe, which has had a general impact on demand for green products, and the small supply base for organic produce, with the associated difficulties and cost of marketing small amounts in niche markets. In many countries, the period since 1991 has seen consolidation in many marketing channels, with growth continuing but at slower rates than previously. At current rates of growth, it seems unlikely that many of the projections in Table 2.1 will be achieved. This will, however, depend on two factors which are difficult to assess at present: the impact of legislative changes relating to organic food, and the growth in the supply base (see Chapter 1, Fig. 1.1). In the latter respect, Tate's (1991) projections for Europe's certified organic

Table 2.1. Market demand and supply growth projections for the organic sector, 1990–2000.

Report	Region	Parameter	Unit	1989/90	1995	2000
Henley Centre (1989)	UK	Retail sales:				
		Vegetables	Per cent	1.56	5	n/a
		Other	Per cent	0.2–0.6	0.8–2.0	n/a
Ross (1991)	UK	Retail sales:				
		Vegetables	Per cent	2.5	5.0	10.0
		Other	Per cent	0.3	2.0	5.0
Mintel (1991)	UK	Retail sales	£ million	80	673	n/a
			Per cent			5–10
Landell-Mills (1992)	European Community	Land area	Thousand ha	198	n/a	n/a
			Per cent			2–3
Tate (1991)	Europe	Retail sales	£ million	900	2700	8200
		Land area	Thousand ha	255	776	2362
Marketdata Enterprises (1990)	USA	Retail sales	$ million	1250	3960	n/a
Lampkin (pers. comm.)	World-wide, incl. Japan	Retail sales	£ million	2000	7000	20,000

Source: Lampkin (pers. comm., 1993) based on references cited.

hectarage in 1995 could prove to have been conservative.

Certification and legislation

A key element of the organic produce market is regulation. It is necessary in order to maintain the high ethical standards of the organic movement, to retain consumer confidence in produce, to encourage and support genuine organic farmers, and lastly to provide a basis for traffic in organic produce across frontiers. The symbol schemes that farmers, wholesalers and processors join are a central feature of organic regulation. Once certified by the association running the scheme, they are entitled to use the symbol. The certification cost is met by a levy on the use of the symbol on produce, or on a farm or hectarage basis.

The bio-dynamic movement was the first to introduce self-regulation. Guidelines for cultivation of Demeter symbol products were set in 1928 in Germany and later became the standards for

Demeter symbol schemes in many other countries. The 1970s were a particularly active decade for the establishment of symbol schemes backed by organic standards. In France, Nature et Progrès launched its scheme in 1972; in Britain, the Soil Association followed in 1973, and in Germany, Bioland introduced its symbol in 1978.

Sets of organic standards contain a minimum universal core of rules that transcend national boundaries. What marks one group off from another is the degree to which they go beyond the bare essentials of organic regulation. If a farmer wins a positive response to his application to join a symbol scheme he must expect to enter a conversion period, ranging from one year in the case of Kontroll-foreningen for Alternativ Odling (KRAV) in Sweden up to seven years under certain bio-dynamic schemes. An in-conversion symbol is usually available for the farmer to use when selling his produce, the best known of which is the universal bio-dynamic conversion symbol, Biodyn. Countries vary considerably in their numbers of indigenous certifying schemes. Italy for example had, until 1993, more than 20 associations, each judging candidates' suitability to market produce under their respective logos. The Republic of Ireland and Portugal, on the other hand, had only one symbol scheme each. The more certifying bodies there are operating independently of each other in one country, the greater the uncertainty in consumers' minds and the greater the requirement for the introduction of a common national standard.

Some of the major organic consuming and producing countries have set up national standards for organic certification. France, a leading organic producer, was the first European country to introduce an official label (AB) for organic cereals, fruit and vegetables. AB stands for *agriculture biologique.* A law was passed in 1980 (80-502, 4 June 1980) which permitted official ratification of sets of organic standards. Later an official definition of organic agriculture was made. A decree in 1983 set the numbers of representatives of farmers, consumers and government on the Commission Nationale d'Homologation who would review the sets of standards presented for ratification. After the Commission's recommendation, the ultimate ratification decision rests with government ministers. Once an association presenting a set of standards has received ratification it grants the official AB logo to those of its members who have already qualified to use the group logo. The system is self-regulating. The first set of standards to be ratified were those of France's leading symbol scheme, Nature et Progrès, in 1986. New legislation in 1988 (88-1202, 30 December 1988) introduced a nationally agreed set of basic standards which underpinned the 14 different sets of standards then in operation and outlawed the use of any non-approved standards with effect from 1 January 1990. French consumers recognize and respect the AB logo, and where supermarkets have begun to sell fresh organic fruit and vegetables, their packaging only carries the AB logo.

Britain followed France in national organic certification. The United Kingdom Register of Organic Food Standards (UKROFS) was set up by the Food from Britain organization in 1987, and is funded by the Ministry of Agriculture. Taking advice from producers and processors, UKROFS has compiled an official set of organic standards (UKROFS, 1993), covering both arable and livestock production, as well as horticulture and the processing of organic products. Voluntary certification schemes need to match UKROFS standards in order to be registered as approved organic sector bodies with UKROFS, and for their symbol holders to be able to use the UKROFS certification mark. UKROFS maintains a public register of all approved producers. As part of the overall policing of the scheme, UKROFS inspectors check a random sample of 10% of the approved producers in the schemes at least once a year. The UKROFS certification mark on produce is not intended to replace the symbols of organic bodies participating in the UKROFS scheme. The UKROFS symbol can be placed alongside other symbols if required to underline their organic credentials. By 1993 the UKROFS Board had registered five approved organic sector bodies: Organic Farmers and Growers Ltd; the Bio-Dynamic Agricultural Association; the Organic Food Federation; the Scottish Organic Producers Association; and the Soil Association.

The EC Regulation

From 1 January 1993 all fresh and processed produce of plant origin sold in the European Community (EC) as 'organic' (or the 'biological/ecological' equivalent in eight other European languages) must meet the organic standards laid down by the European Commission (EC, 1991). Produce of animal origin is not yet included. For the first time trading standards officers now have the power to pursue organic cheats operating in the so-called 'grey market'. A minimum two-year conversion period is required. Each EC country has an appointed Designated Inspection Authority (DIA) which will implement and police the scheme either alone or in conjunction with approved organic certifying bodies. In Britain, the previously voluntary scheme of UKROFS is now compulsory, with UKROFS as the DIA. Other EC countries have had to establish bodies to be DIAs specifically to meet EC legislation.

The introduction of the EC legislation has had a major impact on the number of approved certification bodies: in France and Italy the number has fallen to five, in Ireland it has risen to three and in Germany it has increased from six to over 50, although the situation is complicated by the existence of separate approved inspection bodies for each of the 16 federal states. In some cases, such as Denmark, France and the Netherlands, certification is now carried out partly or entirely by governmental or quasi-governmental agencies.

Organic imports into the EC now have to be from a country on the EC's approved organic source list. Those on the list must follow the EC scheme's organic production rules and will need equivalent inspection measures. The local inspection body will have to issue inspection certificates to accompany produce entering the EC.

There is no EC organic logo as such. National logos will continue to be used as long as the standards they represent are up to EC standards. Concern has been expressed that in certain southern European Community states with no organic tradition, the inspection authority will be delegated to local government which will be too lax, preferring to encourage quasi-organic production for social, economic and political reasons rather than to protect the good name of organic produce. The criticism has also been voiced that the EC organic standards are basically undemanding and as such they may lower standards rather than raise them.

United States legislation

Legislation defining organic food production has also been introduced in the United States (US, 1990). This provided for the establishment of a National Organic Standards Board with powers to set detailed standards for organic crop and livestock schemes and to regulate the certification process. The law was due to come into force in 1993, although the process has been delayed due to funding problems.

International certification and accreditation

At the widest international level initial moves towards global organic regulation have been made by the International Federation of Organic Agricultural Movements (IFOAM). IFOAM was founded in 1972 and has its headquarters at Tholey-Theley in Germany. Nearly all the major national European organic certifying associations and many European non-certifying organic associations and institutes belong to IFOAM, which means that European members dominate the Federation, although there is a rapidly increasing representation from developing countries and the former socialist economies of central and eastern Europe (Table 2.2).

IFOAM has drawn up a set of minimum production standards for primary organic produce to which all its members must subscribe in principle (IFOAM, 1992). The IFOAM set of minimum standards has

Table 2.2. IFOAM membership, 1986–1992.

Continent	1986	1990	1992	1992 (%)
European Community	60	151	158	41
Europe (non-EC)	19	34	53	14
North America	14	23	37	10
South America	3	56	79	20
Africa	0	13	22	6
Asia	7	17	30	8
Australasia	11	9	13	3
Total	114	303	386	100

Source: International Federation of Organic Agriculture Movements.

been used as the basis for the EC scheme and others. IFOAM's objective is to produce one global 'organic seal' free of any national or corporate ties. The process of endorsing national organic seals by IFOAM is under way but still in its early stages.

On the international level, the Food and Agriculture Organization (FAO) has responded to the growth of the organic sector and the need for certification by tabling proposals for international organic standards as part of the FAO's *Codex Alimentarius* (FAO, 1991). The proposals are still under development, but draw heavily on EC and United States legislation, as well as national codes from Australia, Canada and Austria, and the IFOAM standards.

International trade in organic products

The growing demand for organic produce has led to the development of international trade. Countries which in some cases have no internal demand, but a favourable climate, are producing to satisfy export markets.

Europe is the world's leading market for organic produce, a considerable amount of which is imported from outside. There is a large inflow of cereals and soya from North America into Europe, notably from producers in the Farm Verified Organic and the Organic Crop Improvement Association (OCIA) schemes of the USA and Canada. Israel is another important supplier. Israeli organic exports include citrus fruits, melons and early vegetables. Other sources are Turkey (figs, nuts and dried fruits), Morocco (oranges), the West Indies and Canary Islands (bananas) and India (tea).

Within Europe there is considerable cross-frontier trade in cereals, wine and fresh organic fruit and vegetables. The most obvious direction is from south to north because northern Europe needs Mediterranean citrus fruit, grapes and other delicate fruits and early vegetables. The countries with the largest deficit in fresh organic produce as a whole are Germany and the UK, while the two with the greatest surplus are France and the Netherlands.

Longer-life organic commodities also tend to go from the south of Europe to the north. These include Italian and Spanish organic rice and olive oil, as well as wine. Processed branded organic food is increasingly traded; the best-known international organic food processor is Lima of Belgium, which has been operating since 1957.

Elsewhere in the world, there is a significant trade in processed and frozen organic produce between the United States and Japan, with some product supplied from New Zealand. Southern hemisphere countries are also supplying increasing quantities of out-of-season fresh produce to the US and European markets.

Countries with a production shortfall need to vet organic imports. To this end a German national regulatory body called Bundesverband Naturkost was established. Formed in 1983, it is an organization of wholesalers, processors and retailers. Bundesverband Naturkost aims to ensure transparency and openness in the organic market, to provide a guarantee of organic quality for the consumer and to develop a country-wide awareness of organic quality. German and foreign symbol schemes to which Bundersverband Naturkost gives its approval win a QK (Qualitäts-Kürzel) number. Organic produce without a QK number is unwelcome in the German market.

In Britain the Soil Association has established a Soil Association Positive List which is supplied to participants in its Processing, Packing and Distribution Scheme. It is a list of national and international sources of supply that are compatible with Soil Association organic standards and hence can be sold under the Soil Association logo for the usually levied fee. The absence of a certifying body from the Positive List does not necessarily mean low organic standards. The produce of certain reputable schemes is not exported and so no request has been made for the Soil Association to verify their standards.

Within the European Community, these vetting processes for third-country schemes are now heavily influenced by the implementation of the EC Regulation.

Marketing channels

Specialist outlets

A wider and more sophisticated demand for organic produce led to the development of a dedicated supply chain connecting producer to consumer. Specialist organic wholesalers have arisen, collecting produce and redistributing it to processors and retailers. Natouring, Germany's leading organic wholesaler, is a good example, acting as a wholesaler of German fresh organic fruit and vegetables, as well as being the major importer of that produce. Natouring also imports and distributes dry and processed organic produce, as well as handling domestic production. German organic cheeses are exported. Natouring has a sister company, Natouring France, which is its principal supplier of fresh imported fruit and vegetables, and makes France the major supply source. Italy and Spain are Natouring's next most important sources, followed by the Netherlands and Israel. The German QK number system restricts Natouring's choice of organic suppliers from abroad. Winter is the most active season for the fresh import trade, as in summer local produce is plentiful and often cheaper than imports. Also the German product is fresher than the imported equivalent.

A variety of different types of retail outlet for organic produce have developed as demand has grown. The earliest organic sales were direct from producer to consumer in the form of farm-gate sales, producer–consumer cooperatives and markets. All three forms survive today in differing degrees in different countries. In Germanic countries, where the organic philosophy is deeply rooted, cooperatives flourish with consumers providing an input into 'extended farm communities', guaranteeing a market in return for a voice in farm enterprise planning. Consumer groups usually appoint one of their number to handle delivery. Some, like the Swiss cooperative BIO Gemüse, operate a mailing service of fresh vegetables on a weekly and fortnightly basis. Organic market (*les marchés bio*) are a particularly strong feature of the French organic scene, organized by the 50 regional Nature et Progrès groups. They are held weekly or fortnightly and enable consumers to buy direct from producers without paying the high organic premium charged by conventional retailers. Nature et Progrès also regularly publishes lists of its farmers to encourage the public to buy direct at the farm gate.

The growing demand for organic produce by the young in the 1970s and 1980s led to the creation of specialist organic retailers, the best known of which are the German *Naturkostläden.* There are about 1800 *Naturkostläden* in Germany, most of which are independent. A complete range of organic groceries is usually offered, ranging from long-life items such as organic wines and grains, to short-life milk, cheeses, bread and fresh fruit and vegetables. Many consumers are not simply looking for organic, they want bio-dynamic produce. Produce bins carry a slate with the name of the certification body, QK number and price per kilo. Produce is unwashed and unwrapped and as such is considered ecologically sound. An older generation of German health-food consumers frequent the *Reformhäuser,* which sell predominantly non-organic Neuform logo products. What organic produce is sold tends to be processed; fresh organic produce is regarded as too expensive by retailers and consumers alike.

The French have no equivalent of the *Naturkostläden* but sell most of their organic produce that is not sold at markets or at the farm gate through *magazins diététiques.* Here organic produce is sold alongside *naturel* products (no chemical treatment after harvest) and *diététique* products (with added vitamins, sugar-free or part of a slimming regime). *Naturel* products are cheap, which can lead to their purchase by those who are unaware of the extent to which they differ from genuinely organic produce. Of the estimated 2200 *magazins diététiques,* about 100 belong to cooperatives which buy jointly and 300 are part of the La Vie Claire chain.

Supermarkets

Supermarkets dominate food retailing in general but have not yet captured the organic produce market. The position varies from country to country. Supermarkets have taken up the idea of organic supply and persisted with it to the greatest extent in the United Kingdom. Safeway began to stock fresh organic fruit and vegetables as early as 1981, Asda followed in 1986, Sainsbury in 1987, and Tesco in 1988. Organic flour and bread are also stocked in most cases, milk in some instances, but meat is rare. Marks & Spencer tried fresh organic produce in 30 of its stores in 1989 but dropped the idea in 1990, except for milk which is now available in most stores. Marks & Spencer blamed the poor quality, unreliability and high price of fresh organic fruit and vegetables for its decision to drop them.

British consumers do not in general have the choice of outlets offering fresh organic produce in particular that their counterparts on the Continent enjoy. Supermarkets are often the only retailers offering these lines. Organic markets are virtually non-existent, health-food shops selling fresh organic produce are few and far between, and direct sales by organic farmers are unrealistic for the urban majority. Britain's organic farmers, growers, wholesalers and processors have entered into a relationship with supermarkets not because they necessarily relish the prospect, but because they have little alternative. Safeway has pioneered the link, and worked hard to strengthen it, also supporting research and development work and sponsoring annual events such as the National Organic Food and Wine Fair and the Organic Food and Farming Centre at the Royal Show.

In the USA supermarkets have not taken over organic retailing any more than they have in continental Europe. In the 1980s natural food stores met the small but growing requirement for organic food. However, concern over the health risks to children from a growth regulator for apples called alar, and discovery of a consignment of Chilean grapes laced with cyanide, caused an upsurge in organic demand in 1989. Supermarkets, particularly in California, hurried to stock organic fruit and vegetables, but then as the controversy began to die away, consumers began to flinch at the 40–50% premium asked for organic produce, and a large amount of produce had to be disposed of. The regularity of supply was just as much a problem in the USA as in Europe, and some supermarkets dropped organic supply as quickly as they had picked it up. Others worked to consolidate the organic market, with the emphasis shifting to processed and frozen products which did not face the same appearance and shelf-life problems as fresh produce. The market for other products once again reverted to the natural food stores.

The ethical basis of organic agriculture has meant that producers have often been at odds with supermarket management. Historically the organic movement has suspected that supermarkets' pursuit of profit is inconsistent with their own environmental concerns. It is felt that the supermarkets' insistence on expensive, wasteful packaging, uniform produce size and quality, and regularity of supply, mean that they inhabit a different and incompatible world to that of organic growers. Mistrust of the middleman is a recurrent theme in organic agriculture; supermarkets, thriving on what are perceived as excessive profit margins, are often cast as the ultimate bogey men of organic demonology.

The Significance of the Organic Industry

Because of the amount of publicity that organic farming receives from the media, it is possible to lose sight of the movement's actual importance. Certified organic land comprised less than 1% of Europe's agricultural land in 1990, and much less than that elsewhere in the world (Tate, 1991). There were thought to be 15,000 organic holdings in Europe, compared to total farm holdings in the EC alone of nearly 9 million. Some ambitious campaigns have been launched in recent years with the aim of drastically extending organic farmland. The Bavarian state government has a target of 25% of its agriculture to be organic by the year 2000. The Soil Association wants to turn 20% of Britain's farmland organic by the same year. To meet these targets the organic consumer would have to metamorphose from current statistical insignificance into one in four or five of the population.

The great mass of people care little and know even less about organic produce. They want cheap convenient food. The still relatively scarce committed organic consumer, in contrast, is prepared to pay a good deal more to buy organic produce on grounds of taste, health, safety or protection of the environment. There is no guarantee that markets can be found for large extra quantities of organic produce. The experience of the Swedish organic market in 1990 is illuminating. A trebled hectarage of organic land resulting from an official conversion scheme led to a swamping of the organic cereal market, and disappearance of the organic premium. Much lower prices for organic produce may be the only long-term way of significantly expanding the market, but this would damage the economics of conversion in the absence of governmental financial assistance.

References

Balfour, E.B. (1976) *The Living Soil and the Haughley Experiment.* Universe Books, New York.

Davy, J. (1980) *Rudolf Steiner, A Sketch of his Life and Work.* Rudolf Steiner Press, London.

EC (1991) Council Regulation (EEC) No. 2092/91 of 24 June 1991 on organic production of agricultural products and indications referring thereto on agricultural products and foodstuffs. *Official Journal of the European Communities* 91 (L198), 1–15.

FAO (1991) *Draft Guidelines for the Production, Processing, Labelling and Marketing of Organic/Bio-dynamic Foods.* Codex Alimentarius Commission, Food and Agriculture Organization, Rome.

Henley Centre (1989) *The Market for Organic Food.* Report by the Henley Centre for Forecasting for Mid Wales Development, Newtown.

IFOAM (1992) *Basic Standards of Organic Agriculture.* International Federation of Organic Agriculture Movements, Tholey-Theley, Germany, pp. 25–64.

Landell-Mills (1992) *Organic Farming in Seven European Countries.* Report for the European Crop Protection Association. Landell Mills Commodities Studies, Brussels.

Marketdata Enterprises (1990) Cited in: Haest, C. (1990) Growing as partners into the 21st century. In: Geier, B., Haest, C. and Pons, A. (eds) *Trade in Organic Foods.* IFOAM, Tholey-Theley, Germany

Mintel (1989) *The Green Consumer.* Mintel, London.

Mintel (1991) *Organic Food.* Mintel, London.

Peter, D. and Ghesquiere, P. (1988) *Bilan des connaissances et des applications de l'agriculture biologique et interet pour l'agriculture Communautaire.* Commission of the European Communities, Brussels.

Ross, C.P. (1991) *Going Organic – the Future for Organic Food and Drink Products in the UK – an Update.* Price Waterhouse, Birmingham.

Rusch, H.-P. (1968) *Bodenfruchtbarkeit: eine Studie biologischen Denkens.* Verlag K.F. Haug, Heidelberg.

SA (1993) *Standards for Organic Food and Farming.* Soil Association Organic Marketing Co. Ltd, Bristol.

Steiner, R. (1924) *Agriculture: A Course of Eight Lectures.* Rudolf Steiner Press/Bio Dynamic Agricultural Association, London.

Tate, W.B. (1991) *Organic Produce in Europe.* The Economist Intelligence Unit, London.

UKROFS (1993) UKROFS *Standards for Organic Food Production.* UK Register of Organic Food Standards, Food From Britain, London.

US (1990) *Organic Foods Production Act of 1990.* United States Senate Committee on Agriculture, Nutrition and Forestry, Washington, DC.

3 Researching Organic Farming Systems

NICOLAS H. LAMPKIN

Introduction

It can be argued that two broad approaches to researching organic farming systems have developed in recent years. The first, comparing organic farming against existing conventional systems, is the one most frequently adopted by researchers and institutions new to the organic farming concept. The second, researching and developing organic farming systems in their own right, is more often adopted by researchers and institutions who regard the concept of ecologically-based systems as valid and are interested in developing and exploring the potential of such systems.

The two approaches reflect the differing objectives of researchers and institutions. In the first case, researchers may be seeking to establish that there are no (or negligible) differences between organic and conventional systems, for example in terms of food quality or environmental impact (CAST, 1980; Buttel *et al.*, 1985). Alternatively, the research may be part of a learning process, of benefit to the researcher or institution, but which frequently brings little direct benefit to the organic producer in terms of improved techniques, reduced costs or improved environmental impact and food quality, although in the long term this type of research may lead to specific proposals for developmental type work. A third reason for adopting the comparative approach is that researchers may be seeking to establish what the differences are in order to determine future research and policy programmes, in which case the results are of importance to policymakers and may have longer-term impacts on organic producers.

The second, more developmental approach, aims to derive results which provide some direct benefits to existing and potential producers.

In this context, comparisons with conventional systems are relatively meaningless – what is important is the performance of the improved organic system compared with existing organic systems.

This dichotomy has tended to lead to a view among established organic farming researchers that the research focus should be on developing and improving organic/biological systems, and not on attempts at systems comparison (Dlouhy and Nilsson, 1983; Woodward, 1989). Given the severely restricted funding of organic farming research, at least until recently, and the desire of the organic sector to put its research and development priorities ahead of the learning objectives of established research institutions, this is not surprising.

However, the needs of farmers are such that in certain circumstances, comparisons can be justified, particularly in the area of economics and environmental impact assessment. Existing organic producers, clearly, will be more interested in developmental work and in comparisons between their own holdings and those of other organic producers. Conventional producers, however, need conventional versus organic comparative data in order to make soundly-based decisions on the implications of conversion, although the most important issue for the conventional farmer is the likely impact on his or her own farm, rather than the generality of farms. All farmers, organic and conventional, are influenced by the decisions of policy-makers and governments in the fields of agricultural, economic and environmental policy. The success or otherwise of policy-making depends critically on the quality (often even more than the quantity) of data available to the policy-maker, and this must include soundly-based, comparative data on different farming systems.

The issues relating to how comparisons between farming systems should be made underpin many of the studies reported in this book. This chapter is therefore devoted to exploring these issues, in order to provide a framework for interpreting previous studies and to act as a basis for the design of future research programmes.

Comparing Farming Systems

The comparison of farming systems is an area of research which is problematic even in situations where the results are not likely to be controversial. Problems may arise due to definitions, the objectives and design of the study, the methods and standards used for comparison, and the need to isolate factors which affect performance but are not necessarily related to the systems being compared (Koepf, 1983; Lockeretz, 1989; Vogtmann, 1990).

The nature of the researcher

Although science is intended to be a means for arriving at an objective 'truth', the way in which the scientific method is implemented is critically dependent on the researcher and institution concerned (MacRae *et al.*, 1989). No researcher is completely free of normative values and prejudices, and these are likely to colour the questions asked, particularly if the researcher is not very well informed about the subject under study. The researcher will also be influenced by an institutional ethic which expresses itself in terms of career opportunities and the availability of research funding. In industrialized countries, it can be argued that the institutional agricultural research ethic has been one of progress defined in terms of increasing output through the increasing application of technology, thus reducing unit costs and improving 'efficiency' (Dahlberg, 1988). It is only recently that research objectives have shifted sufficiently to allow different questions to be asked.

Thus it may be possible to reach very different conclusions using the same scientific method on exactly the same situation or set of data – a situation analogous to the proverbial question of whether a bottle is half empty or half full. The answer depends entirely on the viewpoint of the observer, an issue complicated by the methodological issues discussed below.

The nature of the farming system

Organic farmers, and particularly bio-dynamic farmers, set great store by the concept of the farm as an organism (Koepf *et al.*, 1976; Scofield, 1986). In a biological sense, an organism usually functions as an independent entity, with the capacity for self-regulation, growth, development and reproduction. The organism will consist of a number of different organs which fulfil different, often unique, functions, but are interdependent on each other and are incapable of independent existence. The idea of a farm as a living, organic whole may be considered to be analogous to that of an organism, in that the soil, the naturally occurring fauna and flora, the crop and livestock enterprises, and the human elements of management and labour, interact with each other in the context of the prevailing ecological, economic, political and social environment.

An alternative but related view is that of the farm as a system. In general terms, a system may be defined as 'a group of interacting components, operating together for a common purpose, capable of reacting as a whole to external stimuli: it is unaffected directly by its own outputs and has a specified boundary based on the inclusion of all significant feedbacks' (Spedding, 1988). Whether a farm can really be considered to be an 'organism' in a real sense, as the bio-dynamic farmers argue, or whether the idea of an organism is simply a model for describing a more mechanistic system, is a matter for debate (Dabbert, 1990a; Schaumann, 1990). However, the concept of the farm as an interactive, independent (bounded) system has long been recognized by agriculturists and agricultural economists and can be traced in the literature of the twentieth century (Lloyd, 1970; Dabbert, 1990a). It forms the basis of much of the agricultural systems literature (Dent and Anderson, 1971; Dalton, 1975; Spedding, 1988; Jones and Street, 1990) as well as the 'farming systems research' approach which has been popularized in developing countries (Gilbert *et al.*, 1980; Hildebrand, 1986).

The systems concept, however, sits uneasily with the emphasis on specialization and commodity-oriented research which has been perpetuated by the research institutional framework and policy mechanisms (e.g. commodity programmes) of the industrialized countries, as well as 'green revolution' type approaches in developing countries (MacRae *et al.*, 1989; Dent, 1990). The need to simplify research to the level of subsystems and system components, on grounds of cost as well as manageability, has sometimes led to the whole-system perspective being lost, in particular the economic and social consequences of technological developments. The interactions between elements of a system such as a farm are such that the 'whole *can* be greater than the sum of its parts'. Thus it may be difficult to reach definitive conclusions about the functioning of a system based on an analysis (and theoretical combination) of its components without some form of study of the system as a whole (Koepf, 1983).

As a consequence, there is now considerable discussion about the development of appropriate research methodologies for ecologically-based agricultural systems in both developed and less-developed countries (Conway, 1985; Chambers *et al.*, 1989; MacRae *et al.*, 1989; Patriquin, 1989; Youngberg, 1990; Lockeretz, 1991; Vereijken, 1991). These emphasize a multi-disciplinary approach, as well as the needs and potential contribution of producers, including their previous experience, indigenous knowledge and traditions, and their role in determining research priorities, innovation and the carrying out of research. Not all research is necessarily suited to this 'farmer first' (Chambers *et al.*, 1989) approach, a point which is relevant to issues such as comparisons between farming systems, to the extent that researchers need to reconcile the objectives of different groups of individuals, including farmers, policy-makers and society (Lockeretz, 1991; Lockeretz and Anderson 1993). However, the problem of generalizing results from any research situation which is inevitably site-specific is such that research on farms, with results specific to the individual farm, is likely to have much greater validity in a developmental and applied research context (Patriquin, 1989).

In the context of comparing organic and conventional farming systems, the key issue is that we are dealing with different *systems*, not modifications to individual *practices*. Thus traditional agricultural research methodologies, involving the manipulation and comparison of a restricted number of variables, is inappropriate in a systems context where a large number of variables are involved, and where restrictions on individual variables might impact on overall system performance. Organic farming *does not* represent the zero-input treatment in a conventional fertilizer or pesticide input trial. It *does* involve the interaction between a number of different system components, such as the inclusion of legumes for biological nitrogen fixation and the creation of habitats for natural predators of crop pests.

Defining the difference between farming systems

The issue of the farming 'system' goes beyond the question of appropriate research methodologies to include the definition of the farming system itself. This has been and continues to be a major problem with comparisons between conventional and organic farming systems.

The first problem is in terms of what is meant by organic (or ecological) farming (Lockeretz, 1990). In Chapter 1, a broad, conceptual definition of organic farming was presented, but the problem of determining which farms come within this category remains. Many researchers choose to define organic farms either in terms of self-definition (by the farmer) or in terms of certification by a particular standards organization. Farmers defining themselves as organic may not actually meet accepted production standards. On the other hand, standards contain both recommendations and requirements or prohibitions. Organic farmers may not necessarily implement the recommendations, so it has been argued that the only clear-cut criteria on which a distinction may be based are the requirements and prohibitions (Hermanowski, 1989).

It is sometimes argued that there is a spectrum of farming systems, with intensive conventional at one end and 'pure' organic at the other (e.g. Klepper *et al.*, 1977). This immediately raises the question: which point on the spectrum should be compared with another? For example, in Fig. 3.1, should point G, an extensive organic farm, be compared with point A, an intensive conventional farm, with point C, an average conventional farm, or with a point just the other side of the organic–conventional divide at E?

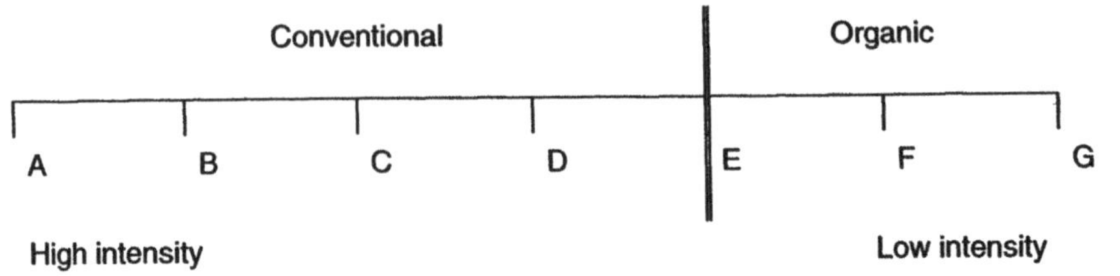

Fig. 3.1. Schematic representation of the spectrum of conventional and organic farming systems with respect to intensity of external input use.

The problem is even more complex, because the spectrum, if it is an appropriate model, is multi-dimensional. Organic and conventional farming systems can not only be differentiated *within* each group in terms of intensity, but also in terms of farm type (arable, dairy, mixed etc.), location and size, as well as a range of physical and financial constraints, the most important of which is the management ability of the individual farmer.

The Development of Farming Systems experiment at Nagele in the Netherlands (Zadoks, 1989) illustrates the problems which may arise when attempting a systems comparison, in this case conventional, integrated and bio-dynamic. The soil types, land area and management are consistent, but the bio-dynamic unit has dairying as its main enterprise while the other two comparative systems are all arable. This is in an arable region on a land area which is arguably insufficient to carry the fixed costs of a dairy enterprise. Is it therefore surprising that the financial returns from the organic system at Nagele are much worse than the other two systems? Or is it a valid premise that an organic system must always have livestock present and that dairying is the livestock enterprise which has most likelihood of being profitable (Zadoks, 1989)?

Objectives and performance measures

A further question which arises is how performance can be measured and against which criteria. Whether one system is 'better' than another depends on the objectives of (i) the researchers; (ii) the operators of the system; and (iii) society or other interested parties. If one farmer aims to maximize yield per unit area and another aims to minimize the use of purchased inputs, both may be equally successful in achieving their objectives, and thus equally 'good', even if yields per unit area are very different. The 'success [of agricultural systems] cannot be judged ... except in relation to the particular purposes for which they were carried out' (Spedding, 1975).

Yet very few studies, including many of those reported in this book, have attempted to assess farmers' objectives, assuming that the desired objectives (and hence the appropriate performance measures) are common to both systems being compared. This is a valid approach if the ability of different systems to meet certain social or policy objectives (e.g. increasing agricultural output or reducing energy consumption) is being tested. However, if these objectives are not also those of the operators of the system(s) being analysed, then the interpretation of the results must be treated with caution. In particular, the results cannot be interpreted as an indication of the potential of a system to meet such social or policy objectives under these conditions.

Time period

Another issue is the time period over which a comparison is conducted. Many studies only cover one year, which means that the results may be strongly influenced by particular climatic, economic or other exogenous factors prevailing in that year. For example, Murphy's (1992) study of organic farming in Britain (discussed in Chapter 5) covers the 1989 crop year, which in some parts of the country was a drought year, as a result of which yields were low, particularly on organic farms that were not able to stimulate growth with fertilizers in the early spring before the drought conditions set in. Would a similar comparison hold in other years?

System- and non-system-determined factors

A number of different approaches have been used in making comparisons between organic and conventional systems, including:

- controlled experimental trials comparing organic and conventional plots varying in size from small replicated plots (e.g. Besson *et al.*, 1978) to adjacent farm units (e.g. Zadoks, 1989);
- individual organic farms compared with individual (paired) conventional farms similar in terms of size, location and type (e.g. Klepper *et al.*, 1977; Mühlebach and Näf, 1990);
- individual organic farms compared with regional averages differentiated by type and size of farm (e.g. Vine and Bateman, 1981; Lampkin, 1993);
- individual organic farms compared with undifferentiated regional or national averages (e.g. Budig *et al.*, 1988; Kaffka and Koepf, 1989);
- averages for groups of organic farms compared with averages for conventional farms differentiated by region, type, size and other characteristics (e.g. Böckenhoff *et al.*, 1986; Murphy, 1992);

- averages for groups of organic farms compared with undifferentiated national or regional averages (e.g. BMELF, 1983; Murphy, 1992);
- averages for undifferentiated groups of organic farms compared with averages for conventional farms differentiated by type and size (e.g. BMELF, 1984–1993).

The question of which approach is appropriate depends partly on which factors are determined by the organic or conventional management of the system. Those which are not should be excluded from the comparison as far as possible, or, if this is not possible, they should be considered explicitly in any interpretation of the results obtained. Non-system-determined factors can include the following, but the issues are not always clear-cut (Dabbert, 1990b):

- location (soil type, climate, topography, altitude);
- farm characteristics (size, layout, land tenure, buildings);
- financial circumstances (capital availability, interest payments, rents, cash flows);
- production (breeds, varieties);
- marketing (processing, distribution, distance to markets, etc.);
- managerial influences (experience, ability, education, motivation).

To what extent, for example, is the enterprise structure of the farm due to organic management? A diverse arable rotation in a predominantly monocultural arable region may well be, but a specialist organic dairy farm in an arable region could be due to other, historical factors and may be very similar to other (isolated) conventional dairy farms in that region. It has been argued that the comparison of an organic livestock farm with a conventional arable farm in a particular region is justifiable if the main alternative to organic livestock farming in that region is arable farming (Hermanowski, 1989). However, the organic farmer's main alternative is conventional livestock production, at least in the short run, because of the fixed asset structure of the farm business. In addition, the priorities of the farmer in terms of allocating scarce resources to grassland or cereal crops will be different in each case, so that comparisons between cereal crops on the (organic) livestock farm and the (conventional) arable farm may say more about farm type than system of management.

A similar issue arises in relation to farm size. To what extent are enterprise structure and intensity, as well as financial performance, determined more by differences in farm size than by the nature of the management system? This issue is clearly illustrated by the study of Murphy (1992) discussed in Chapter 5, where the failure to differentiate farms on the basis of farm or business size resulted in the inclusion of a high proportion of very small organic farms, whereas farms of similar size were excluded from the conventional comparison groups, with the result that the reported relative performance of the organic farms appeared to be very poor.

The use of groups of paired or closely matched groups of organic and conventional farms (Klepper *et al.*, 1977; Mühlebach and Näf, 1990) allows for many factors which are not due to the management system to be averaged out. But the selection of the group can itself lead to bias where the enterprise structure or farm type differences *are* due to differences in the management system. This danger is even greater when individual farms are considered in isolation or paired with single conventional farms. The DOK (Dynamisch Organisch Konventionell) experimental comparison of organic, bio-dynamic and conventional (integrated) farming in Switzerland (Besson *et al.*, 1978) succeeded in holding a number of variables (e.g. variety, soil type) constant so that the influences of manuring and pest control practices could be evaluated, but in order for comparisons to be obtained, the rotations were also identical. Is this valid?

Attempts have been made to pair for management ability as well (Jochimsen, 1982), a notoriously difficult factor to assess objectively, but one which may be critical to the end result, especially if the argument that organic farmers need to be better managers to make their systems work is accepted (Lockeretz, 1989). Of the studies reported in this book, Wynen (Chapter 11) asked all farmers to rate their management ability and includes this issue in her assessment.

Even comparisons of organic and conventional crops under the same management can be suspect if the cultivars are different. Spring wheats are typically lower yielding than winter wheats, but are spring wheats a characteristic of an organic management system? Bread-making (milling) wheats are typically lower yielding than feed wheats. Is the use of the former part of the organic system, or determined primarily by marketing considerations? Or do the marketing considerations themselves form part of the characteristics of the organic system? The same questions can be applied to livestock breeds, for example milk yields from organic Jersey cattle compared with yields from conventional Friesian cattle (Murphy, 1992).

Similar questions can be asked when looking at fixed costs such as labour. Are higher labour costs a characteristic of organic systems, or a consequence of restricted capital availability which prevents adequate mechanization? Is restricted capital availability a consequence of newcomers to farming putting most of their capital into land, or of poor financial returns to organic management? Which is cause and which is effect?

Finally, it is also argued that organic systems are relatively undeveloped when compared with conventional systems which have benefited from considerable research funding (Woodward, 1989). To what extent is the comparison of the *potential* of an immature system with the *achievements* of a mature one (like comparing the performance of a child against an adult) possible or appropriate?

Generalizability of results

Comparisons of individual organic farms with averaged group data for conventional farms can be helpful in terms of illustrating potential, but they do not provide a basis for generalization, because the individual farm cannot be considered in any way representative of the population of organic farmers. Group averages for organic farms compared with group averages for conventional farms may in principle be more generalizable, but issues of representativeness remain a problem and such comparisons often ignore the variations which occur *within* the groups, which may be much more significant than any differences *between* the groups.

Given a sufficiently large sample which is representative of the total population, and factors which are not system-determined such as management, farm size may be assumed to be normally distributed if the sample is randomly selected from a known population and therefore statistical comparisons of averages become to some extent meaningful. While it has been possible in industrialized countries to identify the total farming population, this is only now being achieved for organic farmers so that the samples of organic farmers selected cannot usually be assumed to be representative of the population. The problem of representativeness has been exacerbated in the past by the unwillingness of organic farmers to cooperate due to a mistrust of conventional research institutions. In addition, the sample size of the organic group is often so small that it is not possible to ensure that the group has the same locational, type, size or other characteristics as the conventional group (e.g. BMELF, 1983–1993; Murphy, 1992). For these reasons, many of the studies reported in this book do not contain statistical analyses of the results.

Criteria for conducting comparisons of farming systems

If comparisons are to be carried out, particularly on a controlled, experimental basis, then the following criteria are important (Dlouhy and Nilsson, 1983):

- Comparisons should be conducted on a sufficiently large scale, preferably on a farm scale, with the appropriate human input (recognizing management and social as well as labour influences), so that the experiment is not divorced from the environmental and management 'context', including farmers objectives, which is critical to the functioning of the farming system.

- Comparisons should only take place after the organic farm has been through a conversion process, unless the conversion period itself is under study, as there are major differences in system structure and performance during the conversion phase (see Chapters 13–17).
- Comparisons should be conducted over several years, preferably encompassing one or two rotations, in order to remove the impact of variability in climatic and other exogenous variables.
- Researchers involved in such comparisons should have a thorough introduction into the organic approach to agriculture and be convinced of the value of such a trial, in order to ensure that the interpretation of findings is based on a sound, informed and unprejudiced approach.
- Comparisons should be dynamic (i.e. flexible), because static comparisons will bear little or no relation to reality after a period of time; accompanying research tools (case studies, single component or subsystem trials, on-farm monitoring, etc.) can be used to develop and evaluate all innovations before their introduction into the comparison.
- Enough farms with suitably matched characteristics are required to give such trials more than local significance.

Alternatives to Conventional versus Organic Comparisons

The use of scarce resources for comparisons of organic and conventional farming systems which produce results of limited validity has justifiably been criticized by those researchers who are more interested in the potential of well-managed, successful, organic farming and the causes of variability between organic farms (Vogtmann, 1990). A number of different approaches have been adopted to overcome this problem.

Experimental units

Dedicated organic experimental units have been established to conduct developmental work on organic systems, particularly in Europe (Besson, 1990). In the United States and Canada, the preference has been for on-farm trials (e.g. Patriquin, 1989). The common characteristic of these approaches is that they involve comparisons between alternative options *within* the organic system, not with conventional systems. Much development-type research in conventional agriculture is also concerned with differences within the system, rather than between systems; it has seldom been argued that research in conventional agriculture needs to involve comparisons with organic farming in order to be scientifically valid.

Surveys

An alternative to the experimental approach is to analyse data from groups of organic farms to determine factors which might lead to yield variability within the group. Factors such as soil quality and the length of time since conversion have been examined on this basis using regression and contingency analysis (see Chapters 6 and 16). The impact of time period since conversion can be analysed on a cross-sectional (Chapter 16) or a time-series basis (Chapter 25). One advantage of the cross-sectional approach is that factors such as weather and price variability can be excluded.

Case studies

A further approach is the use of case studies of 'well-functioning' organic farms to determine what are the factors which contribute to their 'success' and to what extent these factors may be replicated or transferred to other farms (Granstedt, 1990; see also Chapters 13 and 14, this volume). This is very much in the spirit of the farmer, not the researcher, as innovator espoused by Chambers *et al.* (1989). The case-study approach can also be used as a diagnostic tool to analyse why problems occur and what potential exists for improvement, and can provide longitudinal data to illuminate change processes (Maxwell, 1986). The major advantage of the case-study approach is that it allows a detailed understanding of a particular situation to be obtained, so that interrelationships and causality, the why and how, can be better understood in the context of the *individual* farm (Maxwell, 1986). This is, however, at the expense of comparability, statistical analysis and generalizability.

In principle, the case-study approach should be multi-disciplinary so that the interactions between soil management and nutrient cycling, pest management, weed control, rotation design, manure management and animal husbandry, as well as farm organization and financial, management, sociological and environmental factors, can be fully appreciated (Maxwell, 1986; Lampkin, 1988).

> Case studies provide insights into how the real world works. They help formulate and test hypotheses, but cannot substitute for other forms of scientific research. In complicated areas of human endeavour, however, case studies can provide useful observations that go beyond the range of controlled experiments. They can indicate promising directions for further research and help demonstrate how many different factors – economics, biology, policy and tradition – interact.
>
> (NRC, 1989)

Modelling

A final approach is that of modelling organic farming systems. This approach allows for non-system-determined factors (exogenous variables) such as weather, management, etc. to be excluded, so that the effect of endogenous variables can be determined, and for alternative scenarios to be investigated. The weakness of the approach is that modelling always involves a considerable abstraction from and simplification of reality and is very dependent on the underlying assumptions (Olson *et al.*, 1982; Jaep *et al.*, 1985; Hermanowski, 1989; Dabbert, 1990b; see also Chapters 17–22, this volume). A range of techniques can be applied, from standard budgeting (e.g. Jaep *et al.*, 1985) and linear programming based optimization techniques to more sophisticated techniques of systems simulation (e.g. Bossel *et al.*, 1988) and expert systems (e.g. Jones and Street, 1990). The latter have not yet been used for researching organic farming systems to any great extent because of the lack of data of suitable quality.

Conclusions

Comparisons between farming systems form the basis for most of the studies reported in this book. These studies all, in one way or another, come up against the methodological problems described in this chapter. Yet the similarities between many of the findings from the different studies, conducted in different countries and in different contexts, do allow some general conclusions to be drawn. We have attempted to highlight these in the discussion chapters at the end of each section in this book.

Despite the increasing availability of information on the financial and physical performance of organic farming and the conversion process, there are still some areas where research is almost non-existent and urgently needed. In the European context, further research is needed to continue monitoring aspects of the relative performance of different farming systems in order to evaluate the impacts of agricultural policy reform, the support for organic farming under the new agri-environment programme (see Chapter 26), and the impact of the GATT settlement.

It should also be possible, as the quantity and quality of the available data improves, to develop better models of organic systems and of the conversion process. Some of the existing attempts to develop models for conversion planning are discussed in Chapter 17. Further work is clearly needed on the impact of widespread conversion to organic farming, including changes in physical output and land use, as well as the related changes in consumption patterns, prices, employment and incomes. The studies reported in Chapters 18–22 are at best preliminary studies in this respect.

Research is also needed on the much neglected issue of risk and uncertainty in organic farming systems, aspects of which are considered in Lampkin (1993). Perceptions that organic farming is more risky may act as a significant barrier to its adoption by many farmers, although actual risk levels have seldom been assessed. Related human, sociological and institutional factors affecting the adoption of organic farming (Padel, 1994), including the role of appropriate information and extension services as well as institutional 'credibility', rather than direct financial support, would also merit further research.

Finally, the description and quantification of the non-monetized environmental and social costs and benefits of organic farming, i.e. the economic as opposed to just the financial issues, require considerable additional research, as discussed in the next chapter.

References

Besson, J.-M. (ed.) (1990) *Biological Farming in Europe – An Expert Consultation.* REUR Technical Series, 12. Food and Agriculture Organization, Rome.

Besson, J.-M., Vogtmann, H., Lehmann, V. and Augstburger, F. (1978) DOK: Versuchsplan und erste Ergebnisse eines Projektes zum Vergleich von drei verschiedenen Anbaumethoden. *Schweizerische landwirtschaftliche Forschung* 17(3/4), 191–209.

BMELF (1983–1993) *Agrarberichte der Bundesregierung.* Bundesmimisterium für Ernährung, Landwirtschaft und Forsten, Bonn.

Böckenhoff, E., Hamm, U. and Umhau, M. (1986) Analyse der Betriebs- und Produktionsstrukturen sowie der Naturalerträge im alternativen Landbau. *Berichte uber Landwirtschaft* 64(1), 1–39.

Bossel, H., Kramer, P., Krieger, H., Schuller, U., Trost, N. and Zerger, U. (1988) *Dokumentation zum Fruchtfolgeberatungssystem FELDSIM.* AG Resourcen- und Systemforschung RSF 88-1, Gesamthochschule, Kassel.

Budig, M., Völkel, G. and Redelberger, H. (1988) *Alternativer Landbau: Erhe-*

bungen und Untersuchungen auf Flächen alternativ wirtschaftender Betriebe in Hessen, 1981–1986. IfB – Information für Beratung und Verwaltung, 56/88. Hessiches Landesamt für Ernährung, Landwirtschaft und Landentwicklung, Kassel.

Buttel, F.H., Gillespie, G.W., Janke, R., Caldwell, B. and Sarrantonio, M. (1985) Reduced-input agricultural systems: a critique. *The Rural Sociologist* 6(5), 350–370.

CAST (1980) *Organic and Conventional Farming Compared.* Report, 84. Council for Agricultural Science and Technology, Ames, Iowa.

Chambers, R., Pacey, A. and Thrupp, L.A. (eds) (1989) *Farmer First: Farmer Innovation and Agricultural Research.* Intermediate Technology Publications, London.

Conway, G.R. (1985) Agroecosystem analysis. *Agricultural Administration* 20, 31–55.

Dabbert, S. (1990a) Der Begriff des Betriebsorganismus. *Lebendige Erde* 90(5), 333–337.

Dabbert, S. (1990b) *Zur optimalen Organisation alternativer landwirtschaftlicher Betriebe. Untersucht am Beispiel organisch-biologischer Haupterwerbsbetriebe in Baden-Württemberg.* Agrarwirtschaft Sonderheft, 124. Verlag Alfred Strothe, Frankfurt am Main.

Dahlberg, K.A. (1988) Toward a typology for evaluating conventional and alternative agricultural systems and research strategies. In: Allen, P. and Dusen, D.v. (eds) *Global Perspectives on Agroecology and Sustainable Agricultural Systems.* Proceedings of the 6th IFOAM International Scientific Conference, Santa Cruz, 1986, 1. Agroecology Program, University of California, Santa Cruz, pp. 103–112.

Dalton, G.E. (ed.) (1975) *Study of Agricultural Systems.* Applied Science Publishers, London.

Dent, J.B. (1990) Optimising the mixture of enterprises in a farming systems. In: Jones, J.G.W. and Street, P.R. (eds) *Systems Theory Applied to Agriculture and the Food Chain.* Elsevier Applied Science, London, pp. 113–130.

Dent, J.B. and Anderson, J.R. (eds) (1971) *Systems Analysis in Agricultural Management.* John Wiley Australasia, Sydney.

Dlouhy, J. and Nilsson, G. (eds) (1983) *Comparisons between Farming Systems.* Report 124. Swedish University of Agricultural Sciences, Uppsala.

Gilbert, E.H., Norman, D.W. and Winch, F.E. (1980) *Farming Systems Research: A Critical Appraisal.* MSU Rural Development Paper. Department of Agricultural Economics, Michigan State University, East Lansing.

Granstedt, A. (1990) *Fallstudier av kväveförsörjning i alternativ odling.* Alternativ Odling, 4. Research Committee for Alternative Agriculture, Swedish University of Agricultural Sciences, Uppsala.

Hermanowski, R. (1989) *Vergleich alternativer und konventioneller landwirtschaftlicher Betriebe in Hessen.* Regionalpolitik und Umweltschutz im ländlichen Raum. Wissenschaftlicher Fachverlag Dr Fleck, Niederkleen, Germany.

Hildebrand, P.E. (ed.) (1986) *Perspectives on Farming Systems Research and Extension.* Lynne Rienner, Boulder, Colorado.

Jaep, A., Scharf, A., Budig, M., Voelkl, G. and Weiss, J. (1985) *Konventioneller und alternativer Landbau – ein betriebswirtschaftlicher Vergleich.* IfB

Landwirtschaftliche Fachinformation, 143/85. Hessisches Landesamt für Ernährung, Landwirtschaft und Landentwicklung, Kassel.
Jochimsen, H. (1982) *Alternativer Landbau im Vergleich: bessere Produkte, höhere Preise, mehr Gewinn?* Betriebswirtschaftliche Mitteilungen, 325, Landwirtschaftskammer Schleswig-Holstein, Kiel.
Jones, J.G.W. and Street, P.R. (eds) (1990) *Systems Theory Applied to Agriculture and the Food Chain.* Elsevier Applied Science, London.
Kaffka, S. and Koepf, H.H. (1989) A case study on the nutrient regime in sustainable farming. *Biological Agriculture and Horticulture* 6, 89–106.
Klepper, R., Lockeretz, W., Commoner, B., Gertler, M., Fast, S., O'Leary, D. and Blobaum, R. (1977) Economic performance and energy intensiveness on organic and conventional farms in the Corn Belt: a preliminary comparison. *American Journal of Agricultural Economics* (February), 1–9.
Koepf, H.H. (1983) Comparisons between farming systems. In: Dlouhy, J. and Nilsson, G. (eds) *Comparisons between Farming Systems.* Report 124. Swedish University of Agricultural Sciences, Uppsala, pp. 67–72.
Koepf, H.H., Pettersson, B.D. and Schaumann, W. (1976) *Bio-dynamic Agriculture – An Introduction.* Anthroposophic Press, Spring Valley, New York.
Lampkin, N.H. (1988) A research concept for investigating organic farming systems: case studies. In: Allen, P. and Dusen, D.v. (eds) *Global Perspectives on Agroecology and Sustainable Agricultural Systems.* Proceedings of the 6th IFOAM International Scientific Conference, Santa Cruz, 1986, 1. Agroecology Program, University of California, Santa Cruz, pp. 121–134.
Lampkin, N.H. (1993) The economic implications of conversion from conventional to organic farming systems. PhD thesis, Department of Agricultural Sciences, University of Wales, Aberystwyth.
Lloyd, D.H. (1970) *The Development of Farm Business Analysis and Planning in Britain: A Methodological Review and Appraisal.* Department of Agriculture, University of Reading, Reading.
Lockeretz, W. (1989) Problems in evaluating the economics of ecological agriculture. *Agriculture, Ecosystems and Environment* 27, 67–75.
Lockeretz, W. (1990) Major issues confronting sustainable agriculture. In: Francis, C.A., Flora, C.B. and King, L.D. (eds) *Sustainable Agriculture in Temperate Zones.* John Wiley, New York, pp. 423–438.
Lockeretz, W. (1991) Multi-disciplinary research and sustainable agriculture. *Biological Agriculture and Horticulture* 8(2), 101–122.
Lockeretz, W. and Anderson, M.D. (1993) *Agricultural Research Alternatives,* University of Nebraska Press, London.
MacRae, R.J., Hill, S.B., Henning, J. and Mehuys, G.R. (1989) Agricultural science and sustainable agriculture: a review of the existing scientific barriers to sustainable food production and potential solutions. *Biological Agriculture and Horticulture* 6, 173–219.
Maxwell, S. (1986) The role of case studies in farm systems research. *Agricultural Administration* 21, 147–180.
Mühlebach, J. and Näf, E. (1990) *Die Wettbewerbsfähigkeit des biologischen Landbaus.* Schriftenreihe der Eidg. Forschungsanstalt für Betriebswirtschaft und Landtechnik, 33. Tänikon, Switzerland.
Murphy, M.C. (1992) *Organic Farming as a Business in Great Britain.* Agricultural

Economics Unit, University of Cambridge, Cambridge.

NRC (1989) *Alternative Agriculture.* National Research Council, Board on Agriculture, Committee on the Role of Alternative Farming Methods in Modern Production Agriculture, National Academy Press, Washington, DC.

Olson, K.D., Langley, J. and Heady, E.O. (1982) Widespread adoption of organic farming practices: estimated impacts on U.S. agriculture. *Journal of Soil and Water Conservation* 37, 41–45.

Padel, S. (1994) Adoption of Organic Farming as an Example of the Diffusion of an Innovation. Discussion Paper 94/1. Centre for Organic Husbandry and Agroecology, University of Wales, Aberystwyth.

Patriquin, D. (1989) Collaborative on-farm research. Paper presented at the 'Research in Sustainable Agriculture' conference, MacDonald College, McGill University, Quebec.

Schaumann, W. (1990) Der Hinweis auf Aereboe. *Lebendige Erde* 90(5), 337–338.

Scofield, A.M. (1986) Organic farming – the origin of the name (Editorial). *Biological Agriculture and Horticulture* 4(1), 1–5.

Spedding, C.R.W. (1975) *The Biology of Agricultural Systems.* Academic Press, London.

Spedding, C.R.W. (1988) *An Introduction to Agricultural Systems.* Elsevier Applied Science, London.

Vereijken, P. (1991) Methodology of farming systems research and development. *Netherlands Journal of Agricultural Science* 34, 387–393.

Vine, A. and Bateman, D.I. (1981) *Organic Farming Systems in England and Wales: Practice, Performance and Implications.* Department of Agricultural Economics, University College of Wales, Aberystwyth.

Vogtmann, H. (1990) Research structure in biological farming: general view, justification, development. In: Besson, J.-M. (ed.) *Biological Farming in Europe.* REUR Technical Series, 12. Food and Agriculture Organization, Rome, pp. 41–47.

Woodward, L. (1989) Considerations for research in organic agriculture. In: Williamson, C. (ed.) *Workshop on Priorities for Organic R&D.* Report from Workshop, Perth, November 1989. Scottish Agricultural College, Edinburgh, pp. 16–20.

Youngberg, G. (ed.) (1990) *Learning from Each Other: New Models for Sustainable Agriculture Research and Information.* Proceedings of the Symposium, Washington DC, 1990. American Journal of Alternative Agriculture Special Issue No. 5(4). Institute for Alternative Agriculture, Greenbelt, Maryland.

Zadoks, J.C. (ed.) (1989) *Development of Farming Systems – Evaluation of the Five-year Period 1980–1984.* Pudoc, Wageningen.

4 Organic Farming and Society: An Economic Perspective

David I. Bateman

Economics and Finance

Much of this book is concerned with questions of *finance* – is organic farming profitable? – and this question has indeed given rise to much controversy and misunderstanding (Bateman, 1993).

There is, however, a second question. It is at least possible that, even if it were true that organic farming was *financially* unsound, it might still be *economically* justified. Sound finance and sound economics are *not* the same thing. A *financial* assessment is concerned with how organic farming performs from the viewpoint of a profit-seeking business person; an *economic* assessment is concerned with its performance as seen from the viewpoint of society as a whole. There are at least three reasons why organic farming might be justified economically (even if it *were* unprofitable):

- Its impact on the environment may be preferred to that of conventional farming.
- Its impact on rural employment may be greater than conventional farming, giving it an important role in the field of pluri-activity.
- Since its output is likely to be lower than that of conventional farming, the costs to the taxpayer arising from the need to dispose of surpluses may be lower.

These arguments will be assessed in this chapter. Because they relate to how society *ought* to organize its affairs (because, in other words, they are, unlike most scientific effort, *explicitly* normative) anyone who ventures into this field has to set out the assumptions with particular care.

Economics and the Environment

Introduction

Any society has a limited supply of resources – natural resources such as oil or good land, man-made resources such as roads or factories, and man-supplied resources such as intelligence and manual labour. The economic problem is to decide how to use these limited resources to best meet the needs of society – needs for intellectual stimulation and personal development, for leisure, for consumer goods and so on.

The simplest and the most common solution to this problem is to rely on markets – on the price mechanism. This approach is so well established that the onus must be on those who dislike its consequences to make the case against it. It is one purpose of this chapter to do just that.

The first step is to identify what the strengths and weaknesses of the price mechanism are. This section tries to do this by contrasting the stylized views of three straw men: the naïve politician (a statement of the arguments for a completely free market); the green environmentalist (a critique of the market); and the conciliatory economist (a defence of the value of a modified market).

The conclusion will be that the economist's position, although only a partial response to the environmentalist's critique, is much more sophisticated than environmentalists believe or than politicians understand. It offers the environmentalist who wishes to move the world a place to stand – a place, moreover, that has a foundation of over 200 years of sustained intellectual effort. Those who demand a *new* economics (Ekins, 1986), it will be suggested, are overstating the very strong case that really exists for a modified economics.

The politician's view: The naïve case for the market

Allocating resources according to the price mechanism means that any member of society who has needs can express them through his or her willingness to pay. At the same time, members of society who own resources will generally prefer to sell them to those who offer most and so (usually via numerous intermediaries and often in a highly processed form) the marginal unit of each resource will be allocated in a way that takes full account of individual tastes and preferences as indicated through each person's expressed willingness to pay.

Compared with alternative systems of allocating resources (such as centralized decision-making), the market has two major advantages – advantages that its opponents often underestimate.

First, there is an *economic* advantage: the intuitive idea that the market acts as a voting system, giving each consumer the opportunity to express his or her own choices and preferences in determining how resources are to be allocated, suggests that such an allocation may, in some sense, be the most efficient.

Second, there is the *political* advantage that the market is highly democratic: *all* consumers in society can express their preferences directly and on a day-by-day basis through their willingness to pay, and (except in providing a basic framework of law) there is no reliance on governments to mishandle the myriad of decisions that have to be made. People do not have to be cajoled to behave in a socially desirable way: all they have to do is to pursue their own self-interest, a natural enough instinct anyway.

Adam Smith's potent metaphor of the 'invisible hand' (Smith, 1954) so neatly encapsulates these two advantages that the degree of influence that they have had is hardly surprising.

The environmentalist's view: A green critique of the market

The arguments presented in this section represent a critique of the economic, rather than the political, case for the market. The environment, though not the only example of 'market failure', is undoubtedly the leading one. It is not surprising then that it is environmentalists above all who have wished to refute the supposed benefits of the invisible hand.

The ethical issue

A fundamental problem that concerns some environmentalists is the ethical position implicit in market economics. First, the model assumes that trade-offs are always possible – everything has its price. The environmentalist, concerned with conservation of wildlife, or habitats, or the ozone layer or fertile soil, may regard all these as beyond price. Trade-offs should simply not be considered.

Second, the market model not only assumes that all individuals *will* pursue their own self-interest, but requires that they *should* do so: altruism is not a virtue – it may even be a sin. Policies are designed on the assumption that self-interest is not merely to be tolerated but is the norm and, where absent, is to be encouraged.

Third, the model contrasts, on the one hand, 'utility' (to be derived from consumer goods and services) with, on the other, 'disutility' (associated with work). Work is a necessary evil: it must be done if we are to have consumer goods, but its performance requires adequate financial incentives. Work satisfaction is absent from the model.

Fourth, asceticism, widely admired in many cultures, again assumes the quality of a sin rather than a virtue: only if everyone consumes as much as possible will everyone else be able to have a job and so also consume as much as possible.

In all these respects, the market model appears to assume either that the underlying form of behaviour that it presupposes is inherent and therefore not worth discussing, or is ethically neutral. Those who disagree may feel that the market model is not just amoral but immoral, not just lacking in humanity but inhuman.

Preferences and welfare

The market allocates resources according to people's choices. But before we can regard this as a reflection of *social* welfare, we need to be certain that choices reflect preferences; that individual preferences reflect individual welfare; and that an increase in individual welfare increases the welfare of society.

For most ordinary consumer goods, these three conditions are probably met. Typically, expository examples have consumers choosing between such goods as oranges and apples where the case for the market seems entirely plausible. The triviality of the example passes as an unimportant aspect of the pedagogic device.

Unfortunately, as one tries to generalize to more serious examples the argument loses this plausibility. Cases where *choices may not reflect preferences* include many of those where there is an altruistic or ascetic element. If, for altruistic reasons, a farmer chooses to farm organically or a consumer joins a conservation body or buys organic food, then there is a separation between choice and preference. In such a case, choice may be a better indicator of welfare than is preference, but what is not known is whether it is a perfect indicator or how many other consumers identify a gap between preference and welfare but do not express it in their choice.

Cases where *preferences may not reflect individual welfare* are extremely common. One group is associated with lack of knowledge, e.g. how to judge the effect of food quality on health. A less obvious group relates to the difficulties of making judgements over time – how far is our preference for consumption *now* a genuine reflection of welfare or a misjudgement which we may subsequently regret? If, for either of these reasons, choice cannot be regarded as a reliable indicator of preferences, the market model loses much of its attraction.

And cases where there is a *divergence between individual welfare and social welfare* are also common. Examples include:

- Exogeneity of preferences: the idea that the satisfaction of consumer preferences is the only target for an economy needs justification. That justification would be more convincing if it could be shown that preferences were innate and immutable, but common observation suggests that they are not. Advertising and fashion are so influential that (at least in the richer societies of the world) the satisfaction of consumer tastes may be a trivial target (Mishan, 1971). In some fields, custom may be an equally fickle former of tastes, e.g. changes over time in the preference for open landscapes versus hedged fields, or (to take an anti-organic example), the 'passing fad' for organic food (Tregear, 1993).
- 'Snake value': few people as individuals would place a high value on the conservation of snakes, poisonous spiders or viruses, compared at least with 'cuddly' forms of wildlife such as pandas, beautiful ones such as red kites or ones with nationalistic significance such as bald eagles. Can this mean that there is no case for conserving snakes and the like? Are environmentalists wrong when they exploit the public's liking for cuddliness by treating the panda, etc. as 'umbrella' species, 'surrogates for a larger ecological entity worthy of preservation' (Gould, 1993)?
- Independence of individual welfares: in order that we can be sure that an increase in individual welfare always increases social welfare, we have to be certain that each individual's welfare is independent of everyone else's – otherwise, considerations such as emulation and envy become relevant. There is good evidence that it is our relative rather than our absolute position that determines welfare – a conclusion with implications for the whole debate about the desirability of growth (Daly, 1992).
- Anthropocentrism: some environmentalists would argue that an entirely anthropocentric approach is inadequate. Wildlife, for instance, is important not only because it contributes to human welfare but *also* for its 'own sake': animals (and plants?) have rights, too. This is an emotive and difficult area. In the sense that all normal people oppose cruelty to animals (and would be offended if it were suggested that this opposition is really only self-interest), everyone believes in animal rights. Many people might go further: given a choice between a world from which all forms of life were removed, and one from which only human life were removed, they might feel that the second would be in some sense preferable to the first. Contrary arguments could be based on the view that 99% of all life forms that have ever existed are already extinct (Gould, 1993), and, whatever human beings may do, at least some of the remainder will

in any case continue (the Gaia hypothesis (Lovelock, 1979)).

The situations that have been described relate not to choices between apples and oranges, but to ones where most people, whatever their particular views, would at least agree that the issues at stake are important. In none of the situations can individual choices be relied on to represent social welfare.

It appears that the market may produce the right answer to trivial problems but the wrong answer to the ones that really matter.

Future generations

Leading on directly from the previous discussion is a more fundamental issue: when we talk of the choices, preferences and welfare of individuals and society, *which* individuals and society are we talking about – just those currently living, or all those who may live in the future? If 'society' comprises, in some sense, all generations, then the superficially plausible argument that the market is a kind of ideal because it provides everyone with an opportunity to exercise, through purchase, his or her individual 'vote' as to which goods and services should be produced loses its plausibility, and it does so for at least two reasons. First, it is not apparent how unborn generations exercise their vote. Second, to the extent that the present generation must act on behalf of the future, the difficulties of knowledge and judgement about choices and preferences (difficult enough within one's own generation) are magnified manifold. Many, perhaps most, of the leading environmental problems of the day are ones in which these issues relating to time are inescapable.

Income distribution

The market allocates resources according to *willingness to pay* as expressed in actual market behaviour. To conclude that the consequential pattern of resource allocation is desirable is to ignore at least three important considerations:

- If incomes are badly distributed (whatever that may mean) then the pattern of resource allocation will be correspondingly distorted.
- Income distribution determines not only *consumer* behaviour but also *producer* behaviour: thus, problems of environmental degradation in Third World countries may be the direct result of low incomes forcing producers to act in a short-term fashion (less plausibly, it is sometimes argued that European farmers would, with higher incomes, be able to afford to behave in a more environmentally friendly way).

- The assumption that resource allocation depends on income distribution is only half the story since, over the long term, incomes themselves will depend on earlier decisions about the use of resources.

As far as the environment is concerned, the consequences of these arguments are unpredictable. Often, environmental *damage* is associated with the activities of the rich (with their motor cars, refrigerators and washing machines). But at other times it appears equally obvious that environmental *concern* is also a predilection of the rich: if so, an unequal distribution of income will strengthen demand for conservation and environmentally friendly goods – organic food is one such example.

Whichever way it works, it is indisputable that the pattern of resource allocation cannot be separated from that of income distribution. To talk of an 'efficient' pattern of resource allocation is no more than verbal prestidigitation unless we have a means of identifying and achieving the 'efficient' pattern of income distribution on which it must be based.

Pervasiveness of competitive markets

The argument that the market can allocate resources efficiently only stands up if all relevant goods and services are in fact sold through markets, and through competitive markets at that.

A moment's consideration shows that many of the goods and services of concern to environmentalists are *not* sold through markets at all: wildlife, landscape, unpolluted rainwater, the atmosphere and so on. The reason is that it is difficult or impossible for anyone to assert property rights in such goods – hence, they cannot be bought and sold.

For conventional goods (that *are* bought and sold), a different problem arises: if markets are not competitive, they will not reflect the preferences that consumers try to express through the choices that they make.

Overview

Five strong reasons have been presented for doubting the value of markets as a solution to the problems of resource allocation – either because the system is morally repugnant, or because the result is not (or may not be) efficient after all, or because, if it is efficient, it only is so by virtue of the fact that efficiency is defined in a narrow and inappropriate way.

The economist's compromise: A defence of a modified market

Have the sources of market failure, identified in the previous section, completely demolished the case for the market as a means of resource allocation? If so, economists must abandon any explicitly normative position and confine themselves to the more positive aspects of their subject (e.g. estimating the effect of a carbon tax on income distribution or on usage of fuels). But economists have a vested interest in defending their right to make normative statements and hence in defending the market. How successful have they been in establishing a compromise position between the two extremes outlined?

In the 200-odd years since Smith put forward the idea of the invisible hand (doing so with a large number of caveats that are often forgotten), the intellectual effort that economists have put into the study of that idea has been immense. They have demonstrated that if the rates at which each pair of consumers are willing to substitute goods for one another are everywhere equal, there can be no gain from further trade between them. Similar conditions apply to producers, and to trade between producers and consumers. In ideal conditions, the market ensures that all these conditions are met since producers and consumers all adjust their behaviour in response to the same prices. In addition, the profit incentive offers dynamic benefits in terms of technological development, etc. The outcome is that it is established that, *provided specified conditions are met,* the market is, indeed, efficient: no alternative way of allocating resources could improve economic welfare. The objections of the environmentalist have not been ignored by the economic theorists: they are covered by the conditions. The issue that has to be examined, then, is whether or how far the conditions are actually met in the real world.

Pervasiveness of competitive markets

The most significant aspect here is the absence of property rights, and this has been one of the most active areas of economics in recent years. An *externality* exists when there is an unpriced interdependency between two economic agents (e.g. a farmer who shoots rabbits on his own farm confers a non-market benefit on his neighbour, while a householder who builds an extension to his house and so blocks his neighbour's view imposes a non-market disbenefit). If a property right could be established in these benefits and disbenefits, not only could a market solution be reached, but also that solution would be 'optimal'. An important and non-intuitive consequence of this way of thinking is the concept of an *optimal amount of pollution.* Pollution is not inherently bad; it should be produced as long as the marginal benefits from its production outweigh the marginal disbenefits.

Externalities are particularly common in the context of *open access resources* such as the atmosphere or the oceans – resources, that is, where it seems to be in the nature of things that no one can own them. In such contexts, the problem is not that pollution (or fishing) occurs, but that it goes beyond the optimum that would be indicated by the market if there were an appropriate charge.

These examples of 'missing property rights' can be compared with two other possibilities: *private property rights* (often conceived to be the normal case) and *common property rights* (where the rights are owned by a group of people) (Stevenson, 1991).

This problem of missing property rights is well recognized by economists. Can they provide practical solutions? In the past, and, indeed, currently, such problems have generally been handled by physical controls (e.g. planning legislation, emission controls on pollution) – policies whose extension is often advocated by environmentalists. Economists point out, though, that such policies override the market entirely and, as a result, are likely to lead to non-optimal solutions (e.g. the householder who wishes to extend his house may or may not get benefit from doing so which more than outweighs the disbenefit imposed on his neighbour, but there is no way in which planning regulations can decide which). More important, physical controls generally do not give the developer an incentive to tailor his proposal to respond to the benefits and disbenefits associated with it.

The corrective policies advocated by economists, therefore, try to preserve the benefits of the market without its disadvantages. Four proposals will be considered here.

Improved bargaining

Some economists (Coase, 1960) have taken the view that, wherever externalities are large, an incentive will exist for someone to find a way to create property rights where they did not previously exist or to modify them where the old system has become efficient.

Two environmental examples may be given in support of this. First, those concerned with conservation have created bodies (such as, in Britain, the Royal Society for the Protection of Birds) to buy wildlife habitats on their behalf, thereby introducing a common property regime to express the preferences which they had not been able to express individually through the market. Second, historically, the open field system in England (which had been a common property regime) became inefficient and gave way to private land ownership.

Taken to its extreme, however, the argument suggests that no intervention is required at all – we simply rely on the market in all cases. More realistically, governments can take positive steps to encourage better bargaining: by legal changes (such as facilitating the creation of environmental covenants); by financial incentives (such as matching grants to aid land purchase by conservation bodies); or by acting itself as a collective agent to own or lease specified property rights (examples vary from country to country, but in Britain, for instance, there are Management Agreements on Sites of Special Scientific Interest and in Environmentally Sensitive Areas) (Bateman, 1989; Bateman *et al.*, 1991; O'Carroll, 1993).

Taxes and subsidies

Wherever an economic activity gives rise to a benefit or disbenefit not reflected in market prices, the prices can, at least in principle, be adjusted by the imposition of taxes and subsidies to whatever would be the 'proper' level. As an example, taxes on motor fuel could be justified as environmental taxes with lower rates on unleaded fuel representing the lower level of pollution associated with it. Such taxes represent the leading example of the politically acceptable 'polluter pays principle' but there are several problems with them.

First, to impose them with exactness requires that the size of the externalities should be measurable, though in the absence of exact measurement a rough tax or subsidy can be used, as is done in the motor fuel case with considerable success.

Second, and more serious, interdependency between many different products means that devising a satisfactory system of taxes can be extremely difficult. The problems of devising a satisfactory carbon tax provide one example. Possibly even more difficult would be the development of an appropriate system of taxes to represent the externalities associated with different varieties of biocides.

A third problem with this solution relates to non-point externalities. An example would be nitrate pollution. Because of the way it occurs, it is, in practice, virtually impossible to tax the pollution itself. But the alternative, to tax nitrogen, would affect resource allocation by farmers in areas where nitrate pollution is not a problem.

Finally, and perhaps most serious of all, is the effect that taxing externalities may have on income distribution. Although the 'polluter pays principle' has the virtue of apparently combining the objectives of justice and resource allocation, it has to be recognized that justice does not necessarily lead to an acceptable pattern of income distribution. In farming, for precisely this reason, it is not unknown for the principle to be reversed, with compensation being paid for income forgone from *not* polluting. This is only an example of the more general problem that the level at which the tax needs to be imposed in order to secure a given reduction in usage depends on the price elasticity of demand and, wherever this is low, the rate of tax will need to be correspondingly high. The effects on the income of those taxed may be politically unacceptable, rendering the taxation solution impracticable.

Marketable permits

An alternative to taxation is to specify the level of pollution (or fishing, or whatever) that is to be permitted and to issue (or sell) marketable permits. Such techniques are already well known to farmers in the form of milk quotas. The advantage in the environmental area is that they enable the government to decide on the level of pollution and then, through the price that the market establishes for permits, to be sure that those firms that can most cheaply cut their pollution will be the ones to do so. Some of the problems associated with taxes apply here, too.

Cost–benefit analysis

Many decisions affecting the environment are, in the end, made by government diktat. But such decisions could still be made in such a way as to try to emulate what the market *would* achieve if it worked properly. This is the basic principle of cost–benefit analysis. There are two main stages.

First, it is necessary to identify and to quantify in physical terms (together with any risk assessment where there are uncertainties) all the consequences (market or non-market, immediate or long-term) of each alternative decision. These will include effects on habitats and wildlife, on atmosphere and water supplies, on job opportunities, on costs to taxpayers, and so on. These are, indeed, the same items that would go into an environmental impact assessment (which can be thought of as the first stage of cost–benefit analysis).

Second, as far as possible, these consequences should be evaluated in money terms. Methods for such evaluations have been developed and widely tested in recent years (Barde and Pearce, 1991). It is probably fair to say that the techniques have a greater degree of sophistication than most environmentalists are aware of. It is also fair to say that the techniques are, quite deliberately, based on willingness to pay – on the premise, in other words, that individual preferences count. They are therefore subject to the comments about preferences and welfare (see Preferences, welfare and ethical issues, p. 59).

Although this is not the place for a review of the techniques for measuring externalities, there is one important conclusion. The values found by such techniques almost certainly *do* exist – if the techniques have any deficiency it is that they *under*value by omitting considerations that lie beyond mere willingness to pay. Nevertheless, the values that are included, limited as they may be, are often enough to swing a decision. To deride the techniques, as some do, is to throw away a tool that can make a critical contribution to environmental conservation – critical both because of its quantitative importance and because of its ability to be understood by politicians and incorporated into their calculations in a way that qualitative arguments are not. This argument is strengthened by current attempts to incorporate such techniques into national accounting frameworks.

The conclusion to this section, then, is that the problem of property rights is one that economists have recognized and tackled, and where they are increasingly developing relevant techniques that go at least some of the way to recognizing and to correcting for the weaknesses that the environmentalist sees in the naïve market approach.

Income distribution

Three ways of answering the criticisms relating to income distribution have been proposed by economists. First, there is the suggestion that income distribution be treated separately from resource allocation: given that some appropriate branch of government can ensure that incomes are 'properly' distributed, the economist can confine his attention to resource allocation *alone*. A variant on this is to say that the economist can recommend all those changes in resource allocation that make some people better off and no one worse off ('Pareto-optimal changes'). Unfortunately, neither of these will do: recommendations about resource allocation invariably affect income distribution, too. The two issues are inseparable.

A second approach is to recommend changes in resource allocation whenever the gains to the gainers exceed the losses to the losers ('*potentially* Pareto-optimal changes'). What is unattractive about this approach is that it can (and probably often does) lead to the recommendation of policies that benefit the rich and harm the poor. A possible justification for it is that if it were applied many times then, in the long run, each individual could expect to benefit. No empirical evidence for such an assertion is known, however, and intuition suggests the contrary: that, in the environmental field at least, many developments continually harm the *same* group, those on low incomes, without cars, etc. This intuition is strengthened beyond recognition when the problem of future generations is taken into account. Although very widely used, therefore, this approach also appears to be without intellectual foundation. The best that can be salvaged from it would be a rule along the following lines: a change can be recommended whenever it is shown (i) that it is potentially Pareto-optimal; and (ii) that the income effects are identified as negligible or beneficial. A change that is potentially Pareto-optimal but has harmful income effects remains undecided.

A third approach, far more satisfactory and relatively new, is to take income effects into account quite explicitly (Blackorby and Donaldson, 1990).

Future generations

The argument that economics necessarily ignores completely the needs of future generations (since those not yet born cannot have a vote) is a misunderstanding. If the market had no concern for the future, it would *never* justify the planting of trees, the fattening of sheep for market or the storage of wine to maturity. Clearly, the future does get a vote of some kind.

How? The owner of a resource (an oil well, good soil or whatever) can profit from it in two ways. He can *exploit* it and benefit now; or he can *preserve* it and profit later. The return from *exploitation* is determined by the current rate of interest (since the owner would either invest the proceeds, or, if he consumes them, would otherwise have had to borrow in order to do so). The return from *conservation* is determined by three things: the biological rate of change of the product (rate of growth of the trees or the sheep, or improvement in quality of the wine); the expected change in price of the product (possibly caused by future shortages if some resource owners are exploiting now); and the expected change in costs of exploitation (affected either by technological changes, or by the early exploitation of the more accessible supplies). The sum of these three items determines the net benefit from conservation.

The resource will be exploited *now* only up to the point where the interest rate exceeds the sum of the net marginal benefits from conservation. This is the mechanism through which unborn consumers exercise their vote in a market system – the market conservation rule. Is it adequate, and, if not, what can be done about it?

Various objections to the market conservation rule have been discussed. Only one will be referred to here – the role of interest rates (Price, 1993). The effect of the market conservation rule is that benefits or costs that occur in the future are *discounted* at the market rate of interest: the weight given to future consumption is lower than that given to current consumption. Even if we make such a judgement (perhaps mistakenly) for ourselves, what can justify us in doing so between generations? Few other economists have gone as far as Price (1990) – 'Changes in value over time are subject to so many complex and contrary forces that to attempt to represent them by a single, simple negative exponential must be considered the work of a madman' – but, equally, few would defend the market's *unmodified* use of discounting.

Proposals that have been discussed include the use of a lower 'social' discount rate for some (or all) projects, the use of different rates for different projects, and the idea that the conventional market conservation rule should be complemented by some other criterion such as sustainable development (or, in some contexts, safe minimum standards).

'Sustainable development' has become, like the 'invisible hand', a phrase for all seasons. Taking its most widely used definition – 'development that meets the needs of the present generation without compromising the ability of future generations to meet their own needs' (World Commission on Environment and Development, 1987) – three comments are worth making:

- The economic problem, by definition, is concerned with making choices between competing ends. The definition of sustainable development quoted here does not tackle this problem but simply spirits it away: in the world of sustainable development, hard choices, apparently, do not have to be made.
- Sustainable development implies that we should not undertake developments that benefit one generation at the expense of another, i.e. *between generations* we should adopt the criterion of actual Pareto-optimality even though this is almost never proposed as a rule for adoption *within* any generation. The contradiction is not explained.
- Attempts to defend such an apparently attractive idea as sustainable development have been made by modifying the definition, but one commentary on such attempts concludes: 'Defensible sustainability is infeasible, but feasible sustainability is indefensible' (Price, personal communication).

What then are the conclusions on the economist's attempts to suggest rules for allocating resources over time? First, the market, even unmodified, does take rather more account of the future than the environmentalist's critique assumed. Second, most economists recognize the need, nevertheless, to modify the market. Third, they are not agreed on how it should be done.

For environmentalists there is a choice: whether to contribute to an intellectually stimulating debate as to how the market can be made to work better in this area; or whether to dismiss the market as a philistine interference with areas of such importance that they transcend quantification.

Preferences, welfare and ethical issues

The environmentalist critique of the market that was made under these headings is one that is seldom discussed explicitly among economists beyond the statement, usually taken as self-evidently plausible, that 'individual preferences count'. This economist, at least, accepts most of the earlier critique and offers only two defences – both, it may be thought, poor ones.

First, the difference between the economist and the environmentalist is one of degree rather than of principle, for the environmentalist position itself is not so far removed from market considerations as its exponents themselves believe. It is not that the economist says that trade-offs are always possible and the environmentalist says never; rather, the economist says *generally,* and the environmentalist says *seldom.* The economist emphasizes that we *do* (some of us) buy Rembrandts and (most of us) Beethoven CDs and he believes that, in the environmental arena as in the arts, at least some areas of trade-off can exist. He notes, for instance, that environmental scientists are sometimes heard to scorn the market mechanism on the grounds that, for example, *every* Site of Special Scientific Interest (SSSI) is unique and beyond value and ought to be inviolable. But the same scientists, faced with the question, 'which of several SSSIs would you be least unwilling to lose assuming that you must lose one?' do not find it impossible to answer: concepts of choice and trade-offs are relevant after all.

Second, can you, critic, offer anything better?

Economics and the environment: Some conclusions

Few economists, then, would defend the *uncorrected* market. Although the corrections that can be accommodated *within* an economic framework fall far short of meeting the environmental critique, they are both subtle and substantial.

Apart from the challenge just issued ('Don't throw out the economic model unless you're sure you've got a better one'), there is another reason why environmentalists should look at economics with sympathy rather than disdain. It is this. Those self-styled hard-nosed people who deride environmental concerns on the grounds that decisions must reflect economic realities miss the point rather neatly: economic realities *do* take full account of the environment. Economics, defined as being concerned with the allocation of scarce resources between competing ends, could hardly do otherwise. Soft-hearted environmentalists who see economics as an unsympathetic discipline are equally missing the point: their target should be those who attach undue importance to *financial* considerations. The market, duly modified, provides a framework for articulating the environmentalist's concerns and should therefore be seen as a friend rather than an enemy. Rather than a negative and unhelpful rejection of the market, the environmentalist can meet the hard-nosed men on their own ground, refute their simplistic conclusions and replace them with arguments that can be based on a proper understanding of both the strengths and the weaknesses of the market.

Finally, there were *two* reasons for the pervasiveness of markets – the political and the economic. The criticisms that have been made relate only to the economic case – the political one remains and will ensure that, whatever their economic efficiency or inefficiency, markets will continue to be the normal way of allocating resources. For this reason, if for no other, those who wish to influence resource allocation can do so best by arguing from *within* a market framework.

Organic Farming and the Modified Market

The previous section has argued that the market is not an infallible guide to the way that resources should be allocated, and, specifically, that where the environment is concerned, there is every reason to suppose that it is a *bad* guide. Traditional economic thinking, apparently to the surprise of many environmentalists (and organic enthusiasts), *is on their side*. It provides a strong and carefully constructed platform from which to argue that, whether or not organic farming is profitable, it *may* offer advantages that would justify government support – and such support might then make it profitable.

This section asks what kind of a research programme would be necessary to establish whether such a case could be made, and examines what research has already been done. Reference will mainly be to work undertaken in Britain though comparable needs exist in other countries, especially in the Third World.

A research programme

To develop a research programme to apply the idea of the modified market, there are five requirements:

- to identify the external (non-market) benefits that might be associated with organic farming;
- to quantify those benefits in physical terms;
- to value the benefits in economic terms as far as may be possible and to suggest a preferred outcome;

- to identify and choose between the policy options for achieving the preferred outcome;
- to state as clearly and forcefully as possible the limitations of economic analysis of this kind.

Identification and quantification

This section discusses the first two of these requirements – necessary preconditions before the *economic* analysis can begin. External benefits usually associated with organic farming include the following:

- The environment in its capacity as a supplier of resources (raw materials for subsequent processing). In this context, organic farming is said to have at least two advantages: better capacity for soil conservation, and lower usage of fossil fuels. Whether such benefits really exist, and if so how large they are, is likely to vary by region, by farm type, by farm size, by farm enterprise (including new ones such as energy coppicing that may be just as suitable on non-organic farms as on organic ones); furthermore, there is likely to be variation between *all* types of farming system, not just between the 'organic' and 'conventional' stereotypes. Simple generalizations will not do. Information on soil aspects is now becoming available but is still incomplete (e.g. Arden-Clarke and Hodges, 1987, 1988), and in other areas the situation is even worse – e.g. the scanty and very dated information to suggest that organic farming has advantages in terms of lower usage of fossil fuels (Vine and Bateman, 1981). In this area, then, research by environmental scientists is necessary before the economist can begin to contribute.
- The environment in its capacity as a direct source of services (landscape and/or wildlife). It is sometimes claimed that organic farming offers advantages in terms of landscape (Redman, 1992). More often, wildlife benefits are claimed, but these do not appear to be fully or carefully articulated (e.g. NCC, 1990), although there is an increasing amount of recent research which supports some of these claims – e.g. studies reviewed in Lampkin (1990), Greenpeace (1992) and Redman (1992) or new empirical studies, e.g. Wilson (1993). There are likely to be the same kind of variations as those listed in the previous paragraph. Until more conclusive results are available from environmental scientists, the possible contribution of the economist is again limited.
- The environment in its capacity as an assimilator of waste. It seems plausible to suppose that organic farming offers benefits here, too, in terms of reduced waste disposal or pollution from nitrate leaching,

for example (EFRC, 1992a,b). Again, in the absence of sufficient raw physical data, there is little that the economist can contribute.

- Job opportunities. Even in a period of widespread unemployment such as the time of writing, governments in many countries spend resources creating jobs in particular areas, including rural ones. If it is true, as is often claimed, that organic farming offers more jobs (per hectare, or per unit of food) than conventional farming, then this constitutes an external benefit. Some evidence is reviewed by Lampkin (1993), but again it is fairly scanty.
- Agricultural policy. Intervention in agricultural policy is the norm, not the exception, and any case for supporting organic farming needs to be judged against this background. Objectives of agricultural policy have included increasing the level of food output (either to achieve a degree of *self-sufficiency* or to reduce *balance of payments* costs), and minimizing *costs to taxpayers.*

  As far as output is concerned, it was always doubtful how far agricultural policy really contributed either to genuine self-sufficiency (since conventional output is heavily dependent on imported inputs; Sturgess, 1992), or even the balance of payments. In both respects, organic farming has possible advantages which could be quantified. But in any case, the emphasis has now shifted towards limitation rather than expansion because of the high cost of surplus disposal to the taxpayer. Superficially, the lower output generally associated with organic farming seems to be in line with current ideas as to the magnitude of output reduction required – not, as some critics have suggested, an indicator of likely mass starvation (see Chapter 20). Given the objective of output reduction, it is reasonable to ask how organic farming compares with alternatives such as set-aside.

The conclusion of this section is that a major research effort is needed by environmental scientists before economists can begin to make reliable assessments of the case for organic farming or of the policies that might be most appropriate.

Valuation

A collation of what is currently known about some of the issues relating to choices, preferences and welfare in this field (Cuttle, 1989) confirms the need for the whole area of attitudes and preferences of conservationists and voters to be seriously addressed by sociologists and social psychologists.

As far as economists are concerned, the general area of the valuation of externalities has become a major research field (see, for instance, Hanley and Splash, 1993). To evaluate the external benefits of organic farming, an assessment would need to be made of: the current use value of the system (i.e. the financial costs and benefits); the indirect values (e.g. ecological benefits); the option values (e.g. the value which might be placed on having access to some of the techniques of organic farming at some point in the future, even if they are not necessarily justified at present); and an existence value (which may be present even if the goods are not used) (Lampkin, 1993). The available valuation techniques depend on willingness to pay (or to accept compensation), which may be assessed:

- *directly* using contingent valuation techniques (based on surveys of public attitudes and willingness to pay premiums for organic food) and household production functions (how much do households currently spend on membership of organic organizations, purchasing books or consuming organic food);
- *indirectly* using hedonic pricing (e.g. house values in areas with different environmental characteristics) or travel-cost techniques (where the cost of travel to a particular environmental site or organic farm is taken as an indicator of the value of that site to the individual).

Since they are based on willingness to pay, these techniques are subject to the criticisms made earlier, in particular: the appropriateness of individual preferences as a basis for judging the environmental and other values associated with a particular site or environmental benefit; the availability of information concerning the actual environmental benefits of a particular site or farming system; and the assumption that values can be measured on the basis of current income distributions.

Part of the value which society places on organic farming might be estimated in terms of consumer willingness to pay more for organically produced food, or the willingness of farmers to incur costs or forgo income to achieve environmental goals. To estimate this accurately, it would be necessary to isolate the environmental quality component from health and other quality characteristics of the product. Market surveys of consumer demand for organically produced food consistently overstate willingness to pay when compared with actual buying behaviour. Whether this is because limited availability prevents consumers from fulfilling their actual willingness to pay, or because willingness to pay data generated by surveys where the individual is not actually having to part with money generally overstate actual willingness to pay, is an open question.

While some of the benefits of organic farming fall readily into the willingness to pay category, not all of them do. Reductions in costs arising from environmental damage (such as soil erosion or flooding) and the costs of damage regulation and abatement (e.g. the costs of research, monitoring and subsidizing reduction of agricultural pollutants, or disposal of surplus output) can also be considered as potential benefits of organic farming. Although the costs may be more readily quantifiable, this approach implies that society is willing to accept the costs currently incurred, and therefore that the environmental benefits are worth at least that amount (always assuming that society at large accepts that these costs are justified).

Even in the most obvious cases, no thorough analysis of the costs and benefits of alternative farming systems is available.

Identification and choice of policy options

Many of the issues relating to policy choices were referred to in the discussion of competitive markets earlier in this chapter. There is, for instance, need for research relating to the different general types of intervention to achieve environmental benefits in agriculture (e.g. changes in property rights versus taxes) as well as on the details of alternative policies such as Management Agreements or Environmentally Sensitive Areas. Organic farming is only one option among many, and all have to be assessed.

To determine the effects of different policies, we need to know how farmers will respond to alternative incentives. In the first instance, this means knowing the relative profitabilities of each farm system including organic farming – a subject on which some work has been done, not all of it uncontroversial (Murphy, 1992; Bateman, 1993; also Chapters 5 and 12, this volume). There is room for much new work, using a variety of techniques – spreadsheets, linear programming and input–output analysis (Lampkin, 1993).

Some evidence

Despite the limited amount of information on the physical environmental effects of organic farming, some work has been done by economists that can be used to illustrate some applications of the ideas discussed in earlier sections. Three examples are noteworthy, although in each case they avoid the problem of evaluating benefits by using the techniques of cost *effectiveness* rather than cost–*benefit* analysis.

The first approach compares the costs of a series of alternative agricultural policies selected by the Nature Conservancy Council as offering unquantified environmental benefits (Jenkins and Midmore, 1992). An extension of organic farming is the only one of the options that brings about a significant reduction in agricultural output and hence budgetary costs (see Chapter 21, Table 21.1). The consequence is that organic conversion emerges as a policy that not only offers environmental benefits but simultaneously reduces, rather than increases, net budgetary costs.

A second approach to an economic assessment of organic farming focuses more narrowly on agricultural output alone (Midmore and Lampkin, 1988): what is the cheapest way of reducing output? The results of their analysis (see also Chapter 26) can be summarized as follows:

- in some, though not all, cases the organic options are cheaper;
- the organic options have a greater degree of permanence in that the reduction in output is likely to continue beyond the period when support ceases;
- the organic options also have the other advantages discussed above;
- but, against these advantages, an expansion in organic output may tend to undermine organic premiums.

A third approach takes a different objective again – the creation of jobs in the rural economy (see Chapter 21) reaching the conclusion that, although organic farming involves a reduction in inputs and hence input–output-related employment, this is more than offset by increases in on-farm employment and in processing activities.

Conclusions

The conclusions of this review of the social value of organic farming are as follows.

- An economic assessment is not the same as a financial one: even if organic farming were financially unviable, there may still be reasons why it should be encouraged.
- The economic case is based on an acceptance of the concept of a modified market – a concept which takes account of some, not all, of the criticisms that environmentalists would make of a completely free market system.
- Environmentalists can best advance their case by arguing from within the concept of a modified market, rather than by opposing it.
- Such quantitative evidence as there is confirms a case for supporting organic farming, but a great deal of further research is necessary – from environmental scientists as much as from economists.

References

Arden-Clarke, C. and Hodges, R.D. (1987) The environmental effects of conventional and organic/biological farming systems. I. Soil erosion with special reference to Britain. *Biological Agriculture and Horticulture* 4, 309–357.

Arden-Clarke, C. and Hodges, R.D. (1988) The environmental effects of conventional and organic/biological farming systems. II. Soil ecology, soil fertility and nutrient cycles. *Biological Agriculture and Horticulture* 5, 223–287.

Barde, J.-P. and Pearce, D.W. (1991) *Valuing the Environment.* Earthscan Publications, London.

Bateman, D.I. (1989) Heroes for present purposes? A look at the changing idea of communal land ownership in Britain. *Journal of Agricultural Economics* 40(3), 269–289.

Bateman, D.I. (1993) Financial and economic issues in organic farming: a case study in pluriactivity. *Journal of the University of Wales Agricultural Society,* 73, 4–31.

Bateman, D.I., Charles, S. and Jenkins, T.N. (1991) Environmental economics and nature conservation. Unpublished report prepared for the Nature Conservancy Council. Department of Agricultural Sciences, University of Wales, Aberystwyth.

Blackorby, C. and Donaldson, D. (1990) The case against the use of the sum of compensating variations in cost–benefit analysis. *Canadian Journal of Economics* 23, 471–494.

Coase, R. (1960) The problem of social cost. *Journal of Law and Economics* III, 1–44.

Cuttle, S.P. (1989) *Public Perceptions of Agriculture and the Environment.* Report prepared for AFRC Institute for Grassland and Animal Production (Welsh Plant Breeding Station, Aberystwyth).

Daly, H.E. (1992) The economic growth debate: what some economists have learned but many have not. In: Markyanda, A. and Richardson, J. (eds) *The Earthscan Reader in Environmental Economics.* Earthscan Publications, London.

EFRC (1992a) *Nitrate Reduction for Protection Zones: The Role of Alternative Farming Systems.* Report on research by Elm Farm Research Centre, R&D Note 108. National Rivers Authority, Bristol.

EFRC (1992b) *Assessment of Nitrate Leaching Losses from Organic Farms.* Final report to MAFF. Elm Farm Research Centre, Newbury.

Ekins, P. (ed.) (1986) *The Living Economy: A New Economics in the Making.* Routledge and Kegan Paul, London.

Gould, S.J. (1993) *Eight Little Piggies.* Jonathan Cape, London.

Greenpeace (1992) *Green Fields – Grey Future. EC Agricultural Policy at the Crossroads.* Greenpeace, Amsterdam.

Hanley, N. and Splash, C. (1993) *Cost Benefit Analysis and the Environment.* Edward Elgar, Aldershot.

Jenkins, T.N. and Midmore, P. (1992) The economic implications of selected nature conservation proposals for agriculture. Paper presented at the 30th European Association of Agricultural Economists Seminar, Château-d'Oex, Switzerland, November 1992.

Lampkin, N.H. (1990) *Organic Farming.* Farming Press, Ipswich.

Lampkin, N.H. (1993) The economic implications of conversion from conventional to organic farming systems. PhD thesis, Department of Agricultural Sciences, University of Wales, Aberystwyth.

Lovelock, J. (1979) *Gaia: A New Look at Life on Earth.* Oxford University Press, London.

Midmore, P. and Lampkin, N. (1988) Organic farming as an alternative to set-aside and an option for extensification. Paper presented to the European Association of Agricultural Economists, Copenhagen.

Mishan, E.J. (1971) Expenditure on commercial advertising is justified. In: *21 Popular Economic Fallacies.* Pelican, Harmondsworth.

Murphy, M.C. (1992) *Organic Farming as a Business in Great Britain.* Agricultural Economics Unit, University of Cambridge, Cambridge.

NCC (1990) *Nature Conservation and Agricultural Change.* Focus on Nature Conservation, 25. Nature Conservancy Council, Peterborough.

O'Carroll, L. (1993) Public land purchase and conservation. PhD thesis, Department of Agricultural Economics, University of Manchester, Manchester.

Price, C. (1990) Social discounting: a game played with meaningless marks on paper. In: Whitby, M.C. and Dawson, P.J. (eds) *Land Use for Agriculture, Forestry and Rural Development.* Department of Agricultural Economics and Food Marketing, University of Newcastle, Newcastle upon Tyne.

Price, C. (1993) *Time, Discounting and Value.* Blackwell, Oxford.

Redman, M. (ed.) (1992) *Organic Farming and the Countryside.* Report for the Countryside Commission. British Organic Farmers, Bristol.

Smith, A. (1954) *The Wealth of Nations.* J. M. Dent and Sons, London. (First published 1776–78).

Stevenson, G.G. (1991) *Common Property Economics: A General Theory and Land Use Applications.* Cambridge University Press, Cambridge.

Sturgess, I. (1992) Self sufficiency and food security in the UK and EC. *Journal of Agricultural Economics* 43(3), 311–326.

Tregear, A. (1993) An investigation of the demand for organically grown produce. MPhil thesis, University of Edinburgh.

Vine, A. and Bateman, D.I. (1981) *Organic Farming Systems in England and Wales: Practice, Performance and Implications.* Department of Agricultural Economics, University College of Wales, Aberystwyth.

Wilson, J. (1993) The BTO birds and organic farming project: 1 year on. *New Farmer and Grower* 38, 31–33.

World Commission on Environment and Development (1987) *Our Common Future.* Oxford University Press, Oxford.

Part 2

Organic Farming as a Business

5 Economics of Organic Farming in Britain

Nicolas H. Lampkin

Introduction

The first, limited attempt to make a financial assessment of organic farming in Britain was in 1974–1975 (Mayall, 1977). This was followed three years later by a more comprehensive study (Vine and Bateman, 1981), commissioned by the Ministry of Agriculture, Fisheries and Food (MAFF) at a time when information on the subject was still very restricted. Ten years later, MAFF commissioned a further study (MAFF, 1991; Murphy, 1992) based on a postal questionnaire and a detailed follow-up survey of 238 farms in 1989–1990 by the Universities of Cambridge, Aberystwyth and Edinburgh. The aim of this chapter is to review these and other studies in order to answer the following questions: What characterizes organic farms and farmers? To what extent are yields and costs reduced? Are labour costs higher? What is the role of premium prices? How do incomes from organic farming compare with conventional farming?

The Characteristics of Organic Farms and Farmers

A postal questionnaire in 1989 yielded information on the characteristics of 557 holdings which at that time were organically managed or had land in conversion (MAFF, 1991; Murphy, 1992). Of the total land area of 42,000 ha, 15,000 ha were farmed organically and a further 3700 ha were in the process of conversion (the area of organically managed land has since increased to over 35,000 ha in 1993). Nearly 40%

Table 5.1. Distribution of area farmed organically and in conversion by farm size as a percentage of total organic and in-conversion area, Britain, 1989.

	Farm size (ha)							
	< 5	5–10	10–20	20–50	50–100	100–200	> 200	All
England	1.9	2.1	2.6	10.0	19.2	21.2	43.0	100
Scotland	0.8	1.2	2.9	8.3	6.0	28.7	52.1	100
Wales	3.8	4.5	13.5	29.9	37.0	11.2	–	100

Source: MAFF (1991).

of the farms were horticultural holdings, which affected the distribution of farms by farm size. However, the larger farms accounted for the vast majority of the organically managed area (Table 5.1). These findings confirm the results of previous studies (e.g. Vine and Bateman, 1981; Dixon and Holmes, 1987; Cudjoe, 1989), but it is clear from more recent information that the number of very small organic holdings is declining; and that the number of larger holdings, and with them average farm size, is increasing.

The organic farms were located predominantly in areas where cropping is feasible, such as the South and West of England and parts of Scotland and Wales (Cudjoe, 1989; MAFF, 1991; Murphy, 1992). Very few farms were located in areas where production options are restricted by climatic factors to livestock only, due to the lack of marketing opportunities for livestock, although many traditional, extensive farms in these areas might qualify as organic. The location of many organic farms also reflects low land values which were attractive to new entrants from urban backgrounds.

Cudjoe (1989) found that many of the individuals converting to organic farming, particularly in the 1970s, were newcomers to farming, but that an increasing proportion of those converting in the mid-1980s had previous farming experience, supporting anecdotal evidence that the organic sector is attracting an increasing number of established farmers rather than new entrants. Murphy (1992) reported that 50% of the organic farmers surveyed had converted in the five years to 1989, while 52% had previous experience of conventional farming, but the results were not cross-tabulated and therefore do not directly support Cudjoe's findings.

Education to degree level, frequently in non-agricultural subjects, and a lack of agricultural training or experience also characterize a significant proportion of organic producers in the United Kingdom (Brighton *et al.*, 1988; Murphy, 1992), particularly those who started farming organically as new entrants to agriculture. The majority had no specific agricultural qualifications and relied primarily on practical experience gained on their own or other holdings.

Physical Productivity

Crop yields

Vine and Bateman (1981) compared winter wheat and spring barley data from a number of individual farms with the regional group averages for that type and size of farm. Relative yields ranged from 54% to 114% of the conventional average for winter wheat, and from 50% to 123% for spring barley. In both cases, the mean relative yield was 89%, although Vine and Bateman were careful to treat the farms individually and did not attempt to present unrepresentative averaged data in their report. More recent data come from Murphy (1992) (Table 5.2). These are mainly for 1989, a year characterized by dry weather conditions in some areas, resulting in lower than average crop yields, especially for spring-sown cereals and vegetables.

The relative yield differences found by Murphy are greater than those reported by Vine and Bateman, reflecting the continued intensification of conventional systems, with conventional winter wheat yields, for example, rising from around 5 t ha^{-1} in the late 1970s to over 7 t ha^{-1} in the early 1990s. Stanhill (1990) reported a similar increase in the relative cereal yield differences from the long-term Haughley experiment (Balfour, 1976) over the period 1952–1965. Relative yields

Table 5.2. Comparative yields for organic and conventional crops, Britain, 1989.

Crop	Organic[a] (t ha^{-1})	Conventional[b] (t ha^{-1})	Relative (conv=100)
Winter wheat	3.73	6.16	61
Spring wheat	3.24	4.95	65
Winter barley	3.09	5.31	58
Oats	3.59	4.41	81
Field beans	1.97	2.96	66
Potatoes, maincrop	18.98	51.27	37
Carrots	19.19	–	–
Onions	27.41	49.47	55

[a] Organic crops grown on partly and wholly organic farms.
[b] Conventional crops grown on partly organic farms.
Source: Murphy (1992).

for vegetables are also influenced considerably by the degree of mechanization, location and quality specifications for particular market outlets.

Yield variability between farms is a key factor mentioned in many of the British studies (Mayall, 1977; Vine and Bateman, 1981; Dixon and Holmes, 1987; Murphy, 1992). Part of this yield variability may be attributable to differences in soil quality, rotation design and management ability (see Chapter 6), and the period since conversion (see Chapter 16), but there is little evidence on these factors from the British studies. More information is available on varietal influences, for example for winter wheat, which in National Institute of Agricultural Botany trials on an organic farm in Herefordshire produced yields for different varieties ranging from 5.30 to 6.78 $t\,ha^{-1}$ in 1991, and from 5.03 to 7.30 $t\,ha^{-1}$ in 1992 (Bulson, 1993). These results contrast with Murphy's average of 3.73 $t\,ha^{-1}$ in 1989, although such variety trials often give better results than farmers would expect to achieve.

Livestock output

Livestock output is affected in two main ways under organic management: first, in terms of the performance of individual animals due to changes in the management of the stock themselves, such as a shift to a lower use of concentrates for ruminants and to the use of cereals and home-produced grain legumes instead of fish meal and soya in livestock rations; second, in terms of forage output and hence stocking rate, as a result of changing to legume-based pastures and reduced reliance on purchased feeds and fertilizers.

Milk yields per cow are typically 10% below conventional and stocking rates are in the range 1.6–2.0 livestock units (LU) per hectare, or about 80% of the level achieved on conventional farms (Table 5.3). The lower yields per cow combine with lower stocking rates to give milk yields per hectare up to 40% lower than on conventional farms. Breed is an important influence in this context. The British studies indicate a preference for specialist breeds, particularly Channel Island, among organic producers. These breeds are typically lower yielding, but compensate for this with higher butterfat percentages. The analysis by Holden (1989), although effectively in the form of case studies, is the only one to present results differentiated by breed, but the comparison group consists of at least 90% Friesian or Friesian-Holstein cattle. Houghton and Poole (1990) make adjustments to take account of differences in breed structure. Murphy's (1992) results differ significantly from the other studies. The average herd size for the 21 wholly organic farms surveyed was only 29 cows, with four herds having fewer than ten cows. Only eight of these herds were on farms

Table 5.3. Comparative milk yields and stocking rates on organic farms, England and Wales.

Source	Year	Farms	Herd size	Breed[a]	Litres per cow[b]	Cows per hectare[b]	Litres per hectare[b]
Holden (1989)	1988/89	2	n/a	F	5025(91)	1.75(80)	8794(72)
Holden (1989)	1988/89	2	n/a	CI	4041(73)	1.71(78)	6910(57)
Murphy (1992)[c]	1989/90	21	29	N	3723(73)	1.12[d]	n/a
Houghton and Poole (1990)	1989/90	11	57	6 F, 4 CI/A[e]	4589(91)	1.56(66)	7159(60)
Redman (1991)	1990/91	13	73	N	5436(93)	1.79(81)	9730(75)

[a] A – Ayrshire, CI – Channel Island, F – Friesian, N – not identified.
[b] Relative values (conventional = 100) in parentheses.
[c] See text; yields compared with conventional dairy herds on partly organic farms.
[d] Organic herds on wholly organic farms.
[e] Adjusted to compensate for high proportion of Jersey or other Channel Island cattle.

classified as dairy farms. Murphy does not indicate to what extent the smaller herds are characterized by Channel Island breeds (especially Jersey cattle).

Comparative trials of beef finishing systems have been conducted which indicate that similar finishing rates can be achieved for organic beef cattle, helped by the greater feed value of clover-based swards, but 16% lower stocking rates contributed to 21% lower liveweight gains per hectare under organic management (Younie, 1989; Younie *et al.*, 1990).

Variability is again a key factor. Vine and Bateman (1981) found that stocking rates on the organic farms they surveyed ranged from 32% to 132% (typically 80%) of conventional rates. Total livestock output per hectare was generally rather less at about 70% of conventional output. Variations in stocking rate may be a reflection of differences in concentrate use rather than output from clover-based pastures. Newton (1994) found that forage yields on organic farms were correlated with site classes reflecting grass growth potential (Table 5.4). Younie (personal communication cited in Newton, 1994), in a study of ten organic farms, recorded maximum stocking rates of 2.2 LU ha^{-1} and an average of 1.7 LU ha^{-1}. Newton's results are higher than those reported by Thomas *et al.* (1991) for trials with white clover swards receiving no nitrogen fertilizer (see also Fig. 12.1), in part due to differences in methodology, but grazing by animals and the use of livestock manures on grassland in organic systems may account for some of the differences.

Table 5.4. Forage yields (t ha^{-1}) from temporary and permanent grassland on organic farms, Britain, 1992–1993.

	Dry matter		Digestible organic matter
Year	1991	1992[a]	1992
Site class			
2	14.1	15.7 (n/a)	10.2
3	11.9	14.6 (22.6)	9.8
4	10.2	12.8 (21.7)	8.7
5	6.1	12.3 (16.8)	8.5
Number of farms	28	30	30

[a] Maximum yield recorded from permanent grassland in parentheses.
Source: Newton (1994).

Crop and Livestock Enterprise Financial Performance

Prices

In Britain, as in other northern European countries, strong market demand for organic products has led to high premium prices for some crops (Table 5.5). The premiums available for crops such as wheat and oats are also closely linked to quality factors. Other arable crops, such as sugarbeet and oilseed rape, have no ready organic market and therefore do not attract premium prices. Since 1990, prices have generally weakened (Lampkin and Measures, 1994), in part as a response to the large increases in supply elsewhere in Europe (Chapter 25).

Livestock products have not generally attracted significant premium prices in Britain because of the lack of an established market, but account for the majority of the output on organic farms. Premium prices for cash crops should be seen in this context. Since 1989, the gradual development of a market for organic meat and dairy products in the UK has meant that premium prices of between 10% and 30% may be obtained. Processing and direct marketing of meat and dairy products can provide an opportunity to increase the margins for livestock production, but these can be eroded rapidly by increased labour and transport charges.

Table 5.5. Ex-farm prices for organically produced cereals and vegetables in Britain, 1989–1990.

Crop	Average price (£ t^{-1})	Relative (conv=100)[a]
Cereals	195	176
Winter wheat	230	213
Spring wheat	219	192
Winter barley	167	137
Oats	166	155
Field beans	200	114
Vegetables		
Potatoes, maincrop	206	179
Carrots	288	450
Onions	302	282
Leeks	615	122
White cabbage	218	172

[a] Conventional prices received on partly organic farms.
Source: Murphy (1992).

Variable input costs

Slightly higher expenditure on seeds and plants is generally more than compensated for by much lower costs for fertilizers and sprays, which are typically less than 10% of conventional fertilizer and spray costs (Vine and Bateman, 1981; Daw *et al.*, 1991; Murphy, 1992; Lampkin and Measures, 1994). Differences in the costs of spreading manures are minimal as not every course in an organic rotation receives manure or slurry, and lower stocking rates will mean less manure or slurry is available to be spread. However, the management of the manure may be reflected in higher fixed costs such as regular labour. With vegetable and row crops there may be higher expenditure on casual labour for weed control, but this depends on the level of mechanization. There may also be additional costs for specific practices such as flame weeding. Very few studies have analysed these aspects specifically, other than in the general context of labour use on organic farms (see below). Seed-bed and other cultivations are usually treated as fixed costs, although they are often an important aspect of weed control and represent an area where costs may differ. Total variable costs for cereals and grain legumes are typically less than 60% of the costs in conventional cropping (Vine and Bateman, 1981; Murphy, 1992). This may have an impact on interest charges for working capital.

The major cost reductions for livestock are in terms of reduced expenditure on fertilizers and sprays for forage production (Younie, 1989; Houghton and Poole, 1990; Younie *et al.*, 1990). To the extent that reduced reliance is placed on concentrates, these costs will also fall, but the premium price for purchased concentrates may cancel out any benefits, particularly in pig and poultry production. There may also be higher costs associated with sourcing organically reared livestock, although reliance is more often placed on rearing stock on the farm. The organic standards requirement for whole milk and longer periods before weaning can contribute to substantially higher rearing costs, but these may be justified in terms of extended production life and lower replacement costs. Veterinary and medical costs may be lower, but the difference is not significant in terms of the overall costs of livestock production.

Gross margins

The combined effect of lower yields, higher prices and lower variable costs can lead to similar or higher than conventional average gross

Table 5.6. Comparative output, variable costs and gross margins for winter wheat and maincrop potatoes, Britain, 1989–1990.

	Winter wheat			Maincrop potatoes		
	Organic	Conv[a]	Relative[b]	Organic	Conv[a]	Relative[b]
Output						
Yield (t ha^{-1})	3.73	6.16	61	18.98	51.27	37
Price (£ t^{-1})	230	113	204	206	90	229
Total (£ ha^{-1})	857	694	123	3919	4636	85
Variable costs (£ ha^{-1})						
Seeds	59	42	140	443	486	91
Fertilizers	15	80	19	61	208	29
Sprays	6	76	8	12	191	6
Casual labour	30	20	150	201	214	94
Marketing	20	11	181	340	148	230
Total	130	228	57	1057	1247	85
Gross margins (£ ha^{-1})						
With premium (actual)	727	466	156	2862	3389	84
Without premium	291	466	62	658	3389	19

[a] Conventional crops grown on partly organic farms.
[b] Conventional = 100.
Source: Murphy (1992).

Table 5.7. Comparative output, variable costs and gross margins (£ per animal) for milk and beef production, Britain, 1988–1990.

	Milk			Beef finishing		
Source	Houghton and Poole (1990)			Younie *et al.* (1990)		
Country	England and Wales			Scotland		
				1988–1990		
Year(s)	1989–1990					
Base	11 farms			Comparative trial		
System	Organic	Conv[a]	Relative[b]	Organic	Conv[a]	Relative[b]
Herd size	57	102	56	–	–	–
Concentrate use (kg)	930	1448	64	–	–	–
Concentrate cost (£ t^{-1})	155	139	112	–	–	–
Stocking rate (head ha^{-1})	1.56	2.37	66	3.57	4.33	82
Output						
Yield (litres per cow)	4589	5049	91	484[c]	479[c]	101
Price (£ l^{-1})	0.200	0.206	97	2.27[d]	1.96[d]	116
Total (£ $head^{-1}$)	1010	1111	91	364	280	130
Variable costs (£ $head^{-1}$)						
Purchased feeds	187	243	77	123	85	145
Forage	34	86	40	6	40	15
Other	89	95	94	29	29	100
Total	310	425	73	158	154	103
Gross margins						
per head	700	687	102	206	126	163
per hectare	1092	1628	67	735	546	135
per litre	0.153	0.136	113			

[a] Conventional sample weighted for breeds.
[b] Conventional = 100.
[c] kg liveweight per head.
[d] £ per kg liveweight.

margins for individual crops (Table 5.6), but very few studies (e.g. Vine and Bateman, 1981) illustrate the wide variability which can exist between farms and regions. Typical gross margins for a wide variety of crop and livestock enterprises are also illustrated in Lampkin and Measures (1994), but not on a comparative basis.

In milk production, similar gross margins per cow may be achieved (including forage costs but without premium prices), but lower stocking rates result in lower gross margins per hectare (Table 5.7). In a situation where quotas are limiting, and not land, the margin per litre produced (or more accurately margin per litre quota or kilogram fat quota) is a more relevant performance measure (Ashworth, 1989). On this basis, the organic systems generally perform better than the conventional. If an organic premium is available, the advantage on a per cow and per litre basis becomes quite significant, and it also helps to reduce the gap on a per hectare basis. Studies of other livestock systems (Table 5.7; see also Lowman, 1989; Younie, 1989; Murphy, 1992; Lampkin and Bateman, 1993) suggest that premiums of 10–25% would be needed to achieve gross margins per hectare comparable with conventional systems, although similar gross margins per animal can be obtained without premiums.

Whole-farm Financial Performance

Farm gross margins

Gross margins for individual enterprises can be misleading, as not all enterprises perform well in organic systems, particularly those which do not benefit from premium prices, and the enterprise mix can be very different from that found on conventional farms (Dixon and Holmes, 1987; Lampkin, 1990, 1993; Murphy, 1992). Cereals, forage maize, sugarbeet and oilseed rape tend to account for a much lower proportion of organically managed land, while temporary grass, fodder crops, potatoes and vegetables are relatively more important. Pigs and poultry are much less common on organic farms. Some examples of whole-farm gross margins for different enterprises mixes are given by Lampkin and Measures (1994).

Another reason for emphasizing whole-farm performance measures is the importance of within-farm transfers of nutrients and energy in organic farming. For example, ruminant livestock on organic farms provide a means of utilizing the fertility building phase of a rotation, where forage legumes 'fix' nitrogen for the following arable crop. The forage enterprises can therefore be seen as part of the 'costs' of achieving the high returns associated with individual crops such as wheat and potatoes. Similarly, beef cattle may give lower returns per hectare than sheep, but the benefits from interactions between the two species are such that the lower returns from the beef enterprise can be seen as a 'cost' of sheep production.

The importance of intra-farm transfers is illustrated by the use of green manures in organic systems. Red and white clover-based green manures had a marked effect on the financial performance of a subsequent cereal crop, more than doubling the gross margin achieved by cereals following a ryegrass-only control treatment, although the results were also influenced by the duration and the management of the green manure (EFRC, 1993). Conversion from an all-arable conventional system to an organic system with livestock entails major capital investments in fencing, water supplies, buildings and breeding livestock. A stockless system, incorporating a legume-based fallow period, may provide a means of avoiding these investment costs. On-farm and research institute trials indicate that the financial returns from stockless systems may be as high as from systems including livestock, particularly if fixed cost changes are taken into account (Lampkin, 1993; Stopes and Millington, 1993).

Labour expenditure and incomes

Labour is the only fixed cost on organic farms which differs significantly from the conventional average (Table 5.8; Murphy, 1992; Lampkin, 1993). Murphy places a higher value per hectare on unpaid farmer and spouse manual labour on the wholly organic farms, but differences in farm size are a major factor, because similar total farmer and spouse labour inputs per farm are spread over a smaller area. Paid labour per hectare is at similar levels irrespective of farm size or management system for each farm type except horticulture, where it is substantially *lower* on the organic holdings. A major influence on labour requirements is the more diverse enterprise structure, with a loss of benefits from specialization and economies of scale, although the distribution of labour requirements may be better in less specialized systems. Increased labour costs may also be associated with labour-intensive, high-value crops such as vegetables and potatoes, and with the development of marketing and processing activities needed to obtain an organic premium.

The impact of lower yields, lower variable costs and higher labour costs on relative farm incomes varies considerably by farm type, but it is generally the case that premium prices are required to allow organic farmers to achieve incomes comparable to conventional farmers (Vine and Bateman, 1981). Despite the existence of premium prices, Murphy (1992) reports a relatively poor overall performance for organic farms, but the results and comparisons made need closer scrutiny and demonstrate clearly the importance of ensuring comparability between groups of farms. Large differences in farm size (land area), business size (see below), farm enterprise structure (indicated by proportion of land cropped), land quality (indicated by value) and location (less favoured area status) had, in some cases, a greater

Table 5.8. Comparative labour expenditure and income measures on wholly organic and partly organic farms in Britain and conventional farms in England, 1989–1990.

Farm type/system	Number of farms, *n*	Farm size (ha)	Cropped area (%)	Farm output (£ ha^{-1})	Farm variable costs (£ ha^{-1})	Farm gross margin (£ ha^{-1})	Hired labour (£ ha^{-1})	Farmer and spouse labour (£ ha^{-1})	Net farm income (£ ha^{-1})	Net farm income (£ per farm)	Labour income (£ ha^{-1})
Cropping											
Wholly organic	8	42.8	36.2	535	163	373	112	167	−28	−1199	90
Partly organic[a]	25	176.3	74.5	785	248	537	106	36	107	18934	216
All organic	33	143.9	65.2	767	242	525	106	45	98	14053	169
Conventional[b]	33	141.9	78.3	895	307	588	115	48	115	16274	244
Dairying											
Wholly organic	8	52.2	11.0	1000	314	686	159	238	146	7638	322
Partly organic	19	207.0	41.7	966	312	654	200	39	81	16779	283
All organic	27	161.1	32.6	969	312	657	196	58	87	14071	301
Conventional	485	73.9	16.2	1590	695	895	174	125	318	23500	494
Mixed											
Wholly organic	39	53.4	21.9	607	151	456	124	133	49	2613	205
Partly organic	63	128.6	44.9	754	275	480	144	69	39	5057	187
All organic	102	99.8	36.1	724	249	475	140	82	41	4123	191
Conventional[c]	177	76.9	15.6	638	257	381	76	108	40	3076	118
Horticulture											
Wholly organic	61	8.4	36.9	2080	608	1472	450	1303	208	1743	785
Partly organic	12	9.0	27.8	2033	490	1543	193	1561	702	6316	895

All organic	73	8.5	35.4	2072	588	1484	405	1348	294	2495	804
Conventional	41	30.9	–	4779	844	3935	1556	271	480	14832	2041
All farms excluding horticulture											
Wholly organic	55	51.7	22.4	656	176	480	127	153	54	2789	186
Partly organic	107	153.7	51.2	813	276	537	147	53	68	10381	218
All organic	162	119.0	41.5	790	262	528	144	68	66	7804	214
Conventional[b]	162	89.6	28.5	852	333	518	102	91	102	9265	205

[a] Weighted average of specialist cereals and general cropping farms.
[b] Weighted average using organic proportions to reflect farm type composition of organic sample.
[c] Lowland livestock farms (organic may include some hill and upland farms).
Sources: Adapted from Murphy (1992) and Bateman (1993).

influence on income levels on organic and conventional farms than the management system itself.

Net farm income (NFI) represents the return to farmer and spouse's manual labour, management and capital inputs, including borrowed capital. On the mixed (lowland livestock) farms, which are the most closely matched in terms of farm size and structure, NFI is similar to conventional (Table 5.8). This is also true for the partly organic cropping farms, but the wholly organic cropping farms, with their much smaller farm size and proportion of land cropped, performed relatively poorly. In contrast, the larger, partly organic dairy farms have lower NFI per hectare than either the wholly organic or conventional dairy groups, which is consistent with the per hectare performance of smaller, specialist dairy farms relative to larger, mixed dairy/arable farms. The wholly and partly organic horticultural holdings have very similar size and structural characteristics, but NFI per hectare varies considerably due to poorer land quality (as indicated by lower land values) and higher costs, particularly paid labour, on the wholly organic farms. The average farm size of both organic horticultural groups is substantially smaller that the conventional comparison group.

Labour income represents the return to all labour, paid and unpaid, and provides an indicator of the contribution which different systems make to societal income and employment objectives (Bateman, 1993). On this measure, the performance of organic farms is improved relative to conventional on the mixed holdings, but is reduced on the horticultural holdings (Table 5.8).

If the average results for all organic farm types are compared with an average for the conventional groups weighted to reflect the composition of the organic sample, the overall results are very similar on a per hectare basis. Premium prices and lower variable costs compensate for reduced yields to give similar farm gross margins, while similar fixed costs result in similar NFI and labour income. The small average farm size of the wholly organic group results in lower incomes per farm, but total incomes are similar when *all* organic farms are compared with conventional. The horticultural holdings have been excluded, because of their very different characteristics relative to other farm types, and because the comparison group used least accurately reflects their size, structural and locational characteristics.

Despite the different interpretations which can be placed on Murphy's published results, the most significant problem with all these comparisons is that business size was ignored completely. The organic farm sample contained a high proportion of farms smaller than four British size units (derived from standard gross margins), while conventional farms of this size are considered to be part-time and are excluded from the conventional comparison groups (MAFF, 1991). Seventeen of the 38 Welsh farms included in Murphy's study were found by Lampkin and Bateman (1993) to be part-time according to this definition, which corresponded to farm output levels of less than £15,000. The data for the Welsh farms were reanalysed and comparisons made with conventional farms from the Welsh Farm Business Survey (FBS, 1991) on the basis of similar farm type, business size and region (Table 5.9). The distinction between wholly organic and partly organic farms made by Murphy was not made in this case, as all the farms involved managed their land organically, and the main differences in livestock production were associated with medication and to a limited extent purchased feeds.

Full-time organic, mainly dairy, farms performed better on average than their conventional counterparts on a per farm and a per hectare basis. This result was achieved despite below average performance of the organic dairy herds, in part due to the prevalence on these farms of organic horticultural crop production, as well as to the processing of milk into cheese on three of the six farms. However, labour and net farm incomes remain higher on the organic farms even if value added by processing is deducted. Lowland cattle and sheep farms also performed better than their conventional counterparts. Comparable conventional farm data are not available for the full-time horticultural group, but income measures per farm and particularly per hectare compare well with the other farm types. The three other full-time farms were of different type, but achieved similar NFIs, averaging £12,800 per farm.

Part-time organic farms in Wales achieved output and gross margin levels per hectare similar to the full-time organic farms, allowing for differences in farm type. Smaller farm size, however, contributed to lower per farm values than those achieved by the full-time organic farms, and resulted in much poorer income levels, on a per farm and per hectare basis. One other part-time holding (of a different farm type) achieved an NFI of £170. Comparative data are not available for conventional farms of this business size, but full-time conventional farms in the difficult farming conditions of the hills and uplands did not fare much better (Lampkin and Bateman, 1993).

The analysis of the Welsh data suggests that if farms are distinguished into full- and part-time types, the organic mixed and dairy farms perform as well as or better than their conventional counterparts despite smaller than average farm size. This may indicate the potential of the organic market to help maintain smaller farms. Horticulture also provided the opportunity for some very small farms in Wales to make a reasonable income, assisted by the

Table 5.9. Output, cost and income measures for full- and part-time[a] organic and comparable conventional farms in Wales, 1989–1990.

Farm type/system	Number of farms, *n*	Farm size (ha)	Business size (BSU)	Farm output (£ ha^{-1})	Farm variable costs (£ ha^{-1})	Farm gross margin (£ ha^{-1})	Net farm income (£ ha^{-1})	Net farm income (£ farm^{-1})	Labour income (£ ha^{-1})
Mainly dairy									
Organic, full-time	6	36.58	14.48	1126	299	827	406	14870	486
Conventional[b]	14	45.99	15.38	981	356	625	183	8413	304
Lowland cattle and sheep									
Organic, full-time	5	72.95	14.28	435	143	292	73	5305	124
Conventional[c]	17	69.96	18.14	598	227	371	32	2258	81
Livestock, organic part-time									
Lowland cattle and sheep	2	22.31	1.88	393	133	260	–55	–1233	–43
Upland cattle and sheep	4	10.81	1.40	428	96	332	21	224	61
Hill sheep	4	27.68	3.01	318	92	226	5	127	39
Horticulture									
Organic, full-time	7	13.45	9.10	2078	526	1552	828	11138	1206
Organic, part-time	6	2.95	3.03	1794	326	1468	–295	–869	787

[a] Full-time = > 4 BSU, part-time = < 4 BSU (see text).
[b] Welsh FBS mainly dairy farms, 4–15.9 BSU.
[c] Welsh FBS lowland cattle and sheep farms, all sizes > 4 BSU.
Sources: FBS (1991) and Lampkin and Bateman (1993).

availability of a growers' cooperative and a major packhouse based in Lampeter. The analysis of the Welsh data also suggests that organic farms are likely to offer greater on-farm employment in marketing and processing, as well as, in some cases, greater labour income in farming itself, thus strengthening the relative performance of the organic group in terms of contribution to the local economy.

Conclusions

The evidence currently available confirms that organic farming systems are more diverse in terms of enterprise mix; have lower yields and higher labour costs which are not compensated for fully by reduced input costs, and that higher prices are essential if organic farmers are to achieve similar incomes to their conventional counterparts. In practice, some farm types in Britain are able to achieve similar income levels, but this is not the case for specialist arable and horticultural producers, and for the most intensive livestock producers. The somewhat poorer relative performance of organic farming in Britain compared with other countries such as Germany (Chapter 6) is to some extent attributable to a less well-developed market for organic produce and to a higher degree of specialization and intensity of conventional production in the UK. The situation described in this chapter is likely to be affected significantly by the current reforms of the Common Agricultural Policy (Chapter 26), with reductions in commodity price support and the introduction of area-based payments for arable crops which are also available to organic producers. In any case, for the farmer interested in changing over to an organic system, the question of what effect the change-over will have on his or her income can only be answered with reference to the current position and the costs of conversion on the individual's own farm (see Chapter 13).

References

Ashworth, A.W. (1989) The economics of organic milk production. In: *Organic Milk Production.* Scottish Agricultural College, Auchincruive, pp. 18–23.

Balfour, E.B. (1976) *The Living Soil and the Haughley Experiment.* Universe Books, New York.

Bateman, D.I. (1993) Financial and economic issues in organic farming: a case study in pluriactivity. *Journal of the University of Wales Agricultural Society* 73, 4–31.

Brighton, R.B., Cooke, J.G. and Armitage, J.C. (1988) *The Production, Distribution and Marketing of Organic Produce.* Worcestershire College of Agriculture, Hindlip.

Bulson, H. (1993) Organic cereals – what to sow. *Elm Farm Research Centre Bulletin* 9, 2–4.
Cudjoe, F.N. (1989) The nature and role of organic farming in Britain. MPhil thesis, School of Geography, University of Leeds.
Daw, M., Slee, B. and Wynen, E. (1991) *Organic Agriculture: A Review of the Marketing and Economics of Production with Particular Reference to Scotland.* SAC Economic Report, 32. Scottish Agricultural College, Aberdeen.
Dixon, P.L. and Holmes, J.C. (1987) *Organic Farming in Scotland.* School of Agriculture, Edinburgh.
EFRC (1993) *The Use of Legumes for Nitrogen Accumulation by Green Manuring in Organic Systems.* Final report, MAFF Contract CSA14660. Elm Farm Research Centre, Newbury.
FBS (1991) *Farm Business Survey in Wales – Statistical Results for 1989/90.* Department of Economics and Agricultural Economics, University College of Wales, Aberystwyth.
Holden, P. (1989) What price organic milk? *New Farmer and Grower* 24, 28–29.
Houghton, M. and Poole, A.H. (1990) *Organic Milk Production.* Genus Information Unit Report, 70. Genus Management, Wrexham.
Lampkin, N.H. (1990) *Organic Farming.* Farming Press, Ipswich.
Lampkin, N.H. (1993) The economic implications of conversion from conventional to organic farming systems. PhD thesis, Department of Agricultural Sciences, University of Wales, Aberystwyth.
Lampkin, N.H. and Bateman, D.I. (1993) The economics of organic farming in Wales, 1989. Discussion paper 93/3. Centre for Organic Husbandry and Agroecology, University of Wales, Aberystwyth.
Lampkin, N.H. and Measures, M. (eds) (1994) *1994 Organic Farm Management Handbook.* Department of Agricultural Sciences, University of Wales, Aberystwyth and Elm Farm Research Centre, Newbury.
Lowman, B. (1989) Organic beef production. In: Chamberlain, A.T., Walsingham, J.M. and Stark, B. (eds) *Organic Meat Production in the '90s.* Chalcombe Publications, Maidenhead, pp. 19–32.
MAFF (1991) *Farm Incomes in the United Kingdom.* HMSO, London.
Mayall, S. (1977) Costings of organic farms. *Soil Association Quarterly Review* (March), 5–8.
Murphy, M.C. (1992) *Organic Farming as a Business in Great Britain.* Agricultural Economics Unit, University of Cambridge, Cambridge.
Newton, J. (1994) Herbage production on organic farms. *Elm Farm Research Centre Bulletin* 11, 3–4.
Redman, M. (1991) Organic dairy costings: update. *New Farmer and Grower* 33, 20–21.
Stanhill, G. (1990) The comparative productivity of organic agriculture. *Agricultural Ecosystems and Environment* 30, 1–26.
Stopes, C. and Millington, S. (1993) Green manure-based arable rotations. *Elm Farm Research Centre Bulletin* 10, 2–4.
Thomas, G., Reeve, A. and Fisher, G.E.J. (1991) *Milk from Grass* (2nd edn). British Grassland Society, Reading.
Vine, A. and Bateman, D.I. (1981) *Organic Farming Systems in England and Wales:*

Practice, Performance and Implications. Department of Agricultural Economics, University College of Wales, Aberystwyth.

Younie, D. (1989) Eighteen month beef production: organic and intensive systems compared. In: Chamberlain, A.T., Walsingham, J.M. and Stark, B. (eds) *Organic Meat Production in the '90s.* Chalcombe Publications, Maidenhead, pp. 41–54.

Younie, D., Watson, C., Halliday, G., Armstrong, G., Slee, W. and Daw, M. (1990) *Organic Beef in Practice.* Scottish Agricultural College, Aberdeen.

6 Economics of Organic Farming in Germany

SUSANNE PADEL AND ULI ZERGER

Introduction

Germany is one of the countries in the world with the largest numbers of organic farms, with up to 5% of the land managed organically in some states. This may be related to the fact that important methods of organic farming, such as bio-dynamic farming, have been used in Germany since the 1930s, even though only in very small numbers. At the end of the 1970s, 500 farms were managed organically; by the end of the 1980s the number of organic farms had reached 2200, and stood at over 12,500 farms on 360,000 ha or 2.5% of the total area in 1993.

Germany also has one of the most important markets for organic products (including imports). Again there is a long history of organic marketing, starting with the first health food movement (*Reformbewegung*) at the beginning of this century. Until the end of the 1970s marketing of organic produce mainly happened as direct marketing and through health food shops (*Reformhäuser*). The growing green movement and environmental awareness led to the development of special wholefood shops (*Körnerläden*) towards the end of the 1970s, at which time the market was nearly 100% self-sufficient. By 1985, the idea of farmer–consumer cooperatives was developed, but never achieved great importance for the sale of organic products. By the end of the 1980s the demand for organic products had overtaken the supply and led to imports of organic products, mainly from other European countries, which accounted for almost 50% of organic sales in Germany (Hamm, 1991).

Until recently, the main certification organizations refused to cooperate with traditional food marketing channels and, as a result of increased consumer demand, so-called 'grey products' (products with a

Table 6.1. Physical and financial performance of organic and conventional farms, western Germany, 1992/93.

	Organic ('main income' farms)	Conventional (comparable farms)	Conventional (all 'main income' farms)	Relative (conv. compar. farms=100)
Farm characteristics				
Number of farms	101	444	8559	–
Farm size (ha agricultural area)	35.0	34.9	34.9	100
Land comparative value (DM ha^{-1})	1,255	1,273	1,365	99
Land use				
Crops (ha)	19.04	18.76	20.86	101
of which (% of crop area) are:				
Cereals	55.4	60.9	58.1	91
Potatoes	4.7	1.9	2.7	247
Sugarbeet	0.0	2.6	5.7	0
Vegetables and other cash crops	11.8	11.1	13.3	106
Forage maize	1.5	14.3	14.4	10
Other forage and temporary leys	26.7	9.3	5.8	287
Permanent grassland (ha)	15.1	16.1	13.6	98
Livestock stocking rates (livestock units ha^{-1} agricultural area)				
Total of which:	0.955	1.163	1.577	82
Dairy cows	0.442	0.578	0.453	76
Other cattle	0.429	0.523	0.618	82
Pigs	0.025	0.054	0.469	46
Poultry	0.029	0.005	0.026	580
Yields (t ha^{-1})				
Wheat	3.53	6.09	6.49	58
Rye	3.20	4.90	4.91	65
Potatoes	14.50	30.90	29.40	47
Milk (kg per cow)	3,915	4,990	5,088	78
Prices (DM t^{-1})				
Wheat	862	333	349	259
Rye	865	307	333	281
Potatoes	552	165	161	335
Milk (DM kg^{-1})	0.713	0.638	0.637	112

Output (DM ha^{-1} agricultural area)				
Cereals	539	338	406	159
Potatoes	150	41	69	366
Total crops	1,096	576	1,034	190
Milk	1,299	1,859	1,484	70
Other cattle	642	730	945	88
Pigs	71	110	891	65
Total livestock	2,190	2,723	3,412	80
Total output	4,694	4,369	5,636	107
Variable costs (DM ha^{-1} ag. area)				
Seeds/plants	152	73	95	208
Fertilisers	37	176	184	21
Sprays	3	80	110	4
Livestock	152	125	468	122
Feeds	167	327	647	51
Total	772	1,008	1,756	77
Gross margin (DM ha^{-1} ag. area)				
Output less variable costs	3,992	3,361	3,880	117
Fixed costs (DM ha^{-1} ag. area)				
Wages etc.	282	99	177	285
Fuel	81	81	84	100
Machinery repairs	323	281	300	115
Machinery depreciation	457	419	450	109
Rents	235	187	242	126
Interest	139	148	173	94
Total fixed costs	2,680	2,211	2,629	121
Total farm costs and profit				
Total costs (DM ha^{-1})	3,452	3,219	4,385	107
Profit (DM ha^{-1})	1,242	1,150	1,251	108
Profit as percentage of output	26.5	26.3	22.2	101
Profit per farm (DM)	43,444	40,115	43,642	108
Change in net worth (DM ha^{-1})	263	44	117	598
Labour use and returns to labour (full-time equivalents (FTE))				
Family labour (FTE per farm)	1.38	1.47	1.42	94
Profit (DM per FTE family labour)	31,414	27,272	30,630	115
Total labour (FTE per farm)	1.75	1.56	1.60	112
Total labour (FTE per 100 ha)	4.99	4.47	4.60	112
Return to total labour[a] (DM FTE^{-1})	30,460	27,928	31,136	109

[a] Profit plus wages per full-time equivalent family and employed labour.
Source: Adapted from BMELF (1994).

natural image but without a clear set of production standards) found their way into the supermarkets. In addition, attempts to establish a joint label for all organic products in Germany under the umbrella organization (Arbeitsgemeinschaft Ökologischer Landbau: AGÖL) failed, with the result that seven competing trademarks for organic products vied for consumer recognition. The German government was not prepared to create a legal framework for organic production as in some other European countries (Hamm, 1991). Several recent initiatives to market organic products with supermarkets have been started, but sales through these channels are developing slowly.

The increasing importance of organic farming has also led to recognition in research, even though economic research was not the main focus of research activities. However, the first reports on the economic situation of organic farming were published as early as 1973 (Ronnenberg, 1973). Some of these early studies were based on very small samples, e.g. Jochimsen (1982) based on four farms, and some, e.g. Jaep *et al.* (1985), are based on models rather than on survey data from actual farms.

In 1982, the federal government started to incorporate a sample of organic farms into the annual farm business monitoring report (*Agrarbericht*) and from this source 12 years' data are now available (BMELF, 1983–1994). One difficulty with the data is that both organic and conventional samples have changed over the years. Initially a very small sample of 15–20 organic farms from disadvantaged areas in southern Germany were compared with the national average for all farms, whereas the conventional farms were analysed by farm type, size and region. This unfavourable comparison was criticized by several people (e.g. Priebe, 1984; Heilmann, 1985) and subsequently the comparison group was changed. Since 1985–1986 the comparability of the groups has improved. The organic sample is now much larger, but it is still not clear that the group forms a representative sample (Priebe, 1990). The most recent of these annual comparisons is presented in Table 6.1 and discussed throughout this chapter.

In addition, some long-term trends derived from the *Agrarbericht* data are presented, which, in order to have a consistent comparison group, are based on a comparison between the organic sample and the average for all 'main income' farms (more than 50% of income derived from farming) in Germany. These comparisons differ from the published annual comparisons with their specially selected groups of conventional farms, but long-term trends in the development of each sector can be seen. In general, the comparisons with the specially selected group of conventional farms are more favourable than the comparisons used for the long-term trends.

The aim of this chapter is to summarize the main results of newer survey studies and to show trends in the development of the economic situation of organic farms in the western states of Germany in the last decade. The data presented focus on the economic situation of certified organic producers. No special reference is made to the conversion aid scheme introduced by the German government under the European Community's extensification programme that led to the rapid increase in the number of organic farms from 1989. The programme influenced the economic situation for some farms directly through subsidy payments (see Chapter 25), but according to the German standards those farms would not have achieved certification by 1991 and would therefore not have been included in the surveys presented in this chapter. However, established organic farmers faced disadvantages not only because of a lack of payments for already converted farms, but also through a fall in organic cereal prices in 1991 as a result of the increase in supply which might have influenced their performance. Also, no special reference is made to the situation of farms under conversion; this issue is discussed in Chapter 14.

Physical Productivity

Yields of arable crops

Yields of some important crops under organic management are shown in Table 6.2, based on a three-year survey of organic and bio-dynamic farms. Peitzmeier (1990) found strong differences between various regions. He also compared the yields between organic (Bioland) and bio-dynamic (Demeter) farms and found similar results for both groups. Lampkin (1993) compared the variation of organic relative to conventional winter wheat yields in 28 German studies from 1978 to 1992. Typically, the organic yields were 30–40% lower than conventional in the more recent studies, although the differences were much less in the earlier studies. Lampkin suggested that the increasing relative difference over time was a consequence of the continuing intensification of conventional systems.

The evaluation of the long-term *Agrarbericht* data as shown in Fig. 6.1 indicates a slight increase in the yield of winter wheat and potatoes under organic management over time, but a larger increase under conventional management. These results confirm the increased intensification in the conventional system compared to organic management. The most recent *Agrarbericht* data, using the more carefully selected conventional comparison group, are presented in Table 6.1. The yield differences of up to 40% for wheat and potatoes are consistent with the results from other studies.

Table 6.2. Average crop yields (t ha^{-1}), western German states, 1984–1986.

State	Winter wheat	Winter rye	Oats	Potatoes
Baden-Württemberg	4.10	3.63	3.70	18.77
Bayern	3.93	3.34	3.82	18.87
Hessen	4.04	3.21	3.25	18.20
Niedersachsen	3.28	2.80	–	19.91
Nordrhein-Westfalen	3.33	3.21	3.14	19.97
Rheinland-Pfalz	3.26	2.67	3.00	16.83
Schleswig-Holstein	3.81	3.33	3.87	18.77
Average	3.70	3.20	3.50	18.80

Source: Peitzmeier (1990).

The main factors influencing the yields of arable crops are soil quality and climate, crop rotation, stocking rate and time under organic management. In most cases it is difficult to quantify the effect of certain factors exactly, which makes it difficult to use them for forecasts of yield changes under organic management, for example for conversion planning. Certain differences in yields found in surveys of organic farms can only be explained by the influence of the farm management.

Influence of soil quality

A 100-point scale (*Bodenpunkt*) is used to rank the soil quality of every field for valuation and tax purposes. Several studies have found a relationship between these soil points and the yield under organic management (Gekle, 1982; Stöppler, 1989; Dabbert, 1990; Peitzmeier, 1990) but the relationships differ. For some crops, water availability has a stronger effect on yields than the effect of soil quality (Peitzmeier, 1990).

Influence of crop rotation and manuring

The main source of nitrogen for organic crops is nitrogen-fixing legumes, grown as forage crops, green manures or as arable legumes in the rotation, but very few studies have attempted to quantify their influence on yield. Dabbert (1990) was not able to find a relationship between yields and the proportion of legumes and cereals in the rotation. Peitzmeier (1990) found a relationship between wheat yields and ruminant stocking rate and a relationship between rye yield and the proportion of legumes in the rotation. The reason for the low

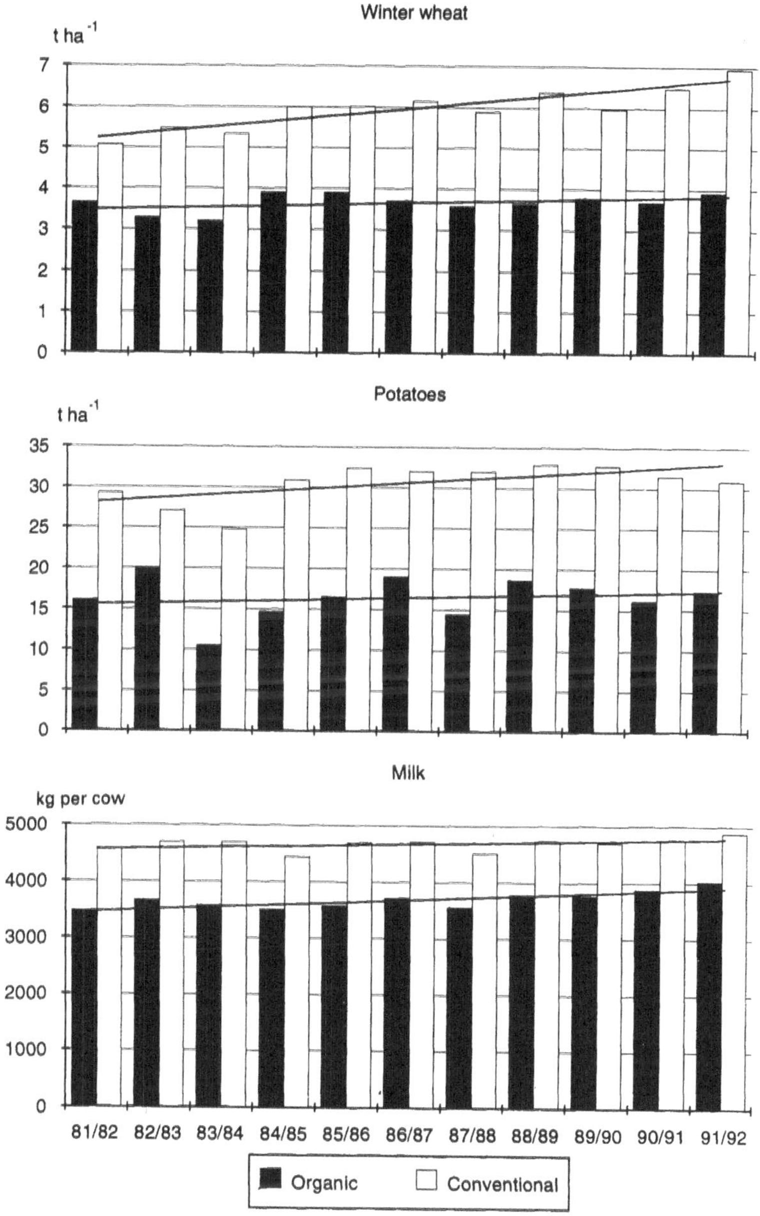

Fig. 6.1. Wheat, potato and milk yields on 'main income' organic and conventional farms, western Germany, 1981/82–1991/92. Source: BMELF (1983–1993).

significance of these relationships may well be the overlapping effect of soil quality and other influences on the cereal yield where survey data from several farms are used.

Time under organic management

Dabbert (1990) and Böckenhoff *et al.* (1986) showed an influence of the duration of organic management on the level of cereal yields on the farm. This might be due to a change in soil structure and biological activity as well as to mistakes in cultivation made by the inexperienced farmer in the early stage of conversion. This issue is discussed in more detail in Chapter 16.

Yields from livestock production

Dairy cows

Less research has been done on livestock productivity under organic management than on crops. Winter (1991) compared a group of 18 organic dairy farms with a conventional group in northern Germany, as shown in Table 6.3. Milk yields per cow were 90% of the conventional yield and stocking rates per hectare were lower under organic management. The organic farms used 73% of the quantity of concentrates used under conventional management and achieved a 14% higher milk yield from forage. Precan (1991) found similar results to Winter (1991) in a detailed three-year survey of 21 organic dairy farms in the state of Niedersachsen, but did not compare them with conventional results. Both reported a wide range for the organic group.

The average milk yields on organic farms reported in the *Agrarbericht* were lower than the results of Winter and Precan (22% lower than conventional in 1992/93 see Table 6.1), but the *Agrarbericht* farms cannot be characterized as specialist milk production farms and are located mainly in other regions of Germany. For all livestock enterprises, feed costs per livestock unit were 49% lower than conventional, while stocking rates were 18% lower in 1992/93. The long-term data from the *Agrarbericht* show a slight increase in the milk yield on the organic farms as well as on the conventional farms over the 11-year period (Fig. 6.1).

Other livestock enterprises

Some detailed data for other livestock enterprises are reported by Lölau and Zerger (1989, 1991) on the basis of a questionnaire survey of 120 organic farms with respect to the situation before and after

Table 6.3. Physical output and gross margins for dairy cows on 18 organic farms compared with conventional averages, northern Germany, 1989/90.

Parameter	Organic	Conventional	Relative (conv=100)
Milk yield (FCM[a]) (kg per cow)	5415	5977	90
Milk price (DM kg^{-1})	0.78	0.76	102
Concentrate use (tonnes per cow)	1.10	1.54	71
FCM[a] from forage (kg per cow)	3241	2832	114
Forage area (ha LU^{-1})	0.70	0.55	127
Costs of forage production (DM per cow)	522	604	86
Gross margin (DM per cow)[b]	3095	3177	97
Gross margin (DM kg^{-1})[b]	0.57	0.55	103
Gross margin (DM ha^{-1})[b]	4677	5805	80

[a] Fat-corrected milk (4.0% butterfat).
[b] Including forage costs.
Source: Winter (1991).

Table 6.4. Physical productivity of livestock enterprises by type, before and after conversion to organic production, Germany, 1989 and 1991.

	Conventional (1989, before conversion)	Organic (1991, after conversion)	Relative (1989=100)
Bull beef finishing (10 farms)			
Liveweight gain (grams per day)	886	819	92
Finishing liveweight (kg per bull)	642	648	101
Farrowing sows (9 farms)			
Number of piglets reared per sow per year	16.9	18.1	107
Finishing pigs (17 farms)			
Liveweight gain (grams per day)	478	391	82
Finishing liveweight (kg per pig)	105	117	111
Egg production (15 farms)			
Number of eggs per hen per year	207	220	106

Source: Lölau and Zerger (1989, 1991).

converting the farm. They have been briefly summarized in Table 6.4, but have to be treated with care because of the small size of the sample for each enterprise. The results demonstrate that beef production is less intensive under organic management than before conversion. It can be assumed that this effect results from the reduced amount of concentrates fed to the cattle.

Until this decade, intensive livestock did not play a major role on organic farms for the following reasons: pigs and poultry were seen as competitors for human food; there was no developed market for their products; feeding costs could not be covered by conventional market prices because of the high price level for organically produced cereals. In recent years, a growing market for these products has developed, so that premium prices can be realized. The production costs, on the other hand, have clearly benefited from the falling prices for organic cereals. The *Agrarbericht* reports that in 1992/93, breeding pigs on organic farms raised 14.2 piglets per year compared to 17.5 on conventional farms. The number of sows on organic farms is small and different results were found in previous years. Lölau and Zerger (1989, 1991) compared the number of piglets raised per year before and after conversion and found an increase. It can be assumed that organic production does not necessarily mean a reduction in the productivity of farrowing sows.

On most organic farms pig finishing is seen as a utilizer of organic by-products, but not as a very intensive enterprise. The results for daily growth rates on those farms before conversion are low compared to specialist farms and it can be concluded that those farms practised a rather extensive form of pig finishing before conversion. This level was reduced even further after converting the farm for the reasons given above (Lölau and Zerger, 1989, 1991). It is possible that pigs fed under organic management with a balanced ration could achieve higher growth rates, and individual examples show the potential (Baldenhöfer, 1992). Another reason for lower growth rates are higher final weights which result from different butcher requirements.

The situation for egg production is similar to that for finishing pigs. In addition, a high variation was found in results from the surveyed organic farms, with a minimum of 120 eggs per hen, while the best farms produced 290 eggs per hen (Lölau and Zerger, 1989, 1991). As for pigs, feeding experiments with well-balanced rations show that higher results could be achieved (Deerberg, 1992).

Enterprise Financial Performance

Prices

German farmers, in general, are not very market-oriented. The importance of marketing for organic farmers, in order to achieve a premium for the organic quality, was recognized at an early stage by Hans Müller, one of the pioneers of organic farming in Switzerland, who encouraged the foundation of a cooperative for organic vegetables in 1946 (Simon, 1990). For many farmers, conversion to organic farming meant getting to know the consumers who buy their products.

Premium prices have an important influence on the financial performance of organic farming and can be achieved in several ways: selling to a specialist organic market outlet or retailer; selling direct to the consumer; or processing on the farm. Traditionally, German organic farms preferred direct selling and/or specialized organic outlets (such as wholefood shops) to supermarkets, but the situation is changing.

The German retail market is reasonably well developed for organic cereals, but oversupply led to a fall in prices and to an increase in quality requirements at the beginning of the 1990s. Milling grain (wheat and rye) had been at a high price level during the 1980s (more than DM 1000 per tonne), but in the early 1990s the price for converting farms just entering the market came down to DM 600 per tonne. For potatoes, the situation is different because of strong annual influences; in general the price for organic potatoes has been twice the conventional price. Despite the increase in supply, the 1992/93 prices reported in the *Agrarbericht* for wheat, rye and potatoes were still close to three times the conventional level (Table 6.1).

Meat and milk were not sold through health-food or wholefood shops and a retail market for animal products is only now slowly developing through the supermarkets. In the beginning, there was no premium price for milk, because no dairies specialized in organic products. In the late 1980s, this situation slowly changed, so that an average premium price of about DM 0.10 per kilogram (15% above conventional) was achieved by the organic farmers in the *Agrarbericht* sample in 1992/93 (Table 6.1). In order to achieve premiums for meat, farmers frequently sell direct to the consumer, often in cooperation with a local butcher to meet the strict hygiene regulations.

The retail market for fresh vegetables was not very well developed in the past. Many growers developed their own local outlet with a stall on the local market or farm shops, whereas a large proportion of the demand for fresh vegetables in the wholefood shops was covered by imports. The limited availability of local fresh vegetables acts as a barrier to the development of supermarket outlets.

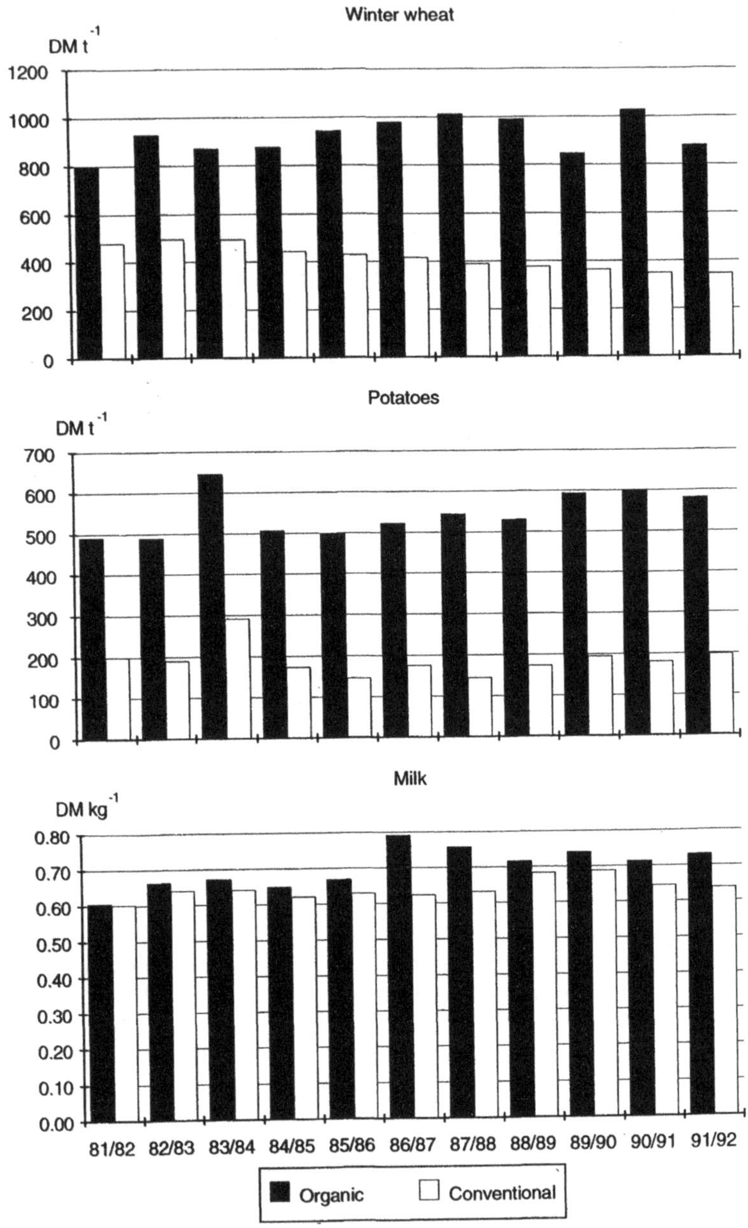

Fig. 6.2. Prices for wheat, potatoes and milk on 'main income' organic and conventional farms, western Germany, 1981/82–1991/92. Source: BMELF (1983–1993).

Long-term price trends from the *Agrarbericht* for wheat, potatoes and milk are illustrated in Fig. 6.2. However, these figures do not give any recognition to the impact of different marketing channels. For the comparison of prices and premiums, these different marketing channels should be differentiated and, for example, farm shops or processing facilities on the farm treated as separate enterprises. A survey by Alvermann and Padel (1991) used this approach and found that in 1989/90, 14% of the whole-farm gross margin resulted from direct marketing activities, even in the rural state of Schleswig-Holstein. A subsequent survey by Holle *et al.* (1993) for the same region, but with a different number of farms, reported only 8% of the gross margin resulting from direct sales for the year 1991/92, with some variation by farm type: mainly arable farms only realized 4% of gross margin from direct marketing activities while livestock producers achieved 17%.

Variable costs (excluding machinery and labour)

Seed costs were 70% higher on organic farms in a three-year comparison of ten organic farms in northern Germany with groups of similar conventional farms (Alvermann and Padel, 1991). This was due to the greater use of more expensive legumes on organic farms and to a greater use of green manures and catch crops. The amount spent on fertilizers was only 30% of the conventional figure, and the concentrate use only 65% of the conventional figure (including home-grown cereals calculated with a notional value). Total variable costs were 45% lower than on conventional farms.

The *Agrarbericht* data for 1992/93 (see Table 6.1) also show higher seed costs and lower other variable costs on the organic farms, especially fertilizer and spray costs which together were 75% lower than conventional costs. The long-term data show a reduction in the

Table 6.5. Crop variable costs[a] on 32 farms, northern Germany, 1990/91.

	Organic (DM ha^{-1})	Conventional (DM ha^{-1})	Relative (conv=100)
Winter wheat	372	870	42
Winter rye	273	594	45
Spring wheat	311	785	39
Oats	240	590	39
Field beans	384	609	63
Potatoes	2082	2004	104

[a] Excluding machinery costs.
Source: Zerger (1994).

level of total variable costs for both organic and conventional groups, with variable costs on organic farms typically 50–60% lower than conventional ones (Fig. 6.3). The main change on the organic farms over the 11-year period has been reduced expenditure on livestock feeds.

Variable costs excluding machinery costs on organic farms have also been reported by Zerger (1994) for 32 farms in the northern part of Germany for 1990/91. These results are compared to average variable costs (minus machinery costs) calculated by the KTBL (1992) for conventional farms (Table 6.5).

These results confirm those of Alvermann and Padel (1991), with cereal variable costs 41% lower on the organic farms. For field beans, the difference is smaller because of higher seed costs for organic farmers and because of less intensive conventional management of this crop. Variable costs for organic and conventional potatoes are almost identical for similar reasons.

Gross margins

Crops

The comparison of farming systems on the basis of enterprise gross margins is not without difficulties. The necessary changes in rotational design, such as the introduction of legumes in order to meet crop nitrogen requirements (see below), are not reflected in gross margins for individual crops. Consequently, very few of the survey studies

Table 6.6. Crop and dairy enterprise gross margins on organic farms compared to conventional standard gross margins, northern Germany, 1991/92.

	Organic (DM ha^{-1})	Conventional (DM ha^{-1})	Relative (conv =100)
Winter wheat	2,812	1,548	182
Winter rye	2,597	999	260
Spring wheat	2,158	1,218	177
Oats	2,304	831	277
All cereals	2,431	–	–
Field beans	1,857	967	192
Potatoes	11,584	4,983	232
Carrots	11,644	–	–
All cash crops	2,585	–	–
Milk (DM per cow)	3,322	2,279	146

Sources: KTBL (1992) and Zerger (1994).

reported here compared organic and conventional gross margins. Some examples of crop gross margins are reported by Zerger (1994), who compares the results of 32 farms for the year 1990/91 with standard gross margins from KTBL (1992) (Table 6.6). This comparison is problematic as it involves a comparison of a limited number of organic farms in northern Germany with data based on national averages, and must therefore be treated with care. Overall, there is a clear advantage for organic farms, with differences of about 150–250% compared to conventional farms.

Milk

Winter (1991) found small differences in the gross margin (including machinery and forage costs) per cow and per kilogram of milk produced (an important measure of the efficiency of quota utilization) which could be accounted for by the difference in milk price obtained (Table 6.3). The margin per hectare of forage area was lower in the organic systems due to the greater forage area required on the organic farms. Winter concluded that in dairy farming, ecological goals, such as the use of more forage instead of concentrates, do not need to be at the expense of financial success.

Whole-farm Financial Performance

Structure of enterprises on organic holdings

Conversion to organic farming means not just changing practices but a change to a different farming system. Organic farming relies mainly on farm-derived resources and within-farm transfers. This usually means changes to the enterprise structure on the farm, and these changes will influence the economic performance and fixed and variable costs. Survey data, including those of the *Agrarbericht* (Table 6.1), as well as case studies of conversion (see, for example, Chapter 14) give an indication of the extent of these changes. The area of forage and grain legumes crops increases to about 26% of the cropping area (three times the conventional proportion), whereas the proportion of cereals is slightly lower (92% of conventional) and sugarbeet and forage maize markedly lower (at 25% and 11% of conventional, respectively). A survey by Böckenhoff *et al.* (1986) in the state of Baden-Württemberg found that organic farmers grow a higher proportion of bread-making cereals (75% compared with 45%). Forty-five per cent of the organic farms, compared with 4% of the conventional farms, grew some vegetables and, in general, more root crops, apart from sugarbeet, were grown on the organic holdings.

Böckenhoff *et al.* (1986) also found a reduction in total stocking rates and on several farms direct marketing and farm processing were introduced as new enterprises following the conversion to organic farming. The *Agrarbericht* data (Table 6.1) confirm the lower stocking rates (18% lower overall), and illustrate the increased importance of ruminant livestock relative to pigs.

Total farm output and gross margins

The long-term *Agrarbericht* data show consistently lower total farm output on the organic farms, with little growth in either group over the 11 years (Fig. 6.3). However, lower variable costs resulted in similar whole-farm gross margins per hectare. In 1992/93 the organic farms achieved 7% higher total output levels than the more closely matched comparison group (Table 6.1), with the 90% higher crop output reflecting the greater importance of crop enterprises and premium prices. Variable costs on the organic farms were 23% lower than conventional, resulting in whole-farm gross margins per hectare 17% higher than conventional.

Alvermann and Padel (1991) found that the total financial output of the organic farms was 20% lower than the conventional farms despite premium prices, reflecting the lower intensity of the organic farms. The share contributed by crop output was about 20% higher on the organic than on the conventional holdings, which reflects the more difficult situation of premium marketing of organic animal products. As a result of lower variable costs the total gross margin was on average 15% higher on the organic holdings than on the conventional farms. A follow-up survey by Holle *et al.* (1993) in the same region found 20% higher total gross margins on organic farms for two of three farm types studied; in the third group the average farm gross margin was similar to conventional.

Fixed costs

The long-term comparison of organic and conventional farms from the *Agrarbericht* data shows a tendency for fixed costs to increase for both organic and conventional farms during the period, but with no clear difference between them (Fig. 6.4). This hides the impact of increased labour and machinery costs arising from the more diverse enterprise structure and the impact of new labour-intensive crop and marketing enterprises such as field vegetables and direct marketing.

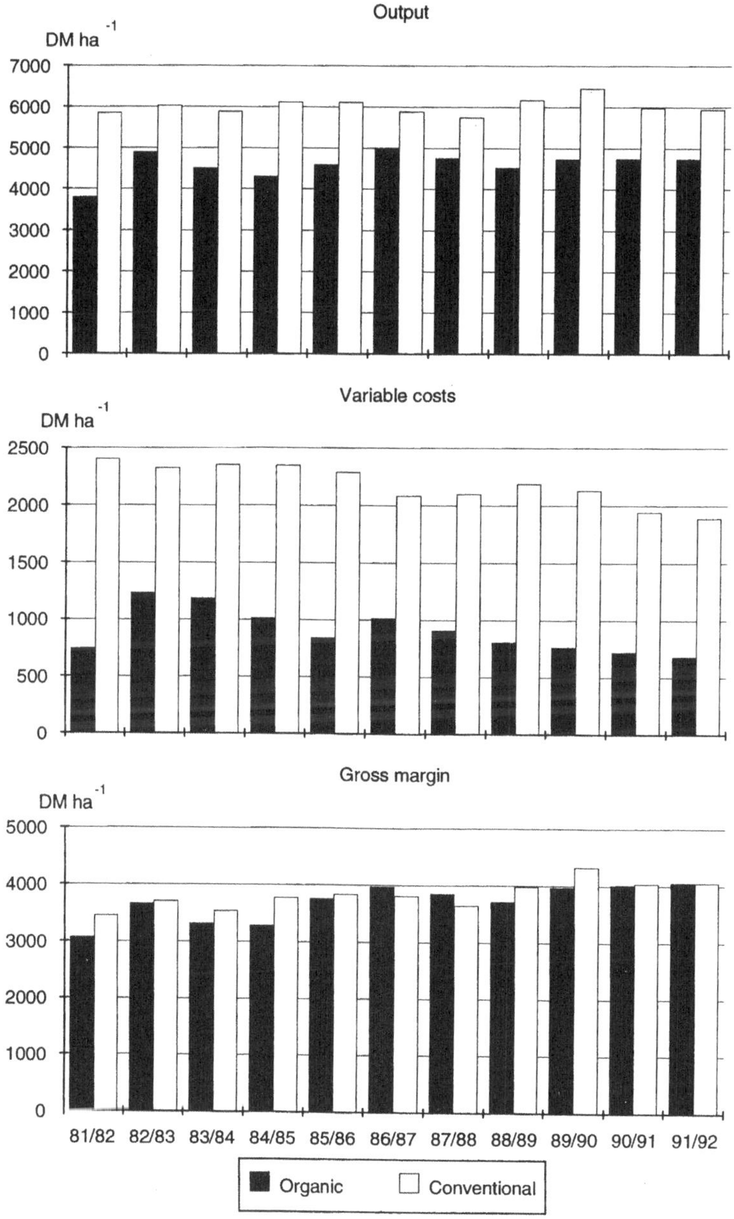

Fig. 6.3. Whole-farm output, variable costs and gross margins on 'main income' organic and conventional farms, western Germany, 1981/82–1991/92. Source: BMELF (1983–1994).

Machinery costs

Although some machinery costs such as fuel vary with the amount of production, they are treated as fixed costs, since most farm accounting systems in Germany do not allocate machinery costs to production enterprises and present only a summary of all machinery costs for the whole farm. Schlüter (1985) found 2–32% higher machinery costs (including diesel, maintenance and depreciation) on bio-dynamic farms in Baden-Württemberg compared with conventional farms. These effects can be explained by the lower level of specialization on the organic farms and possibly by higher machinery use as a result of mechanical cultivation. None of the studies has tried to differentiate these effects. Alvermann and Padel (1991) reported 15% higher costs for machinery and labour on organic farms compared with conventional farms in the state of Schleswig-Holstein, but the study does not differentiate between labour and machinery. Holle *et al.* (1993) found no evidence for higher machinery costs on organic farms in the same region in 1991/92. The machinery costs ranged between DM 800 per hectare for arable farms and DM 900 per hectare for livestock farms. Zerger (1994) found slightly higher costs than Holle *et al.* with arable farms averaging DM 830 per hectare compared to DM 1100 per hectare for livestock farms for the years 1988–1992. Because of smaller farm sizes in the southern parts of Germany higher machinery costs per hectare can be expected. In the *Agrarbericht* data (Table 6.1), fuel costs were identical, repair costs 15% higher and depreciation costs 9% higher on the organic farms than on the comparable conventional farms.

Labour use and costs

Even though additional labour requirements are frequently mentioned as a characteristic of organic farming, not many studies present a detailed analysis of this subject. Schlüter (1985) found 20–200% higher wage costs on bio-dynamic farms in Baden-Württemberg. The main reason for this increased labour input was an increase in the cultivation of root crops and vegetables on the bio-dynamic farms. Little difference was found in the labour requirement for cereal and livestock enterprises between conventional and bio-dynamic farms. Another reason for higher labour requirements after conversion to organic farming is the increased importance of direct marketing and processing on organic farms (Rantzau *et al.*, 1990).

The analysis of the long-term data from the *Agrarbericht* shows an initial increase in the workforce on the organic farms which may be due to the changing structure of the organic sample. This was followed by a decline in the numbers of full-time equivalent (FTE) workers from

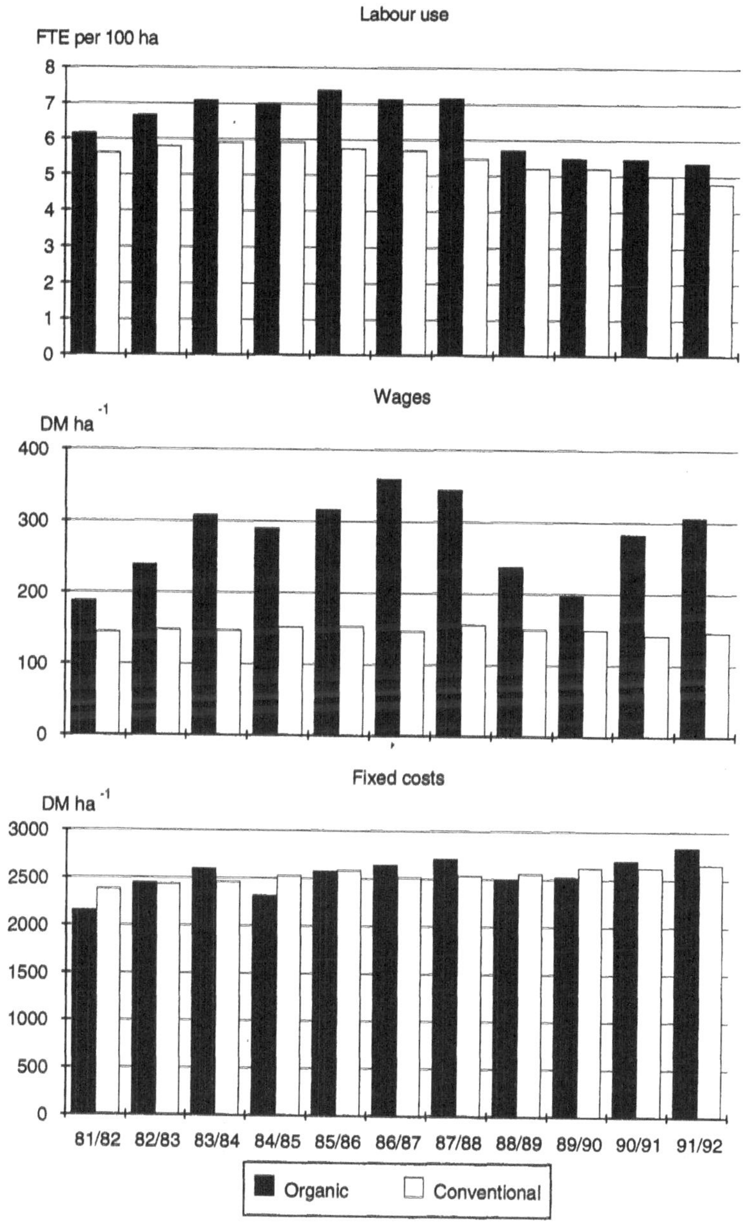

Fig. 6.4. Labour use (FTE per 100 ha), paid labour and fixed costs (DM ha^{-1}) on 'main income' organic and conventional farms, western Germany, 1981/82–1991/92. FTE, full-time equivalents (2200 h year^{-1}). Source: BMELF (1983–1994).

1985/86, so that by 1991/92, the FTE use per 100 ha was only about 10% higher than on all 'main income' conventional farms (Fig. 6.4). The conventional farms were characterized by very stable average wages and salaries cost levels, while there were wide variations in labourcosts on the organic farms. In general, the labour costs were almost twice as high as on the conventional farms. The differences are more marked when the organic farms are compared with the more closely matched conventional farms (Table 6.1). Labour use was almost 12% higher on the organic farms, while wages costs were nearly four times as high.

The study of 250 farms in Baden-Württemberg by Böckenhoff *et al.* (1986) showed that the differences in labour inputs are much higher on small organic farms (60.5 compared with 23.3 FTE per 100 ha on farms under 10 ha), whereas on larger organic farms only a slight increase occurs (4.9 compared with 3.4 FTE per 100 ha on farms over 50 ha). Based on a survey of 58 organic farms, Zerger (1994) also found a decrease in rate of labour use as farm size increases (Table 6.7). Zerger's

Table 6.7. Labour use on organic farms by farm size, Germany, 1988/89–1991/92.

Farm size (ha)	Labour use (FTE[a] per 100 ha)	Number of farms
< 20	13.5	2
20–30	7.4	13
30–40	5.6	11
40–60	4.1	10
60–80	2.7	12
80–100	2.9	4
> 100	2.2	6
Total	4.9	58

[a] Full-time equivalents (2200 h year^{-1}).
Source: Zerger (1994).

Table 6.8. Fixed costs on organic arable and livestock farms, northern Germany.

	Arable farms		Livestock farms	
	DM ha^{-1}	Number	DM ha^{-1}	Number
1988/89	2220	7	2439	17
1989/90	1822	17	2331	17
1990/91	2182	17	2126	13
1991/92	1726	13	2139	7
Average	1963	54	2290	54

Source: Zerger (1994).

results are lower than those of Böckenhoff *et al.* because of regional differences and a shortage of available workers since the late 1980s.

It is likely that there are important differences between the various types of organic farms. Holle *et al.* (1993) found higher wages on organic farms in two of three comparison groups; the third group had 40% lower wages than the conventional comparison group. One of the groups with higher wages was characterized as having a large proportion of field-scale vegetables and seasonal labour. The group with lower wages represents less intensive arable farms.

For the purpose of conversion planning and the planning of organic farms, two collections of data for enterprise costings and labour requirements under organic management have been published (Redelberger *et al.*, 1990; KTBL, 1991). These collections are not based on the detailed observation of organic farms, but on the recalculation of conventional standard data on the basis of reported changes in the cultivation procedures under organic management.

Other fixed costs

Alvermann and Padel (1991) and Holle *et al.* (1993) found between 10% and 40% higher amounts for other fixed costs on organic holdings compared with conventional figures. This increase can be explained by higher general expenditure on marketing, certification, and in some cases higher costs for training and advice. On some farms, investment in marketing or processing facilities may lead to an increase in the depreciation figure for buildings, but the survey data do not cover this aspect in great detail. The results of Zerger (1994) illustrate that there is a clear difference between arable and livestock farms regarding fixed costs (Table 6.8).

Gross income and profit

The published *Agrarbericht* comparisons of organic farms with the more carefully selected comparison group show a higher profit rate for organic farms since 1986/87. In 1992/93, profits per hectare and per farm were 8% higher on the organic farms (Table 6.1). Similar results have been found in other studies. Alvermann and Padel (1991) found 15% higher gross incomes on the organic farms relative to conventional, and Holle *et al.* (1993) found 10–40% higher gross incomes on the organic farms.

The long-term *Agrarbericht* data comparing organic farms with all 'main income' farms in Germany (Fig. 6.5) show a different picture, with the profit levels on the organic farms lower than for the conventional holdings. The contrast between the two comparisons illustrates clearly the importance of the selection of the comparison

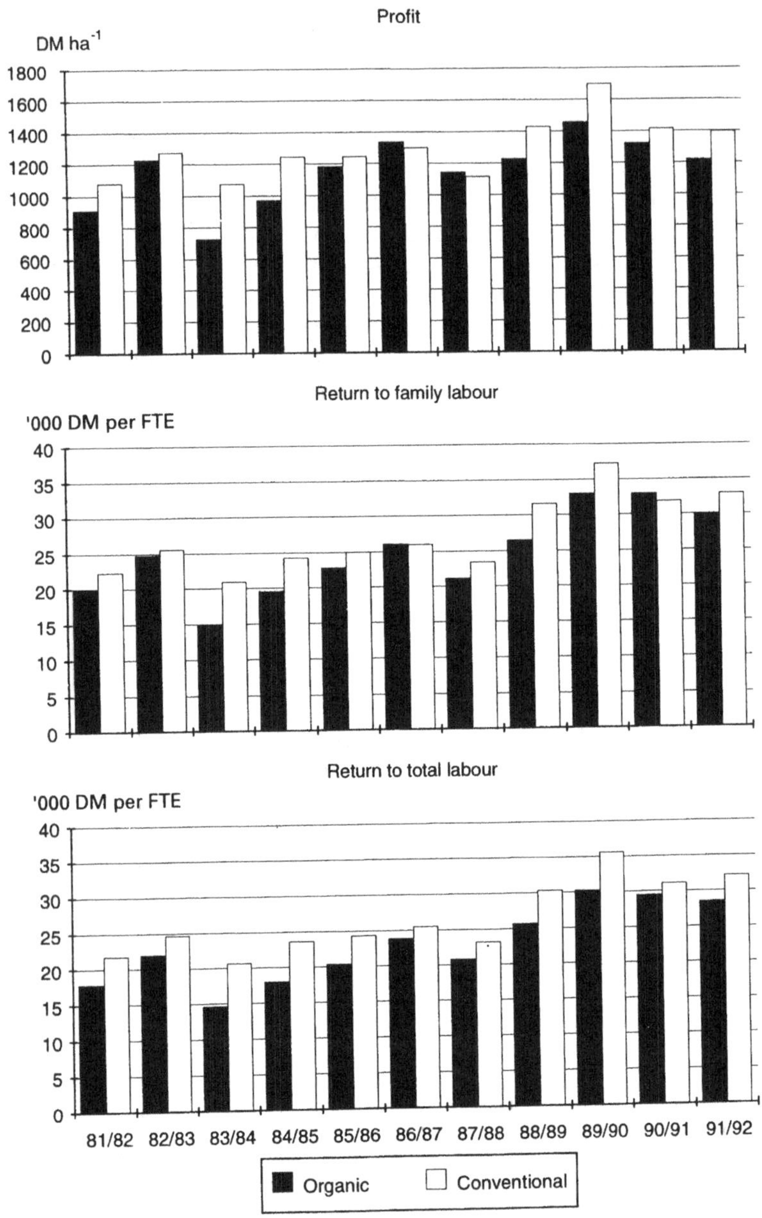

Fig. 6.5. Profit per unit land area (DM ha^{-1}), profit per unit family labour and profit plus wages per unit total labour (DM per FTE) on 'main income' organic and conventional farms, western Germany, 1981/82–1991/92. FTE, full-time equivalents (2200 h $year^{-1}$). Source: BMELF (1983–1994).

group in assessing the relative financial performance of organic systems. Figure 6.5 illustrates the high degree of variation in the results of both organic and conventional farms over time, although the profit levels on the organic farms generally follow those of the conventional farms. This implies that the influence of external factors, such as climate and agricultural policy changes, on profits in both systems is substantial.

Returns to family labour, measured as profit per full-time equivalent (FTE) family labour unit, are closer to conventional levels in the long-term comparison (Fig. 6.5), and were 15% higher than the more carefully matched comparison group in 1992/93 (Table 6.1). Returns to total labour, measured as profit plus wage costs per FTE family and employed labour, are typically 10% lower than conventional levels in the long-term comparison (Fig. 6.5), but when compared with the more closely matched conventional farms, returns to total labour were 9% higher on the organic farms in 1992/93 (Table 6.1). This measure is an important indicator of the potential contribution of organic farming to social employment and income objectives (see Chapter 5). The lower returns per FTE total labour compared with returns per FTE family labour on the organic farms may be due to a higher reliance on student labour, although no direct evidence exists to support this. The *Agrarbericht* data do, however, indicate the potential for organic farming to support both higher employment and higher returns to the family and hired labour employed.

Summary and Conclusions

These data show that under German conditions organic farming was on average equally profitable when compared with conventional farms of similar type. Lower yields for arable crops are compensated by reduced costs for variable inputs and by premium prices which are available for most crops. Moderately reduced yields in milk production are compensated by reduced input costs and dairying is exceptional as an enterprise where in some cases income can be maintained without access to premium prices, but the organic farms have lower stocking rates and therefor lower returns per hectare. Very little is known about the situation for other animal enterprises.

Increases in labour and machinery costs are not only a result of the specific requirements of an organic system, but have to be seen in relation to the different enterprise structure on organic farms. Organic farms are likely to have a lower proportion of cereals, a higher proportion of forage crops and an increased area of vegetables and root crops compared with conventional farms.

Compared with carefully selected conventional comparison groups, organic farms have higher profits per hectare and per farm (BMELF, 1983–1994). The long-term trends show that profits on organic holdings follow the profit trends of agriculture in general, but where these trends are based on comparisons with the national average for 'main income' holdings, organic farms have lower profits per hectare. Similar profitability on average does not, of course, mean that each individual farm achieved a profit. Survey studies like Alvermann and Padel (1991) mention large variation within the organic group. Padel (1992) therefore states that the requirement for management ability usually increases with the conversion to organic farming and conversion cannot be expected to save a farm already in financial difficulties. However, increased profit is not usually mentioned as an exclusive motivation for conversion to organic farming. It can be concluded that conversion to organic farming in the past did not generally mean an economic disadvantage. This may explain the more frequent mentioning of financial stability as a reason for converting to organic farming in the more recent studies.

The picture is very likely to be different in the future. The conversion aid scheme led to a substantial increase in the number of organic farms in Germany and has influenced the market for organic products and the premiums for crops. The absence of an organic certification requirement in order to qualify for the aid payment, and the foundation of further certification bodies, increased confusion on the market and for the consumer. Existing organic farms, which did not receive any payments, may have been disadvantaged through falling prices. The studies providing the material for this chapter did not include farms receiving payments under the conversion aid scheme (for effects of the scheme, see Chapter 25), but future surveys will need to differentiate between those farms which were or were not supported and to show the influence of these payments on the profit of organic farms.

Current changes in European agricultural policy (CAP reform) will influence farming profitability in general and also the performance of organic farms. The introduction of a support scheme for conversion and continuing organic farming as part of the new agri-environment programme (Chapter 26) will also have a significant impact on the profitability of organic farming although payments to existing organic farmers will not be available in all German states.

The future of the organic industry will also depend on the market development for organic products. The aim must be to reach more consumers, potentially through supermarkets, if price premiums are to be maintained in the context of increasing supply. The European Community's organic certification scheme (EC, 1991), the single market and the GATT agreement will increase pressure on the German producers and their organizations and the future of an increasing number of organic trade marks appears uncertain.

Last but not least, German reunification also led to the development of organic farming in the former East Germany, where the general need for change appears to make the decision to convert to organic farming easier for some private farms and cooperatives. The increase of large-scale organic production in the eastern German states will have an influence on the market and on the profitability of organic farming, particularly for the generally smaller-scale organic farmers in the western states of Germany.

References

Alvermann, G. and Padel, S. (1991) Betriebswirtschaftliche Aspekte des alternativen Landbaus. *Betriebswirtschaftliche Mitteilungen der Landwirtschaftskammer Schleswig-Holstein* 433, 3–12.

Baldenhöfer, M. (1992) Schweinehaltung. In: Neuerburg, W. and Padel, S. (eds) *Organisch-biologischer Landbau in der Praxis.* BLV-Verlag, Munich, pp. 262–273.

BMELF (1983–1994) *Agrarberichte der Bundesregierung.* Bundesministerium für Ernährung, Landwirtschaft und Forsten, Bonn.

Böckenhoff, E., Hamm, U. and Umhau, M. (1986) Analyse der Betriebs- und Produktionsstrukturen sowie der Naturalerträge im alternativen Landbau. *Berichte über Landwirtschaft* 64(1), 1–39.

Dabbert, S. (1990) *Zur optimalen Organisation alternativer landwirtschaftlicher Betriebe. Untersucht am Beispiel organisch-biologischer Haupterwerbsbetriebe in Baden-Württemberg.* Agrarwirtschaft Sonderheft 124. Verlag Alfred Strothe, Frankfurt am Main.

Deerberg, F (1992) Hühner und Geflügelhaltung. In: Neuerburg, W. and Padel, S. (eds) *Organisch-biologischer Landbau in der Praxis.* BLV-Verlag, Munich, pp. 276–286.

EC (1991) Council Regulation (EEC) No. 2092/91 of 24 June 1991 on organic production of agricultural products and indications referring thereto on agricultural products and foodstuffs. *Official Journal of the European Communities* 91(L198), 1–15.

Gekle, L. (1982) *Ökonomische Aspekte des alternativen Landbaus.* Schriften der Gesellschaft für Wirtschafts- und Sozialwissenschaften des Landbaus e.V., Münster-Hiltrup.

Hamm, U. (1991) 20 Jahre Vermarktung von Bioland-Produkten aus der Sicht eines Ökonomen. *Bioland* 91(2), 32–34.

Heilmann, H. (1985) Kritik am Vergleich der Buchführungsdaten alternativer und konventioneller Betriebe im Agrarbericht 1985. *Lebendige Erde* 85(6), 285–286.

Holle, R., Winter, R., Alvermann, G. and Zerger, U. (1993) Ökonomische und arbeitswirtschaftliche Kennzahlen ökologisch wirtschaftender Betriebe in

Schleswig-Holstein. Studie im Auftrag des Ministers für Landwirtschaft, Ernährung, Forsten und Fischerei, Schleswig-Holstein, Kiel (unpublished).

Jaep, A., Scharf, A., Budig, M., Voelkl, G. and Weiss, J. (1985) *Konventioneller und alternativer Landbau – ein betriebswirtschaftlicher Vergleich.* IfB Landwirtschaftliche Fachinformation, 143/85. Hessisches Landesamt für Ernährung, Landwirtschaft und Landentwicklung, Kassel.

Jochimsen, H. (1982) *Alternativer Landbau im Vergleich: bessere Produkte, höhere Preise, mehr Gewinn?* Betriebswirtschaftliche Mitteilungen, 325. Landwirtschaftskammer Schleswig-Holstein, Kiel.

KTBL (1991) *Datensammlung alternative Landwirtschaft.* Kuratorium für Technik und Bauwesen in der Landwirtschaft, Darmstadt.

KTBL (1992) *Standarddeckungsbeiträge 1991.* Kuratorium für Technik und Bauwesen in der Landwirtschaft, Landwirtschaftsverlag, Münster-Hiltrup.

Lampkin, N.H. (1993) The economic implications of conversion from conventional to organic farming systems. PhD thesis, Department of Agricultural Sciences, University of Wales, Aberystwyth.

Lölau, S. and Zerger, U. (1989, 1991) Datenauswertung im Rahmen des Agrarkulturpreises der KLS-Stiftung. Munich (unpublished).

Padel, S. (1992) Betriebswirtschaft. In: Neuerburg, W. and Padel, S. (eds) *Organisch-biologischer Landbau in der Praxis.* BLV-Verlag, Munich, pp. 35–48.

Peitzmeier, M. (1990) Bestimmungsgründe pflanzlicher Erträge im ökologischen Landbau. Diplomarbeit am Fachgebiet ökologischer Landbau, Gesamthochschule Kassel, Witzenhausen.

Precan, D. (1991) *Untersuchungsvorhaben Milchviehhaltung und Futterbau im ökologischen Landbau,* Vol. 3. Jahresbericht und Schlußbericht, Ökoring Niedersachsen, Walsrode.

Priebe, H. (1984) Zur Diskussion um die alternative Landwirtschaft. *Lebendige Erde* 84(3), 106–109.

Priebe, H. (1990) Ergebnisse ökologischer Wirtschaftsweise. *Lebendige Erde* 90(3), 162–167.

Rantzau, R., Freyer, B. and Vogtmann, H. (1990) *Umstellung auf ökologischen Landbau.* Angewandte Wissenschaft 389. Landwirtschaftsverlag, Münster-Hiltrup.

Redelberger, H., Scharf, A. and Voelkl, G. (1990) *Datenmaterial für die Betriebsplanung – ökologischer Landbau – Pflanzenproduktion.* IfB: Information für Beratung und Verwaltung, 115/90. Hessisches Landesamt für Ernährung, Landwirtschaft und Landentwicklung, Kassel.

Ronnenberg, A. (1973) *Ökonomische Aspekte der biologisch-dynamischen Wirtschaftsweise – Konsequenzen für den Einzelbetrieb und für den Produktmarkt.* Verlag 'Lebendige Erde', Darmstadt.

Schlüter, C. (1985) *Arbeits- und betriebswirtschaftliche Verhältnisse in Betrieben des alternativen Landbaues.* Agrar- und Umweltforschung in Baden-Württemberg, 10. Eugen Ulmer, Stuttgart.

Simon, B. (1990) Vorwärts zur Natur, Geschichte und Gegenwart des organisch-biologischen Landbaus nach Müller-Rusch. Diplomarbeit am Fachgebiet ökologischer Landbau, Gesamthochschule Kassel, Witzenhausen.

Stöppler, H. (1989) *Weizen im ökologischen Landbau.* KTBL-Arbeitspapier, 138.

Kuratorium für Technik und Bauwesen in der Landwirtschaft, Darmstadt.
Winter, R. (1991) Economic questions of dairy production in ecological agriculture in northern Germany. In: Boehncke, E. and Molkenthin, V. (eds) *Alternatives in Animal Husbandry.* Proceedings of the International Conference, July 1991, University of Kassel, Witzenhausen.
Zerger, U. (1994) Betriebsvergleiche im ökologischen Landbau - vorläufige Auswertungen und Arbeitsergebnisse. Universität Kassel, Witzenhausen (unpublished).

7 Economics of Organic Farming in Denmark

ALEX DUBGAARD

Introduction

Based on empirical findings, this paper attempts to assess to what extent organic agriculture represents an economically viable alternative to conventional agricultural practices. The focus is on the economic consequences of the limitations on inputs in organic farming – primarily the non-acceptance of chemical fertilizers and pesticides and the limitations on the use of non-organically produced feedstuffs. The statistical basis for the study is an agricultural accounts survey describing the economic characteristics and performance of organic farming in Denmark relative to conventional agriculture. The following aspects of economic performance have been analysed and compared: enterprise structure; crop yields; price relations; labour and capital intensity; and factor remuneration.

Characteristics of the Farms Surveyed

The study is based on accounts from 36 organic or bio-dynamic farms selected from the population of about 350 such farms in Denmark in 1988 (Dubgaard *et al.*, 1990). All farms selected had started conversion before 1985 and 90% of the agricultural area of these farms had been converted to organic practices by 1988. Three-quarters of the farms in the sample were fully converted.

The statistical description of organic agriculture in Denmark was facilitated by the existence of an official definition of organic farming and a control system ensuring compliance with the standards set. The selection of organic farms for the accounts analysis was based on a statistical survey recording structural characteristics and time of conversion for 331 of the organic and bio-dynamic farms in Denmark in 1988 (Dubgaard and Sørensen, 1988).

The problems of comparing the two farming systems have been addressed by classifying the conventional farms in groups comparable to the organic sample. Some of the most important structural causes of differing economic performance have been eliminated such as larger average farm size and the much larger importance of pig production in conventional agriculture. Still, there are of course limitations of the representativeness due to the small size of the organic sample.

Structural characteristics of organic agriculture

The enterprise structure in organic farming must be seen in relation to the absence of synthetically compounded fertilizers, pesticides, growth regulators, and livestock feed additives. Due to these restrictions, organic farmers must place considerable emphasis on recycling of plant nutrients, biological nitrogen fixation, hand weeding, and husbandry methods to control pests. In addition, the organic farm sector must maintain a fairly high degree of self-sufficiency with respect to nutrients.[1] These limitations on factor availability affect the (shadow) prices of resources in organic farming, leading to specialization patterns different from those of conventional agriculture.

For example, the nitrogen-fixing ability of clover is a factor alleviating to some extent the yield restrictions due to the absence of chemical fertilizers in organic agriculture. In addition, there are considerable nitrogen carry-over effects in (non-nitrogen-fixing) crops following clover–grass mixtures and other legumes. Also, clover–grass leys are relatively resistant to pests and quite competitive in relation to weeds, i.e. the absence of herbicides does not affect labour requirements significantly.

This is clearly reflected in the cropping patterns reported in Table 7.1. In the organic sector grass and green fodder occupies nearly half the agricultural area, whereas conventional agriculture uses only 20% of the land for this type of crop. Conversely, only one-third of the organic area is being used for cereals and oilseeds. In conventional agriculture 70% of the farmed area is devoted to these crops.

For row crops the absence of herbicides implies that weed control is highly labour-intensive in organic agriculture. The row crops grown

---

[1] Danish standards for organic farming limit the purchase of (non-organically produced) animal feeding stuffs to 15% of total feed consumption (measured in feed units). All purchased feed must be of domestic origin. The use of manure from conventional farms and other forms of conventional off-farm plant nutrients is limited to a maximum transfer of 80 kg ha^{-1} $year^{-1}$ nitrogen and 16 kg ha^{-1} $year^{-1}$ phosphorus.

Table 7.1. Structural characteristics of organic and conventional farm samples, Denmark, 1988.

	Organic[a]	Conventional (comparable[b])	Conventional (whole sector)	Relative[c] (conv=100)
Land area (ha per farm)				
Agricultural	27.7	28.7	33.1	97
Organic	24.6	–	–	–
Livestock density (number per 100 ha)				
Dairy cows	53	37	31	143
Cows for suckling	6	2	3	300
Heifers and heifer calves	43	32	30	134
Bulls and bull calves	9	20	17	45
Pigs	10	43	220	23
Sheep (ewes)	11	–	2	–
Hens and chickens	37	–	524	–
Cropping pattern (% agricultural area)				
Cereals (grain)	31	56	57	55
Pulses (grain)	1	6	5	17
Oilseeds	0	7	8	0
Seeds for sowing	0	2	2	0
Sugarbeets	0	4	2	0
Potatoes	6	2	1	300
Vegetables	5	1	1	500
Fodder beets and swedes	2	4	4	50
Grass and green fodder	49	18	20	272
Green manure and fallow	6	0	0	∞

[a] Converted activities and converted area only.
[b] Conventional sample comparable as far as possible to organic sample with respect to farm size and enterprise structure.
[c] Organic relative to comparable conventional group.
Source: Dubgaard *et al.* (1990).

are mainly high-value cash crops such as potatoes and vegetables. As reported in Table 7.1, approximately 10% of the organic area is used for potatoes and vegetables compared to only 2% in conventional agriculture.

The comparative advantages of grass and green fodder in organic farming – and the limitations on feed purchase – affect the pattern of livestock production. In organic agriculture livestock capable of transforming large amounts of roughage into products for human consumption has a comparative advantage relative to livestock depending primarily on concentrates. As can be seen in Table 7.1, this is reflected in a dominance of ruminants – especially dairy cattle – in organic livestock production. On the other hand, livestock enterprises depending to a high degree on concentrates (grain, oilcakes, etc.) play a rather insignificant role. Thus, pig production is almost absent in the organic sample. In order further to reduce the dependence on concentrates, bull calves (from the dairy herd) are not being fattened to any significant extent on organic farms.

Yields

Organic and conventional yields are reported in Table 7.2. The comparison is biased to some extent due to the fact that the eastern parts of Denmark are over-represented in the organic sample. (Until the mid-1980s conversion had taken place mainly in the eastern parts of the country.) Since soil quality is generally above average in the east, a more even distribution of organic farms throughout the country would probably have reduced estimated organic yield averages.

The organic crop yields reported in Table 7.2 are about 40% below the conventional average. Yield differences are most noticeable for (fertilizer and pesticide) intensive crops such as wheat and potatoes with organic yields around half the conventional averages. Milk

Table 7.2. Yield levels in organic and conventional agriculture, Denmark, 1988.

	Organic[a]	Conventional (whole sector)	Relative (conv=100)
Crop yields (t ha^{-1})			
Cereals (all types)	3.0	5.0	59
Wheat (winter and spring)	3.4	6.8	51
Rye	3.6	4.5	79
Spring barley	3.5	4.6	77
Oats and mixed cereals	1.9	4.7	40
Potatoes	14.8	30.6	48
Forage/livestock yields			
Grass and green fodder (100 feed units per ha)	46[b]	79	58
Milk (kg per cow, 4% fat basis)	4560[b]	6390	85

[a] Converted activities and converted area only.
[b] Based on a subsample of eight farms.
Source: Dubgaard *et al.* (1990).

production per cow is some 15% below the conventional average. The yield differentials found in this study are in line with the findings of similar German investigations (Böckenhoff *et al.*, 1986; AID Informationen, 1988).

Labour and capital intensity

The findings reported in Table 7.3 indicate that the organic sample uses about twice as much labour per hectare as the conventional counterparts. Some of this difference is explained by the considerably larger share of labour-intensive crops – such as vegetables – in organic farming and the greater importance of labour-intensive dairy production. If these structural differences were eliminated, labour requirements of organic farming would probably exceed conventional averages by about one-third 'only' (Dubgaard *et al.*, 1990).

It seems to be the general assumption that organic farming is less

Table 7.3. Labour and capital intensity in organic and conventional agriculture, Denmark, 1988.

	Organic[a]	Conventional (comparable)	Conventional (whole sector)	Relative[b] (conv=100)
Land area (ha per farm)				
Agricultural	27.7	28.7	33.1	97
Organic	24.6	–	–	–
Labour (man years per farm)				
Family labour	1.08	0.77	0.92	140
Hired labour	0.65	0.22	0.23	295
Total per farm	1.73	0.99	1.15	175
Total hours (h ha^{-1} $year^{-1}$)	123	61	60	202
Capital (1000 DK per farm)				
Land and buildings	1000	969	1224	103
Livestock	112	149	208	75
Other agricultural assets	200	240	289	83
Total agricultural assets	1312	1358	1721	97
Total agricultural assets (1000 DK ha^{-1})	53	47	52	113
Turnover/agricultural assets (1000 DK ha^{-1})	0.33	0.29	0.31	114

[a] Converted activities and converted area only.
[b] Organic relative to comparable conventional group.
Source: Dubgaard *et al.* (1990).

capital intensive than conventional agriculture. This assumption is not supported by the findings of the present study. As reported in Table 7.3, capital intensity is about the same in the two sectors of agriculture if measured as capital per hectare or capital in relation to turnover.

Price premiums

In Denmark, price premiums are available for most organically grown plant products and for milk, while most organically produced meat is sold without a premium. In the base year of the accounts survey (1988) organic farmers obtained 75–100% price premiums on bread grain and vegetables, while potatoes carried a price premium of about 200% (Dubgaard *et al*., 1990). Since the mid-1980s a collection and processing system for organically produced milk has been established, resulting in a price premium of about 25% in 1988 – and a considerable improvement of the profitability of organic dairy farming.

Factor remuneration

The economic performance measures (reported in Tables 7.4 and 7.5) indicate that organic farming fares quite well when compared with conventional agriculture. In 1988, labour obtained an average remuneration of about 42 DK per hour in the organic sample, 34 DK per hour in the sample of comparable conventional farms and 40 DK per hour in conventional agriculture as a whole (cf. Table 7.5). Thus, compared with conventional farms of similar size (acreage) and enterprise structure the remuneration of labour was significantly better in the organic sample.

Still, care should be taken with interpreting these results. Most of the holdings in the organic sample are full-time farms while about half the holdings in the 'conventional comparable sample' are part-time farms. Factor remuneration is generally relatively low on part-time farms. To widen the basis for comparisons, a representative sample of full-time farms (from the official accounts statistics) is also presented in Tables 7.4 and 7.5. On full-time conventional farms remuneration of labour amounted to an average of 58 DK per hour, i.e. about 40% more than the organic average.

It should also be noted that labour remuneration in organic farming – and agriculture as whole – is less than half the rate obtained by unskilled labour in off-farm employment. This has a profound effect on the other economic performance measure used in this survey – returns to capital.

When measured by the rate of return to capital, the organic sample performs somewhat poorer than conventional agriculture. In 1988 the

Table 7.4. Output, costs and gross profit in organic and conventional agriculture, Denmark, 1988.

	Organic[a]	Conventional (comparable)	Conventional (whole sector)	Conventional (full-time)	Relative[b] (conv=100)
Land area (ha per farm)					
Agricultural	27.7	28.7	33.1	50.9	97
Organic	24.6	–	–	–	–
Output (DK ha^{-1})					
Crop products	7,720	5,513	4,377	4,290	140
Livestock products	9,657	7,647	11,371	14,265	126
Other agricultural sources	356	782	584	475	46
Total	17,733	13,943	16,332	19,031	127
Costs (DK ha^{-1})					
Raw materials and auxiliaries	5,147	5,844	7,528	8,775	88
Services	3,198	2,091	2,097	2,041	153
Depreciation	1,724	1,778	1,903	2,096	97
Hired labour	2,382	932	819	1,064	256
Property taxes	300	272	422	335	110
Total	12,751	10,917	12,672	14,311	117
Gross profit (DK ha^{-1})	4,982	3,026	3,660	4,720	165

[a] Converted activities and converted area only.
[b] Organic relative to comparable conventional group.
Source: Dubgaard *et al.* (1990).

rate of return was −4.0% in the organic sample compared to −2.7% in the selected sample of conventional farms and −1.3% in conventional agriculture as a whole. Only in the sample of full-time conventional farms was a positive rate of return to capital obtained – equal to 0.9%.

The occurrence of a negative rate of return to capital in agriculture is due to the fact that the gross profit obtained is lower than the imputed value of farm-family labour. This is so in the organic as well as in two of the conventional samples. However, in the organic sample this discrepancy has a much larger impact on the calculated rate of return to capital because of the higher labour intensity in the organic sector.

Table 7.5. Returns to labour and capital in organic and conventional agriculture, Denmark, 1988.

	Organic[a]	Conventional (comparable)	Conventional (whole sector)	Conventional (full-time)	Relative[b] (conv=100)
Returns to labour (DK ha^{-1})					
Gross profit	4982	3026	3660	4720	165
Plus costs of hired labour	2382	932	819	1064	256
Less opportunity costs of capital[d]	2132	1896	2079	2119	112
Returns to family and hired labour	5231	2062	2401	3665	254
Rate of return to labour[e] (DK h^{-1})	42	34	40	58	124
Returns to capital (DK ha^{-1})					
Gross profit	4982	3026	3660	4720	165
Less opportunity cost of family labour	7102	4321	4347	4270	164
Return to capital	−2120	−1295	−686	450	−164
Rate of return to capital (% of total agric. assets)	−4.0	−2.7	−1.3	0.9	−148

[a] Converted activities and converted area only.
[b] Organic relative to comparable conventional group.
[c] Total net income from farming available for remuneration of capital and family labour.
[d] Calculated interest on the cash value of agricultural assets using a (real) rate of interest of 4%.
[e] Returns to labour divided by recorded labour input in the form of manual work as well as management.
[f] Average earnings per hour by unskilled workers (in non-agricultural occupations) multiplied by recorded input of family labour (manual as well as management).
Source: Dubgaard *et al.* (1990).

Cost savings versus price premiums

It was also investigated to what extent cost savings due to the non-use of chemical fertilizer and pesticides compensate for lower yields and higher labour requirements in organic farming. On average, cost savings on fertilizer and chemicals cover about 40% of the losses or extra costs incurred by lower yields and higher labour requirements (Dubgaard *et al.*, 1990). Thus, considerable price premiums on organically produced farm products are needed to obtain a remuneration of labour and capital at about the same level as in conventional agriculture.

Perspectives for Organic Agriculture in Denmark

Since the base year of the statistical survey presented above, organic agriculture has expanded from about 350 farms in 1988 to around 700 farms in 1992. In the same period the area converted or under conversion has increased from 7000 ha to about 18,000 ha – or from 0.2% to 0.6% of the total agricultural area in Denmark. However, as reported by Dubgaard and Holst (Chapter 23) conversion has now come to a standstill. The most important limitation to the expansion of organic agriculture is insufficient demand for organic food products at the price premiums required to make organic farming economically viable.

Organic products are now available on a broad basis at the retail level and consumers' awareness of these products is relatively high in Denmark (Bjerke, 1992). The price premiums obtained are typically around 50% at the retail level. Still, the attempts to create a stable demand for organic products at premium prices have been successful mainly for relatively cheap plant products like potatoes and other vegetables. Where animal products are concerned, stable sales at premium prices have been realized for liquid milk products only. But this market absorbs no more than about half the amount of organic milk currently produced (National Ecological Information Centre, 1992). Processed organic dairy products such as cheese and butter have not acquired market shares large enough to significantly alleviate the 'surplus' problems in the organic dairy sector. Several attempts have been made to market beef and pork under an organic label and bearing a price premium. Until now these attempts have failed (Bjerke, 1992).

In response to the prevailing price relations, organic farmers have allocated most of their land, labour and capital to the production of vegetables, potatoes and milk/forage. Feed grains, on the other hand, and livestock depending on grain, play a minor role in the organic sector. This particular production structure represents an obstacle to the expansion of organic farming beyond the very limited level found at present. In contemporary agriculture little land is allocated to the production of staple food items such as crop products and liquid milk for human consumption. Most resources are employed in the production of meat and relatively expensive dairy products such as cheese and butter. For such products there is limited scope, it seems, for obtaining the price premiums needed to make production economically viable under organic practices (Drake and Holm, 1988).

Conclusions

The results of the accounts survey presented in this chapter indicate that factors of production receive about the same remuneration in conventional and organic agriculture – after a conversion period of some length. Substantial price premiums on output are essential for the economic viability of organic farming. Until now price premiums have been available only on vegetables, potatoes, bread grain, and milk. Consumers' lack of willingness to pay significant price premiums on meat and dairy products (aside from liquid milk) seems to be the most important obstacle to the expansion of organic farming.

It is concluded that organic agriculture is a viable alternative to conventional agriculture only within a relatively narrow range of enterprises – primarily vegetables and to a certain extent dairy production. Meat production, on the other hand, is economically unrealistic under organic practices except as a by-product from milk production. Altogether it seems that organic farming will not expand to much more than the present 0.6% of the agricultural area unless public support for this sector is considerably increased.

References

AID Informationen (1988) *Alternative und konventionelle Betriebe im Vergleich, Buchführungsergebnisse im Agrarbericht.* No. 8. Auswertungs- und Informationsdienst für Ernährung, Landwirtschaft und Forsten, Bonn.

Bjerke, F. (1992) *Forbrugernes interesse for økologiske produkter.* (Consumers' awareness of organic products.) Institut for Samfundsøkonomi og Planlægning, RUC, Roskilde.

Böckenhoff, E., Hamm, U. and Umhau, M. (1986) Analyse der Betriebs- und Produktionsstrukturen sowie der Naturalerträge im alternativen Landbau. *Berichte über Landwirtschaft* 64, 1–39.

Drake, L. and Holm, H. (1988) *Konsumentarnas attityder till alternativt producerat kött – preliminär redovisning av intervju och enkätstudie.* (Consumer attitudes towards alternatively produced meat.) Rapporter Allmänt 147. Sveriges Lantbruksuniversitet, Uppsala.

Dubgaard, A. and Sørensen, S.N. (1988) *Økologisk og biodynamisk jordbrug – en statistisk undersøgelse.* (Organic and bio-dynamic farming – a statistical survey.) Rapport 43. Statens Jordbrugsøkonomiske Institut, Copenhagen.

Dubgaard, A., Olsen, P. and Sørensen, S.N. (1990) *Økonomien i økologisk jordbrug.* (Economics of organic farming.) Rapport 54. Statens Jordbrugsøkonomiske Institut, Copenhagen.

National Ecological Information Centre (1992) *HV Guide.* No. 4. Økologisk Landscenter, Aarhus.

8 Economics of Organic Farming in Switzerland

Irene Mühlebach and Jörg Mühlebach

Introduction

Around 60% of the utilizable agricultural area in Switzerland consists of natural pastures and meadows. A large proportion of this is not suitable for crop production. The output of these areas can therefore only be utilized for livestock production. As a result, the proportion of arable/grassland farms or pure grassland farms is high at almost 50%.

Although the conversion of such farms to organic farming is possible in most cases without significant difficulties, there are still only 1000 farms with 10,000 ha operating according to the principles of organic farming (Table 8.1). This represents just 1% of all agricultural holdings and agricultural land in the country. According to the last agricultural census, 53% follow the organic-biological approach and 26% the bio-dynamic approach. The remaining organic farms gave no indication of their farming methods.

The interest of farmers in organic farming has without doubt increased in recent years. The official farm statistics of the Federation of Swiss Organic Farming Organizations (Vereinigung schweizerischer biologischer Landbau Organisationen: VSBLO) shows 280 new farms entering conversion in 1992. Various cantons currently attempt to minimize the risks of the conversion process with financial aid (see also Chapter 24). The main aim of these area- and farm-related payments is to compensate for the income reductions which occur during the conversion period. At the same time, however, these subsidies are also intended to encourage the adoption of more environmentally friendly farming methods. In the context of the new federal agricultural policy measures, farms which are managed according to the VSBLO standards will, from 1994, be paid SF 750 per hectare for arable or horticultural

Table 8.1. Organically managed farms by size group, Switzerland, 1990.

Size group by crop area (ha)	Number of farms		Agricultural area	
	Total	%	Total	%
< 5.0	349	33.8	567	5.6
5.01–10.0	178	17.3	1,133	11.3
10.01–15.0	193	18.7	2,098	20.8
15.01–30.0	241	23.4	4,046	40.1
> 30	70	6.8	2,235	22.2
Total	1,031	100	10,079	100

Source: Swiss federal agricultural census, 1990.

land and SF 150 per hectare for grassland. In addition, many organic farms can benefit from the recently introduced payments for environmental management (hedges, trees, extensively managed pastures, etc.)

Data Source

The financial results are derived from the farm accounts of farms which are managed according to the VSBLO standards. These standards affect above all the use of yield-increasing or yield-protecting inputs as well as the feeding and housing of livestock. They are compulsory for those farms which market their produce under the common VSBLO trade mark for organic produce ('Knospe').

In order to evaluate the profitability of organic farming methods, the results are compared with those of conventionally managed farms (i.e. those farms which are not managed organically). The allocation of conventional partner (matched) farms is done randomly on the basis of natural locational conditions, farm size, farm type and tenure.

The collection of the data was carried out by the Swiss Federal Research Station for Farm Management and Agricultural Engineering (Eidg. Forschungsanstalt für Betriebswirtschaft und Landtechnik: FAT) in collaboration with the Research Institute for Biological Husbandry (Forschungsinstitut für biologischen Landbau: FIBL) (FAT, 1993).

Farm Resources and Production Structure

The farms analysed operated an average of about 14.6 ha utilizable agricultural area (UAA) (Table 8.2). Of this, at least half was included

Table 8.2. Farm resources and production structure on organic and conventional farms, Switzerland, 1989/91.

	Organic	Matched conventional	Relative[b] (conv=100)
Number of farms, *n*	34	34	
Agricultural land area (ha), of which:	14.58	14.61	100
Permanent grassland (ha)	6.42	6.19	104
Temporary grassland (ha)	3.35	2.46	136
Arable land (ha), of which:	4.50	5.65	80
Bread cereals (ha)	2.01	1.62	124
Feed cereals (ha)	0.79	1.41	56
Potatoes (ha)	0.48	0.56	86
Vegetables (ha)	0.40	0.18	222
Total cattle (LU)[a]	19.8	20.2	98
Dairy cows as proportion of cattle LU (%)	70.0	71.0	99
Total grazing livestock (LU)	20.0	20.3	99
Forage area (ha LU^{-1})	0.53	0.49	108
Stocking rate (manure LU ha^{-1})	1.5	1.6	94
Labour per farm (day), of which:	674	557	121
Family labour (day)	421	431	98
Labour per ha agricultural land (day)	46.2	38.1	121

[a] LU=livestock unit. 1 LU represents the feed consumption or the manure production of a 600 kg dairy cow.
[b] No statistical analysis available.
Source: FAT (1993).

in the crop rotation. The proportion of open crop land on the organically managed farms was just over 30%, a substantially lower proportion than on conventional farms. In contrast, temporary forage crops were of greater significance in the rotation, in view of the need to maintain natural soil fertility.

The larger area of bread-making cereals is characteristic of organically managed farms. Apart from wheat, many organic farmers grow rye and spelt wheat. Feed cereals, particularly barley, are used predominantly within the farm. As a result, the market for organically produced feed cereals is currently still limited. In place of maize, sugarbeet and oilseed rape, which have no ready (organic) markets, more vegetables are produced on organic farms. Alongside crops for processing such as carrots, beetroot and celery, organic farms also produce fresh vegetables for market. The extent of vegetable production is determined primarily by the available labour and marketing opportunities. Apples, pears, cherries and various berry crops serve to complete the product range, particularly on farms marketing direct to consumers.

Livestock production plays a central role on organic farms. As utilizers of forage crops, and as suppliers of on-farm manures, cattle form an essential link in the nutrient cycle of an organic farm. The extent and composition of the livestock enterprises is similar under both organic and conventional management. In comparison with the conventional farms, the organic farms keep a few more pigs and poultry. The stocking rate per hectare UAA is about 1.5 manure livestock units (LU). As a consequence of the extensive management of the forage crops, the organically managed farms require on average about 0.04 ha more forage area per grazing LU.

On the basis of the farm accounts data, organic farms require about 20% more labour. This is supplied from non-family sources. The extra labour is not necessarily due solely to the system-determined labour increase caused by organic production methods. A significant proportion of the increased labour can be attributed to marketing activities. Although direct marketing is characteristic of many organic farms, there is no causal link between the two.

FAT investigations (Mühlebach and Näf, 1990) indicate that there is a 12–13% system-determined increase in labour requirements on

Table 8.3. Increased labour requirements in organic farming, excluding direct marketing, Switzerland, 1985 (17 farms).

	Labour-use increase (%)	Enterprise labour use as % of farm total
Cereals	62	3.2
Other arable crops	49	7.6
Vegetables	72	6.7
Fruit and berries	2	9.7
Forage crops	20	4.7
Manure management	26	2.5
Farm woodlands	0	1.1
Livestock	0	45.5
Other work	19	19.0
Whole farm	12.4	100

Source: Mühlebach and Näf (1990).

Table 8.4. Performance of selected arable enterprises, Switzerland, 1989–1991.

	Wheat			Rye			Barley			Potatoes[a]		
	Org	Conv[b]	Rel[c]	Org	Conv[b]	Rel[c]	Org	Conv[b]	Rel[c]	Org	Conv[b]	Rel[c]
Number of observations, *n*	105	105	–	61	61	–	75	75	–	81	81	–
Area (ha)	1.82	2.49	73	1.28	1.25	102	0.89	1.43	62	0.93	1.18	79
Yield (t ha^{-1})	4.8	6.2	77	4.4	5.4	81	4.6	6.1	75	27.6	37.2	74
Price (SF t^{-1})	1,410	1,010	140	1,410	990	142	780	730	107	540	370	146
Output (SF ha^{-1})	7,255	6,673	109	6,492	5,700	114	5,424	6,299	86	15,313	14,007	109
Variable costs (SF ha^{-1}), of which:	750	1381	54	611	1061	58	425	934	46	2,651	3,551	75
Fertilizers (SF ha^{-1})	112	373	30	70	357	20	61	251	24	219	515	43
Sprays (SF ha^{-1})	1	315	0	0	145	0	13	228	6	158	690	23
Gross margin (SF ha^{-1})	6,505	5,292	123	5,881	4,639	127	4,999	5,365	93	12,662	10,456	121

Conv, conventional; Org, organic; Rel, relative.
[a] Without differentiation by use.
[b] Matched conventional farms.
[c] Relative (conventional = 100), no statistical analysis available.
Source: FAT (1993).

organic farms. There are, however, large differences between production enterprises. From Table 8.3 it can be seen that the increased labour requirement for arable and vegetable crops of between 50% and 70% is particularly high. The main cause of this is labour-intensive weed control. In contrast, livestock production involves practically no additional labour.

Enterprise Results

Yields

Generally, lower yields are achieved in organic farming than in conventional agriculture (Table 8.4). The differences in crop production are clearly greater than in livestock production. The lower yields are primarily due to the reduced use of yield-promoting inputs. Through the conscious avoidance of synthetic fertilizers and plant-protection chemicals, it is often not possible for the genetic potential of the crop to be fully exploited. In addition, mistakes made in organic production cannot be corrected with the help of chemical inputs. Crop yields and therefore the performance of the farm business are far more dependent on the management ability of the farmer.

Yield trends are similar in conventional and organic production, albeit at a lower level (Fig. 8.1). In crop production, the natural climatic factors result in similar annual variation in both management systems. The results for potatoes demonstrate clearly that unfavourable growing conditions have a greater impact on yields on organic farms. Milk yields per cow have increased significantly since the early 1980s. The average milk yield increase is of the order of 70 kg per cow per year.

Prices

The marketing of organic products is conducted via a number of different channels. Alongside private traders and producer cooperatives, direct marketing to consumers plays an important role. Direct marketing in various forms (farm-gate sales, weekly market, local distribution rounds, etc.) is practised on many organic farms. The good image of organic produce may mean that wholesalers also show more interest in this market segment in future.

Certified organic products carrying the 'Knospe' symbol can, as a rule, achieve prices significantly above the price level for conventional products (Table 8.4). For bread-making cereals, the difference between organic and conventional grains is around SF 400 per tonne. This price difference results primarily from the 'organic premium' paid by the

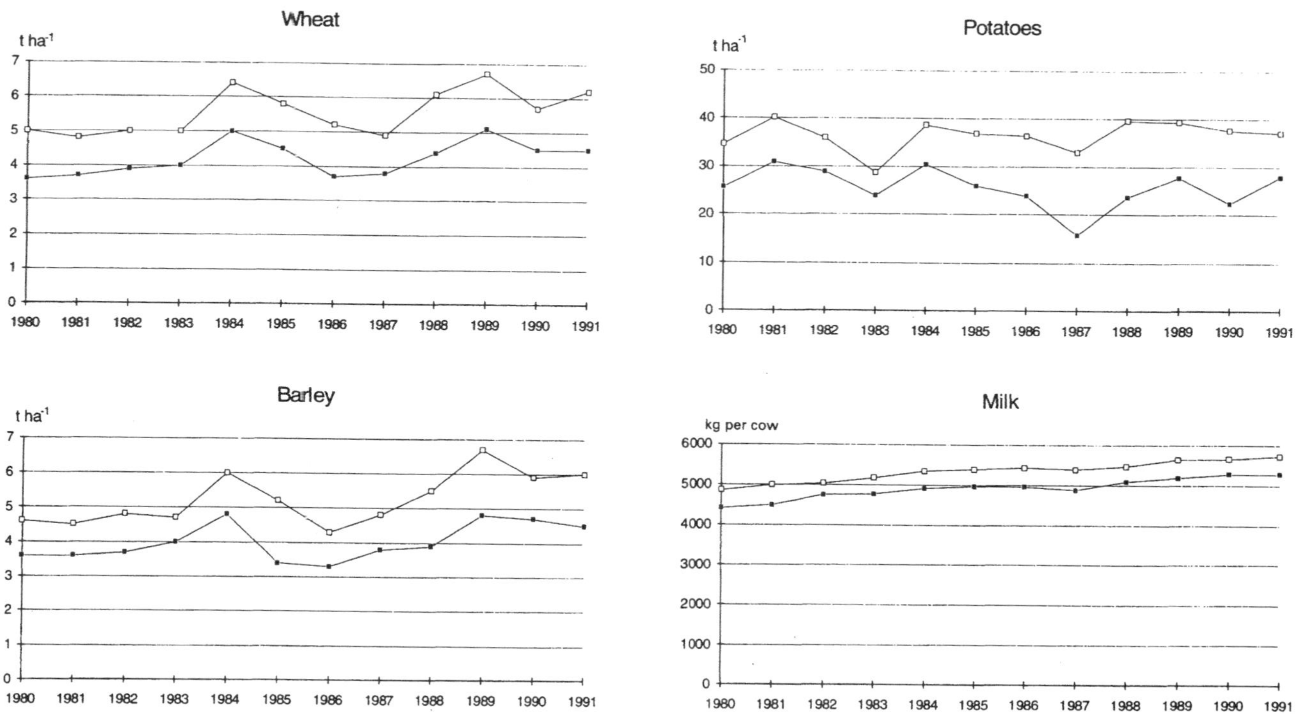

Fig. 8.1. Yield trends in selected organic and conventional enterprises, Switzerland, 1980–1991. ■ Organic farms; □ partner farms.

purchaser ex-farm. In addition, organic farmers are not liable for the levy on surplus bread-making cereals. Feed cereals are used primarily on the farm itself. Since these internal transfers are valued at conventional prices, the price differences are relatively small. Organically produced potatoes are often marketed directly to consumers, allowing high prices to be obtained.

Animal products are at present largely marketed through conventional channels. Specific organic markets for meat, milk and dairy products are still under development, and have initially only regional significance. Higher prices are therefore normally only available for those producers who market their produce directly. Information on the milk price achieved in practice is not available from the farm accounts.

Variable costs (excluding machinery and labour contracting charges)

A sharp reduction in production input use is characteristic of organic farms. Expenditure on these items is consequently also lower (Table 8.4). In crop production, the expenditure on fertilizers and sprays is significantly lower. Depending on the enterprise, savings in variable costs of between 30% and 60% are possible. In interpreting these figures, it needs to be remembered that the parameters shown only include the directly allocatable fertilizer and plant-protection costs. Indirect costs, which result for example from the specific requirements of the rotation design (legumes, green manures, diverse rotation, etc.) are not included.

Table 8.5. Cattle enterprise performance, Switzerland, 1989–1991.

	Organic	Matched conventional	Relative[a] (conv=100)
Number of observations, *n*	141	141	–
Herd size (LU)	21.6	23.4	92
Cow proportion (%)	71	72	99
Milk yield (kg per cow)	5301	5640	99
Output (SF LU^{-1}), of which:	4658	4820	97
Milk (SF LU^{-1})	3625	3767	96
Variable costs (SF LU^{-1}), of which:	617	781	79
Concentrates (SF LU^{-1})	393	531	74
Vet. & med. (SF LU^{-1})	107	127	84
Gross margin (SF LU^{-1})	4041	4039	100

[a] No statistical analysis available.
Source: FAT (1993).

In addition, the rejection of herbicides is often accompanied by increased labour use and hence higher labour costs.

In livestock production, the organic farms' variable costs were 20% lower. The greatest savings of about SF 140 per LU were for concentrated feeds (Table 8.5). However, expenditure on veterinary charges and medicines was also lower on average.

Gross margins (enterprise output less variable costs)

In organic farming, the yields are generally lower than in conventional agriculture. These lower yields may, in part, be offset by higher prices and lower variable costs. These three factors influence the level of the gross margin. Depending on the enterprise, the gross margin results for the two management systems differ correspondingly.

Bread-making cereals, which can be sold at higher prices than in conventional agriculture, achieve higher gross margins per hectare (Table 8.4). The lower yields of these crops can be fully compensated. The price difference for feed cereals is much lower, with the result that the gross margin is about SF 350 lower per hectare than conventional. Potatoes, on the other hand, also achieve better enterprise gross margin results, where, despite considerably lower yields, the gross margin lies some SF 2200 per hectare above the level for the matched conventional farms.

The gross margin for cattle production is a little over SF 4000 per LU (Table 8.5). This is the same as for the comparable conventional farms. As already indicated, livestock products are marketed almost exclusively through conventional channels. Where it is possible to market these products at a higher price, then the results for cattle production would, without doubt, turn out better than for the comparison farms.

Gross margin comparisons, in particular between different management systems, are not, however, completely without problems. Thus for example the better prices achieved on those farms which market their produce directly are reflected in the enterprise gross margins, but the related costs of processing, storage and marketing are not. These are felt first in the whole-farm fixed costs. Further, the gross margin results give no information about the (labour and capital) factor inputs. The more labour-intensive (organic) systems therefore often result in better enterprise performance than the more capital-intensive (conventional) production methods.

Whole-farm Results

The land productivity, measured in gross output terms, reaches SF 12,500 per hectare UAA on the organic farms (Table 8.6). This is 5% less than the conventional farms achieve. Analogous to the production structure of the farms, about 60% of the gross output comes from livestock production. The share of crop production, at around SF 3500 per hectare is similar in both groups. The remaining output includes various output parameters (building rental value, hirework, non-allocatable direct payments, etc.), whose value may hardly be influenced by the production system.

As already indicated in the discussion of the individual enterprise results, cost savings can be achieved in both crop and livestock production due to the lower intensity of production in organic farming. In contrast, little or no difference exists between building and machinery costs. The difference in sundry fixed costs is also small. Together, the production costs show a difference of SF 600 per hectare in favour

Table 8.6. Whole-farm results, Switzerland, 1989–1991.

	Organic	Matched conventional	Relative[a] (conv=100)
Number of farms, *n*	34	34	–
Farm size (ha UAA)	14.58	14.61	100
Output (SF ha^{-1}), of which:	12,500	13,136	95
Crops (SF ha^{-1})	3,517	3,499	101
Livestock (SF ha^{-1})	7,441	8,282	90
Production costs (SF ha^{-1})	5,335	6,111	87
Labour costs (SF ha^{-1})	1,472	753	196
Rent and interest (SF ha^{-1})	749	920	81
Farm income (SF ha^{-1})	4,944	5,352	92
Income as % of target income[b]	78.1	82.3	95
Labour income (SF ha^{-1})	5,354	4,981	108
Labour income (SF day^{-1})	115.85	130.45	89
Return to family labour (SF ha^{-1})	3,882	4,229	92
Return to labour (SF day^{-1})	134.25	142.95	94

[a] No statistical analysis available.
[b] For comparison purposes, a target income is defined, against which the effective income can be measured. This includes notional interest on farmer equity and an appropriate payment for farm family labour. The wage rate used is derived from employee wages and takes into account the actual circumstances of the farm business.
Source: FAT (1993).

of the organically managed farms. Capital costs are also lower. Labour is a different matter. Since the increased labour requirement was met primarily by hired labour on the organic farms studied, the wage costs are about twice as high as for the conventional partner farms.

Farm income is used as a measure of the success of the business. It includes the surplus generated from the farm during the year and represents the return to the farm family for their labour and for their capital invested in the farm business. As a consequence of the high wage costs, the farm income in organic farming is around SF 400 lower per hectare than in conventional agriculture. On average for the three years, the organic farms still achieved 78% of the estimated target income (see footnote to Table 8.6). The use of more labour-intensive production methods results generally in poorer labour productivity. The labour income per day, as well as the return to family labour, lie below the values for the partner farms.

Results for the Agricultural Business

The change in net worth or owner equity (i.e. the farm family's own capital invested in the farm business) is an important indicator of the growth potential of a farming business. By definition, this relates to the share of total income which is not consumed. The annual increase in net worth on the organic farms amounted to over SF 28,000, or just under SF 2000 per hectare UAA (Table 8.7). This amount should be sufficient to cover replacement and new investments, as well as the financing of current liabilities of SF 225,000.

The level of net worth increase depends not only on the farm income, but also on income from other sources (non-agricultural activities, capital investments, social payments, etc.), as well as on

Table 8.7. Income (SF) results for agricultural business, Switzerland, 1989–1991 (34 farms).

	Organic	Matched conventional	Relative[a] (conv=100)
Farm income	72,039	78,116	92
Other income	12,203	11,437	107
Total income	84,242	89,553	94
Family drawings	55,877	58,865	95
Net worth increase	28,365	30,688	92

[a] No statistical analysis available.
Source: FAT (1993).

private expenditures. The type and extent of these secondary activities, as well as the level of private consumption, bears a direct relationship to organic management only in a few cases. A comparison of these values with data from conventional farms is therefore less useful.

Conclusions

In view of the relatively narrow data base, it is difficult to make a final judgement about the economic situation of organic farming in Switzerland. It can, however, be assumed that organically managed farms in total achieve a satisfactory level of income. However, they need to use more labour than their conventional counterparts to achieve this. The current demand for organic produce allows for much higher prices than normal to be achieved for a number of products. These higher prices are essential, since the costs savings alone are insufficient to compensate for the lower yields in organic farming.

In the medium to longer term, the economic performance of organic farming will depend on market development and agricultural policy changes. The recent government proposals to give financial support for environmental outputs from agriculture provide some hope for a positive environment for organic farming in Switzerland.

References

FAT (1993) *Bericht über biologisch bewirtschaftete Betriebe 1991.* Eidg. Forschungsanstalt für Betriebswirtschaft und Landtechnik, Tänikon, Switzerland.

Mühlebach, J. and Näf, E. (1990) *Die Wettbewerbsfähigkeit des biologischen Landbaus.* Schriftenreihe der Eidg. Forschungsanstalt für Betriebswirtschaft und Landtechnik 33. Tänikon, Switzerland.

9 Economics of Organic Farming in Canada

John Henning

Introduction

Canada is a very large country, with a diverse climate and important regional differences in its forms of commercial agriculture. The western Prairie provinces concentrate on grain, oilseed and livestock production, on farms that average nearly 400 ha. In other provinces, farms are smaller and more diversified, but there are regional specializations. British Columbia produces significant amounts of tree fruits and greenhouse products, while dairying represents 36% of farm sales in Quebec (Science Council, 1991). This diversity is also reflected on organic farms, and taken together with the small body of Canadian research, it is indeed a challenge to provide an account of the economics of organic farming.

Overview of the Organic Sector

The market for organic products

No accurate measure of the size of the organic market in Canada exists, although speculative estimates have been published. Nevertheless, all have placed the organic market at the level of a niche market, with the most optimistic giving organic products perhaps 1% of the retail food market. In the late 1980s, Kramer (1989) estimated the organic market share at approximately 0.2% in dollar terms (C$ 95 million), with only about 0.4% of Canadians buying organic on a regular basis. On the other hand, a later estimate of C$ 20 million for the Quebec retail market by Thériault (1991) suggests that the earlier estimate might have been somewhat optimistic.

Based on the survey work carried out since 1988, what is known is that most organic products originate in the USA, are sold in major urban centres, through farmers' markets, health-food stores and other small retailers, to relatively young families with above average incomes, and the market is growing (Hanisch, 1989; Hay, 1989; Canadian Organic Growers (COG), 1990; Weymes, 1990; BFB, 1992). As in other countries, the growth in consumer interest in organic foods has coincided with changing attitudes about food that are being driven primarily by health concerns, and organic foods have benefited from their positive image in this respect. An important part of this phenomenon has been the concern expressed by many consumers over the issue of chemical residues (Hay, 1989; Weersink *et al.*, 1990).

Consumers who at least occasionally consume organic foods tend either to make their purchases direct from farmers or through small retailers, or to grow their own (Hay, 1989). Although sales growth has been encouraging for small retail outlets, organic foods have not been a success for the supermarket chains (Weymes, 1990; Cole, 1991; Hancock, 1991; BFB, 1992). Among problems commonly cited are that organic produce is expensive, requires more handling, has a short shelf life, inferior cosmetic appearance and poor packaging. The supermarkets found that organic foods had relatively slow turnover, and yielded low profit margins.

Price has been identified as a significant obstacle in several surveys. An Agriculture Canada survey of major urban centres (Hay, 1989) suggested that 25% of consumers would purchase organic vegetables if the price premium was no more than 25%, although this rose to 53% for occasional purchasers. However, there appears to be significant resistance to premiums above 30% at the retail level (COG, 1990). This is probably an even more important factor now, as Canada slowly emerges from recession. A recent national survey of consumer attitudes indicated that only 15% of respondents were willing to pay any price premium for 'green products' in general (Rusk, 1993).

Consumers are very price conscious, so given the current economic climate, it is difficult to estimate current or future growth potential for the organic sector, although there are some relatively recent estimates.

In 1989, Montreal wholesalers and retailers expected retail sales growth of 34% (Hanisch, 1989). About the same time, Weymes (1990) reported on a national survey with growth rates among retailers of 11–20% for raw and lightly processed products and 21–50% for processed products. Retailers and processors expected the organic market share to expand to about 10% by the end of the century, while farmers expected a 20% share by the same date. Participants in the COG (1990) survey anticipated strong growth in grains and oilseeds, fruit, vegetables, and meat, but weak growth in dairy products (Table 9.1).

Table 9.1. Characteristics of the market for organic produce, Canada, 1989.

Product	Responses[a]	Growth[b]	US market[c]	Costs[d]	Premium[e]
Grains (and oilseeds)	54	126 (155)	15	0.80	34
Fruits	42	177	12	1.15	34
Vegetables	94	209	7	0.95	71
Meat	51	166	1	0.90	39
Dairy	18	108	0	0.85	13

[a] Based on 236 respondents from various levels of the organic marketing chain, from farmers to retailers.
[b]Projected sales for 1991 as a percentage of 1989.
[c]Respondents reporting sales in the United States.
[d]Farm expenses as a proportion of conventional.
[e]Respondents reporting receipt of a price premium.
Source: COG (1990).

Respondents in the BFB (1992) survey estimated annual retail growth at 15–20%, and looked forward to significant growth in processed products during the late 1990s. At present, at least in the Montreal market, large organic wholesalers and retailers are reporting that sales are good, with annual growth rates in the neighbourhood of 15%.

Numbers of organic producers

Estimating the number of organic farmers in Canada is hampered by the fact that there is no comprehensive registry, some farmers are certified with more than one agency, and many farmers choose to remain uncertified. In spite of this, a reasonable estimate is that there are about 1000 certified farmers, with perhaps as many uncertified, and several thousand in transition. In 1990, Canadian Organic Growers (COG) estimated that the certified area totalled 30,000 ha with another 10,000 ha certifiable; however, there is closer to 50,000 ha certified now with considerable short-run growth potential.

Until fairly recently, most organic farmers were concentrated in central Canada. However, approximately 50% are now located in provinces west of Ontario and these account for most of the recent growth in numbers and certified area. Green (1990) estimated Saskatchewan's organic land base (certified and transition) to be about 80,000 ha. Thériault (1991) reported that of the 38,000 farmers in Quebec, 4300 considered themselves to be farming wholly or partly using organic methods. It would be reasonable to project that the number of farmers and the organic land base will increase quite dramatically for the remainder of the 1990s and perhaps beyond.

Motivation to convert

Farmers are converting to organic methods for a variety of reasons, but the most important have to do with their unease with the potential health and environmental impacts of conventional practices, increasing disease and pest problems, and the expectation that organic methods might be more profitable (Weymes, 1990; Molder *et al.*, 1991). For example, one issue of particular focus has been soil degradation, that is thought to have a cost in the order of C$ 1.3 billion per year (Science Council, 1986). Governments have invested considerable resources to encourage farmers to adopt soil conservation practices.

Another motivating factor for some farmers is the severe financial stress they have faced for nearly a decade. Between 1981 and 1991, consolidation and abandonment reduced the number of farms by 40,000 to under 300,000; farm bankruptcies in western Canada increased 530% from 1979 to 1989 (Science Council, 1991). Frankly, some farmers regard organic practices as a strategy for survival.

Organic standards and certification

There are 37 certification bodies across Canada, and many follow the Organic Crop Improvement Association (OCIA) standards. At the provincial level, unlike the USA where 16 states have legislated standards (Cramer, 1992), in Canada only the provinces of Quebec and British Columbia have done so to date (1994). Although there are no national standards, since 1988 there has been putative federal regulation of the word 'organic' through the Food and Drugs Act. The Act generally prohibits the use of fraudulent product claims, but the industry is expected to self-regulate and respect a voluntary code of behaviour. In practice, this has provided little protection for the word 'organic'.

Meanwhile, national standards are being developed through an industry-led and government-supported consultative process known as the Canadian Organic Unity Project (COUP). To date, standards and an accreditation model have been drafted, and will represent minimum standards for Certification Bodies (CBs) that wish to sanction the use of what will be a national trade mark, 'Canada Organic' (COUP, 1992). The certification of farms will be the responsibility of the CBs, which in turn will be accredited by the Canadian Organic Advisory Board. Unfortunately, this may not provide the comprehensive control of the word 'organic' that the industry wanted. Agriculture is a shared jurisdiction between provincial and federal governments, and the federal government can only effect control when products cross provincial or national boundaries. Thus, it will also be up to each province to apply the standards within its jurisdiction. If they choose not to, uncertified products could be marketed as 'organic', as long as they remain within a province. It is expected that the system will be implemented in 1995, about the same time as national standards come into force in the United States.

Characteristics of Organic Farms and Farmers (Survey Results)

Several recent surveys have been undertaken to learn more about organic farmers in Canada (COG, 1990; Green, 1990; Weymes, 1990; Molder *et al.*, 1991; Saskatchewan Research Council, see Rutherford *et al.*, 1992; C. Van Bers, pers. comm.). Our survey in Quebec (Henning *et al.*, 1990) produced results that are typical and will be used as the basis for a discussion of general characteristics of organic farms.

Our purpose was to provide an overview of organic farmers in Quebec with particular attention to marketing and credit problems (MacRae *et al.*, 1989). Data were gathered in 1990 from 80 certified farmers. In general terms, the results indicated that, compared to other farmers, organic farmers are relatively young, have not been involved in agriculture as long, and tend to operate smaller farms with lower sales levels. We also identified serious difficulties associated with marketing organic products.

Social characteristics

According to the Farm Credit Corporation (1988), the average farmer in Quebec had operated his or her farm for between 17 and 20 years. Meanwhile, the average organic farmer had been operating for eight years, and more than 50% had been operating for four years or less. Organic farmers were also found to be comparatively well educated, with 43% having some post-secondary education. This compares to 30% for Canadian farm operators in general (Statistics Canada, 1987).

These findings can be explained by the fact that the farmers in our sample tended to be relatively young. A large proportion of respondents fell into the 30–34 year age group (38%) compared to 11% for census farmers. At the other end of the distribution, only about 10% of producers were aged 50 years or more compared to about 40% in general. This tendency was also observed in Saskatchewan by Molder *et al.* (1991, p. 9): 'younger producers are now starting to farm by using organic farming practices whereas older producers are in the process of converting'.

Farm size

As might be expected, for a population of recent entrants, organic farms in Quebec tend to be relatively small. Added to the age factor is the fact that, in Quebec, vegetable production is the most important organic enterprise and generally takes place on 1–10 ha (Thériault, 1991), although there are a growing number of dairy operations (generally 40 ha or more). In our sample, only 15% of farmers operated on more than 60 ha and 35% operated on 4 ha or less. In Quebec, only about 20% of all conventional farms are 28 ha or less, compared to 46% for our sample.

However, this is not the case in western Canada where farms tend to be specialized in grain and oilseed production. In Green's (1990) sample of 42 farms in Saskatchewan, organic farms occupied 390 ha on average and ranged in size from 4 ha to nearly 2000 ha. This compares with Molder *et al.* (1991) who found that for their sample of organic farmers in Saskatchewan, 336 ha of the land was owned and 322 ha was rented. The Saskatchewan Research Council reported an average size of 470 ha in its survey (Rutherford *et al.*, 1992).

Financial output

The smaller resource base of organic farmers in Quebec was reflected in terms of their value of sales. Figure 9.1 illustrates the distribution of

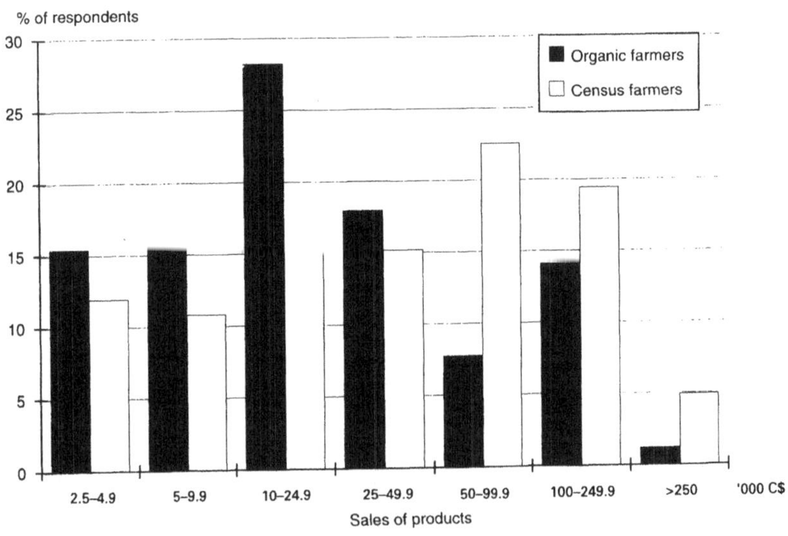

Fig. 9.1. Distribution of sales by value on organic and census farms, Quebec, 1986. Sources: own survey and Statistics Canada (1987).

sales for organic and census farms. In each class up to C$ 50,000, the percentage of organic farmers in each sales class was greater than for farmers in general, with the opposite above C$ 50,000. Similar results were reported for the COG (1990) national survey, with 158 respondents: farm sales were highly variable with over 60% reporting sales of less than C$ 50,000, while sales extended up to C$ 5.3 million.

Yields

In our survey, the yields achieved by organic farmers were only addressed in general terms. However, respondents tended to disagree with the statement that 'yields are lower for organic products', and strongly supported the idea that 'organic yields are more stable'. Weymes (1990) reported that, compared to conventional, 40% of his sample achieved organic yields that were at least as high, while 26% reported yields that were lower by 10% or more. On average, yields were reported to be 6% below conventional. This variance in technical productivity is reflected in the quantitative results reported by other authors. However, all of these results, reported below, involve relatively small samples and should not be taken to be conclusive.

Table 9.2. Crop enterprise physical and financial performance, Ontario, Canada, five-year average, 1986–1990.

	Conventional[a]	Low input[a]	Organic[a]
Tillable land (ha)	234	192	228
Maize			
Yield (t ha^{-1})[b]	5.5	6.5	6.4
Price (C$ t^{-1})	126	126	126
Gross revenue (C$ ha^{-1})	691	816	811
Direct costs (C$ ha^{-1})	606	326	325
Gross margin (C$ ha^{-1})	85	489	485
Winter wheat			
Yield (t ha^{-1})[c]	2.9	3.3	3.3
Price (C$ t^{-1})	160	160	202
Gross revenue (C$ ha^{-1})	460	524	662
Direct costs (C$ ha^{-1})	285	273	209
Gross margin (C$ ha^{-1})	175	251	453

[a] Only two sample points for each category.
[b] Ten-year provincial average yield: 6.4 t ha^{-1}.
[c] Ten-year provincial average yield: 3.8 t ha^{-1}.
Source: Stonehouse (1991).

Table 9.3. Physical and financial results for organic and conventional vegetables, Ontario, Canada, 1991.

	Machinery ($C\$\ ha^{-1}$)	Labour ($C\$\ ha^{-1}$)	Other ($C\$\ ha^{-1}$)	Yield[a] ($t\ ha^{-1}$)	Net return[b] ($C\$\ ha^{-1}$)
Sweetcorn					
Organic	921	926	1182	9.20	−1165
Conventional	921	977	637	15.36 (7.90)	626
Green beans					
Organic	156	2813	2176	7.40	−1971
Conventional	156	1835	420	8.19 (3.92)	1039
Cabbage					
Organic	958	1816	4917	40.21	−2038
Conventional	958	1475	2887	49.82 (27.06)	1659
Tomato					
Organic	988	4801	4894	27.23	−4138
Conventional	988	5478	4839	49.37 (23.47)	590
Spanish onions					
Organic	1643	3055	3626	17.28	−3505
Conventional	1643	432	2292	22.26 (19.31)	1892

[a] Yield is the average for 1991 and 1992; 1991 provincial average yield in parentheses.
[b] Excluding organic premium prices (see text).
Source: Sellen *et al.* (1993).

For various field crops in Saskatchewan, Green's (1990) sample recorded yields that were the same as or slightly below conventional, with some reports that yields were higher than conventional in dry years. Stonehouse (1991) reported five years of on-farm data for grain corn and winter wheat in Ontario for which the organic yields were as good as either conventional or low-input practices (Table 9.2). On the other hand, for a variety of organic vegetables grown under experimental conditions in Ontario, Sellen *et al.* (1993) found that yields were only 55–90% of conventional (Table 9.3).

Van Bers (Environment Canada, Ottawa, pers. comm.) has recently completed a survey of 51 experienced organic farmers from across Canada. Her purpose was to develop estimates of what was possible in

Table 9.4. Yields for organic vegetables and fruits, Canada, 1990.

	Provincial averages (t ha^{-1})[a]				National average[b] (t ha^{-1})	Sample size
	British Columbia	Alberta	Ontario	Quebec		
Vegetables						
Beets	26.7				8.9	3
Broccoli	8.9			12.0	5.2	4
Cabbage	26.7			57.1	28.2	5
Carrots	40.0	22.2	49.4	37.9	23.6	8
Leeks		9.3			3.1	3
Lettuce				48.6	13.2	4
Onions			29.7		9.9	3
Potatoes	22.2	27.4			18.3	6
Squash	22.2			24.7	11.7	4
Tomatoes	32.3				16.1	4
Fruit						
Apples	28.5		32.8		20.2	8
Cherries	13.5				8.1	5
Peaches	11.9		2.8		5.3	5
Strawberries	4.4		11.0	7.8	4.7	5

[a] Data for some provinces and some crops not shown.
[b] National averages reflect data for all provinces.
Source: C. Van Bers (personal communication).

Table 9.5. Average yields (t ha^{-1}) for organic field crops, Canada, 1990.

	Alberta	Saskatch-ewan	Manitoba	Ontario	Quebec	Prince Edward Island	Nova Scotia	Average
Wheat[a]	2.47	1.78	2.47	2.74	2.16	2.78		2.36
Oats	2.99	2.11	3.23	2.80	3.20	2.47	2.90	2.80
Barley	2.39	3.21	3.23	2.43	3.04		3.92	2.89
Rye (fall)	2.68	1.79	2.73	3.12	2.47		4.73	2.78
Buckwheat		1.08		0.79	1.24			1.05
Mixed grain	1.56	2.16			1.85	1.43		1.75
Oilseeds[b]	1.26	1.33	0.99	1.21				1.18
Pulses[c]	3.12	1.28	2.42	3.21		1.85		2.06
Hay	6.80	4.94	3.09	6.51	4.82	6.18		5.49

[a] Wheat: all varieties including spelt.
[b] Oilseeds: flax, canola and sunflower
[c] Pulses: lentils, soyabeans, beans and peas.
Source: C. Van Bers (personal communication).

organic farming rather than what is currently the case. Yet the survey covers a wide range of products, and is the most comprehensive survey completed to date (Tables 9.4 and 9.5).

Marketing and premium prices

For our Quebec survey, we also asked farmers about their experiences when it came to marketing, and the majority expressed some kind of problem. The problems reported are common to other surveys, and regrettably are similar to those reported in the early 1970s (AERC, 1972). The problems generally relate to the structure and performance of the market, with the most important being: lack of market information; regulatory problems; lack of uniform certification standards; scarce supply of organic livestock feed; few slaughtering facilities; and poor consumer understanding of the nature of organic agriculture.

When it came to prices, farmers in our survey expressed the view that organic prices are higher, and on average they reported receiving a premium of about 30%. However, many producers complained that they were unable to get a price premium or of extreme premium instability. Reports of this type of price behaviour are consistent across other surveys (COG, 1990; Green, 1990; Coxworth, 1991) and are symptomatic of what is still a very thin market (see Table 9.1).

Production costs

Although not all farmers are earning price premiums, there appears more of a consensus that production costs are lower. For example, one national sample of organic farmers estimated cost savings compared to conventional to be 18% on average (Weymes, 1990). For the COG (1990) survey, 158 organic farmers were asked to estimate their expenses relative to conventional (see Table 9.1). Except for fruit production, farmers believed that their production expenses were lower.

In spite of their problems, our respondents were convinced that 'organic is as profitable as conventional'; 72% expressed strong agreement with this statement. Again, this sentiment is fairly consistent across other surveys, and holds up for most case studies.

Enterprise Case Studies

Field crop and vegetable production in Ontario

Stonehouse (1991) has reported the results of an on-farm comparison of conventional, reduced-input, and organic weed control systems for several common field crops in southern Ontario. Economic comparisons were made using partial budgets based on five years of data gathered from six farms of comparable size (see Table 9.2). With only two farms of each type, the results should be interpreted as preliminary.

The organic farms achieved yields comparable to conventional, and a lower incidence of weeds to control was also reported. Stonehouse observed that on the organic farms, weeds did not appear to be a serious problem, and attributed this to the practice of composting all manure so as to kill weed seeds. On the other farms, weeds were controlled effectively with herbicides. The organic farms were also distinguished by their tendency to include a wider range of crops, with more complex rotations, including hay, pulses, a variety of small grains, and more frequent use of cover crops.

For both wheat and corn, gross margins for the organic farms were at least as great as the alternatives. For corn, the five-year average price received was the same across farming systems (C$ 126 t^{-1}). For wheat, the organic farms received a premium, but would still have had the highest returns at the conventional price.

Sellen *et al.* (1993) have recently completed a comparison of organic and conventional vegetable production at the Agriculture Canada research station in London, Ontario. The conventional practice was based on soil tests and standard production recommendations, while the organic protocol followed the methods described in Coleman (1989). Data were collected from experimental plots in 1991 and 1992, following a one-year establishment period.

Compared to conventional, organic yields were lower by 8–45%, and input costs were much higher (see Table 9.3), particularly for compost (C$ 937–1875 ha^{-1}) and labour charges for hand hoeing (C$ 200–2900 ha^{-1}). Based on conventional prices, net returns were markedly lower for organic. To earn a zero net return, price premiums of 11–59% would have been required. The authors note that there were a number of problems encountered with the experiment. As such it provides an example of what can happen to yields and net returns during the transition to organic methods.

Field crop production in Saskatchewan

An example of an enterprise analysis in the Prairie region is provided by Rutherford *et al.* (1992). The Prairie provinces of Manitoba, Saskatchewan and Alberta comprise 80% of Canada's farm land and nearly 50% of farms. In the dry land regions, where summer fallowing is practised, soil erosion is of particular concern. The land is left bare during the entire season as a means of storing precipitation, and weeds are

Table 9.6. Physical and financial results for organic and conventional wheat, Saskatchewan, Canada, 1991.

	Yield ($t\ ha^{-1}$)	Price ($C\$\ t^{-1}$)	Income ($C\$\ ha^{-1}$)	Costs ($C\$\ ha^{-1}$)	Net return ($C\$\ ha^{-1}$)	Returns[a] ($C\$\ ha^{-1}$)
Wheat enterprise[b]						
Organic	1.95	86	168	109	59	n/a
Conventional[c]	1.95	86	168	133	34	n/a
Relative (conv=100)	100	100	100	82	174	n/a
Complete rotation						
Organic[d,e]	n/a	n/a	205	92	113	157
Conventional[f]	n/a	n/a	196	156	40	104
Relative (conv=100)	n/a	n/a	105	59	283	151

[a] Including income from government programmes.
[b] 1991 data based on 13 organic farms.
[c] 1991 provincial average yield, 2.15 $t\ ha^{-1}$.
[d] Organic rotation: wheat–fallow–lentil–fallow.
[e] Income based on conventional prices.
[f] Conventional rotation: wheat–lentil–wheat–fallow.
Source: Rutherford *et al.* (1992).

controlled by either mechanical or chemical means. Organic systems are being investigated as an alternative.

Their results are based on partial budgets for wheat, lentils, and flax, for 13 organic farms. The budgets include extra marketing and certification costs (C$ 7.35 t^{-1}). The performance of the organic farms was compared with local average yields, and costs based on a group of conventional farms that participate in the Top Management Program (TMP) operated by the University of Saskatchewan.

Wheat was grown according to various rotations, but generally followed a sweet clover or fall rye green manure, summer fallow, or another crop. For 1991, in ten out of 13 cases, the returns to organic were higher than for conventional. As shown in Table 9.6, wheat yields and income for the organic group were identical to conventional, yet net returns were higher even when conventional prices were used. The results for 1990 were qualitatively the same. The authors also computed economic returns for lentils and flax grown on seven farms; however, these results varied widely.

Rutherford *et al.* also compared the economic returns for a complete rotation on one farm, with and without the inclusion of government programme payments (Table 9.6). The organic rotation incurred lower costs and produced higher average annual returns over cash costs.

Including government programmes caused a reduction in the relative performance of the organic rotation, as shown in the last row of the table.

In spite of the impression given by these results, the authors advise caution. First, the analysis is based on a small sample. Second, while the yields for the organic group are actual yields, the conventional yields are local averages, while at the same time conventional costs were based on farmers from the TMP who may incur higher costs than average, but also have higher yields.

However, Coxworth (1991) suggests that the results achieved at Indian Head, Saskatchewan indicate that organic farmers can make good profits even if lower yields are achieved. To investigate the economics of an organic system, a wheat–wheat–sweet clover green manure rotation was evaluated over a seven-year period (1978–1984). Yields were lower by about 9%, but with a 30% price premium, returns above variable input costs were higher for their organic system (C$ 138 ha^{-1} versus C$ 95 ha^{-1}).

Whole-farm Analyses of Dairy Farms in Quebec

A transition model

The economics of the transition to organic practices for a Quebec dairy farm were investigated by Forest (1992). A six-year transition was analysed using whole-farm, multi-period linear programming, with the primary concern being the optimal choice of a rotational plan among ten potential rotations, under two crop yield assumptions.

The analysis was based on an existing farm of 50 ha of drained land, with 42 milking cows, producing an average of 7566 litres per cow per year. The traditional rotation prior to conversion was three years of hay followed by two years of corn silage and one year of barley. Only one-third of the farm was converted each year, and the farm was constrained to be self-sufficient in livestock feed from year 3 onwards, since the market for organic feed is still poorly developed. In the base model (Model A), yields were assumed to be equivalent to ten-year average conventional yields. In a second model (Model B), a 30% decrease in yields during the first year of the transition was assumed, followed by a linear recovery over three years. Conventional output prices were used, and so the results do not depend on the existence of price premiums.

The general evolution of profit for both models was a decrease followed by a recovery as the herd size adjusted to the constraints imposed. The improved profitability as transition progressed could be traced to lower expenses for external feed, additional revenue from the sale of soyabeans, and revenues from reducing the size of the dairy herd. In Model A, forcing the farm to be self-sufficient in feed caused the dairy herd to shrink to 36 cows, and the sale of some milk quota.

Although there were differences between models, both resulted in optimal rotations that were judged to be economically viable throughout the transition. After allowing for a standard rate of return to capital investment, at worst a C$ 36,092 surplus was available for family living expenses. The effect on profit during the first year of the transition was minimal, since only one-third of the farm was converted. However, a faster conversion would have had a more serious impact on profit.

The results support the view that a conversion to organic agriculture need only have a modest negative effect on economic performance if the transition is done gradually, and with the presence of livestock such as dairy cattle. The presence of livestock is an important factor since it allows the farm to grow crops that are essential in an organic rotation, yet have little value in the local market.

A comparison of organic and conventional dairy farms

To date, the most comprehensive comparison of organic and conventional dairy production in Canada has been undertaken by Burgoyne (1992). Data was sourced from AGRITEL, an extensive data bank on 1500 farms managed by the Fédération des Syndicats de Gestion Agricole du Québec.

The sample included 336 conventional farms and 16 organic. For the purposes of comparison, the conventional farms were stratified along several dimensions according to their farming practices. The most general divisions were made between high- versus low-yield farms, and between farms that tended to use high versus low levels of inputs. These divisions resulted in 12 conventional comparison groups, all based on 1989 data. The sample size was large enough to produce some statistically significant results; however, the sample of organic farms was not considered large enough to be statistically representative.

Table 9.7 summarizes the results for the organic group and two of the conventional groups, and also shows the range of results over all 13 groups. CON-1 is those farms that were identified as intensive in terms of their milk output per unit area of forage production, and CON-2 is the group with high milk yield per cow.

In terms of general characteristics, the three groups were fairly similar in terms of labour availability and number of cows. As might be expected, yield per cow and crop yields for the organic farms were lower than for the intensive farms, but well within the range of all

Table 9.7. Physical and financial performance of organic and conventional dairy farms, Quebec, 1989.

	Organic	CON-1[a]	CON-2[b]	Range[c]
Number of farms, *n*	16	56	61	16–61
Number of cows	45	52	41	32–52
Labour (FTE)[d]	2.0	2.1	2.0	1.6–2.1
Milk yield (kg per cow)	5936	7130	7640	5255–7640
Cereal yields (t ha^{-1})	2.4	3.9	3.0	1.2–4.5
Forage yields (t ha^{-1})	4.0	6.8	5.5	3.4–6.8
Veterinary costs (C$ per cow)	47	73	86	47–90
Feed purchased (C$ per cow)	283	522	562	283–609
Concentrate fed (kg per cow)	1848	2704	2730	1508–3175
Labour cost (C$ per cow)	210	135	153	79–210
TVC[e] dairy (C$ per cow)	450	629	692	450–728
TVC field (C$ ha^{-1})	61	205	152	61–245
SRL[f] (C$ per cow)	976	825	791	416–976
SRL (C$ ha^{-1})	431	688	517	187–431
SRL (C$ l^{-1} milk)	16.4	11.6	10.4	7.0–16.4
Revenue (C$ per cow)	3602	3962	4204	2816–4204
Total costs (C$ per cow)	2857	3273	3561	2511–3698
Net return (C$ per cow)	745	689	643	280–745
Return on assets (%)	3.4	3.1	2.1	−1.4–3.4

All data are recorded on an annual basis for 1989.
[a] CON-1: intensive farms in terms of milk produced per forage hectare.
[b] CON-2: intensive farms in terms of milk yield per cow.
[c] Range of results for 12 conventional and one organic groups.
[d] FTE: full-time labour equivalent (3000 h $year^{-1}$).
[e] TVC: total variable costs.
[f] SRL: standardized return to labour (see text).
Source: Burgoyne (1992).

conventional farms. However, the organic farms recorded the lowest variable costs on a per cow basis, and for their field operations. This reflected lower charges for veterinary services, off-farm feed, and concentrates. Yet labour costs per cow were the highest of any group.

Financial performance was measured in several ways. Standardized return to labour (SRL) is the return over variable costs that is available to labour. It accounts for differences in debt servicing, depreciation and return to capital, so it is useful for comparisons between farms with different capital structures. For all measures using the SRL, the organic farms showed the best performance.

As might be expected, for the organic group, revenue per cow was not as high as for the intensive groups, yet total costs per cow were lower. As a result, the organic group showed the highest net returns per cow and return on assets. It is worth noting that this was achieved without an organic price premium. All farms received approximately the same milk price (C$ 0.50 l^{-1}).

Burgoyne advises some caution in the interpretation of these results since only one year of data was used (1989), and no attempt was made to account for the role of management ability. However, all these farmers belong to the AGRITEL system, and are generally regarded to be the most progressive farmers in the province. Thus, the fact that the organic farms compared so favourably within this group is quite encouraging.

Conclusions

The foregoing summary makes it evident that there has been relatively little scientific research on economic questions related to organic farming in Canada. The same could be said for research in general. This simply reflects the fact that it is only since the late 1980s that public institutions and funding agencies have begun to consider organic farming as a legitimate alternative to conventional. Now that research funding priorities are being reoriented to elevate the rank of environmental issues, one can expect the pace of research on organic farming to accelerate, but results will lag by several years, given the long-term nature of this research.

However, even though the economic studies completed thus far cannot be generally taken to be conclusive, they are encouraging. As in other countries, the farmers who have already made the transition to organic methods are showing themselves to be energetic, opportunistic and more market-oriented than many of their predecessors. In many cases, they are quite successful. The same may not hold for those who follow.

Continued progress in the organic sector will depend on continued development and dissemination of the organic knowledge base. But, it will also depend on a continuing transition of consumer values and attitudes. The current recession has slowed the pace of this transition and consumers' interest in organic still appears to be motivated more by direct personal concerns related to pesticides, than by general environmental concerns related to farming practices. As the loop closes, so the organic sector in Canada will develop.

References

AERC (1972) *Potentials of the Organic Food Market and Implications to Gaspé Farmers.* Research Monograph Agricultural Economics Research Council of Canada, Ottawa.

BFB Consultants (1992) *Canadian Natural and Organic Food Report.* Prepared for Oregon Department of Agriculture and Western United States Agricultural Trade Association, Vancouver, Washington.

Burgoyne, D. (1992) Analyse économique comparée de l'impact du niveau d'intensification, de l'utilisation d'intrants et de la production biologique sur la rentabilité des entreprises laitières au Québec. MSc thesis, FSAA, Université Laval, Quebec.

COG (1990) *A Study of Potential Market Niches for Canadian Organic Producers.* Prepared by Canadian Organic Growers for Agriculture Canada, Ottawa.

Cole, H. (1991) Popularity of organic food seems limited. *Western Producer, Winnipeg,* (6 June), 7.

Coleman, E. (1989) *The New Organic Grower.* Chelsea Green, Chelsea, Vermont.

COUP (1992) *Draft National Standards for Organic Products.* Canadian Organic Unity Project Agriculture Canada, Ottawa.

Coxworth, E. (1991) *Sustainable Agriculture and Regional Rural Processing Centres.* Saskatchewan Research Council, Regina, Saskatchewan.

Cramer, C. (1992) Organic rulemakers have to make hay. *The New Farmer* 14(7), 2–4.

Farm Credit Corporation (1988) *Farm Survey 1988.* Farm Credit Corporation, Statistics Canada, and Agriculture Canada, Ottawa.

Forest, J.F. (1992) The economics of conversion to organic agriculture: a rotational plan. MSc thesis, McGill University, Quebec.

Green, C.O. (1990) *A Needs Assessment Study of the Non-chemical Crop Production Industry.* Saskatchewan Department of Agriculture and Food, Regina, Saskatchewan.

Hancock N. (1991) Organic foods don't check out at registers. *Canadian Grocer* (May), 11.

Hanisch, S. (1989) *The Organic Food Market in Québec.* Department of Agricultural Economics, McGill University, Macdonald Campus, Quebec.

Hay, J. (1989) The consumer's perspective on organic foods. Food Market Commentary. *Agriculture Canada* 10(4), 45–50.

Henning, J., Thomassin, P. and Baker, L. (1990) *Organic Farmers in Québec: Results of a Survey.* Department of Agricultural Economics, McGill University, Macdonald Campus, Quebec.

Kramer, D. (1989) Popular organics: the nuts and bolts of Canada's fastest-growing food industry. *Harrowsmith Magazine* (July), 32–43.

MacRae, R.J., Henning, J.C. and Hill, S.B. (1989) *Financing Sustainable Agriculture: Current Problems and New Strategies.* Ecological Agriculture Projects, McGill University, Quebec.

Molder, P.J., Negrave, P.D. and Schoney, R.A. (1991) A descriptive analysis of Saskatchewan organic producers. Paper presented at the Agricultural Institute of Canada National Conference, Fredericton, New Brunswick, July.

Rusk, J. (1993) Green loses sheen as shoppers try to stay out of red. *Globe and Mail, Toronto* (12 February), 1.

Rutherford, A., Gimby, M., Smith, G., Mair, A. and Coxworth, E. (1992) *Viability of Organic Farm Practice: Second Year Annual Report.* Saskatchewan Research Council, Regina, Saskatchewan.

Science Council of Canada (1986) *A Growing Concern: Soil Degradation in Canada.* Science Council of Canada, Ottawa.

Science Council of Canada (1991) *Sustainable Agriculture: Some Policy Concerns.* Committee on Sustainable Agriculture, Science Council of Canada, Ottawa.

Sellen, D., Tolman, J.H., McLeod, D., Weersink, A. and Yiridoe, E. (1993) *The Economics of Organic and Conventional Horticulture Production.* Working Paper, Department of Agricultural Economics and Business, University of Guelph.

Statistics Canada (1987) *Census Canada 1986: Agriculture: Québec.* Minister of Supply and Services Canada, Ottawa.

Stonehouse, P. (1991) *Economics of Weed Control in Alternative Farming Systems.* Proceedings of the 5th REAP Conference. McGill University, Macdonald Campus, Quebec.

Thériault, J. (1991) *La mise en marché des produits biologiques au Québec.* Ministère de l'Agriculture, des Pecheries et de l'Alimentation du Québec, Quebec.

Weersink, J., Allan, S., Funk, T. and Goddard, E. (1990) *Consumers' Nutritional Attitudes and Perceptions: a Pilot Study.* Department of Agricultural Economics and Business, University of Guelph, WP90/26.

Weymes, E. (1990) *The Market for Organic Foods: a Canada Wide Survey.* Faculty of Administration, University of Regina, Saskatchewan.

10 Economics of Organic and Low-input Farming in the United States of America

Molly D. Anderson

Introduction

Two recent political developments in the United States have had strong impacts on the current economics of organic farming, and should be mentioned before a review of existing studies. First, the US Department of Agriculture established a research and education programme on 'low-input' farming systems in 1988, which has received continued funding each subsequent year. The significance to organic farming is twofold: increased public support has helped to legitimize organic farming, placing it in the agricultural research mainstream. Whereas quite recently organic farming was perceived by most university agricultural scientists as a radical farming system unworthy of serious study, it is now the subject of academic studies in many regions. But at the same time, greater public support for low-input agriculture has diluted the attention organic farming attracts because organic farming is combined in this highly-visible research and education programme with all other farming systems having reduced synthetic chemical inputs. It has also created difficulties in analysing organic farms as a separate entity because most US studies do not report results from organic systems separately from 'low-input' or 'reduced-chemical-input' systems, even when, in fact, the systems being studied include those with no synthetic fertilizers or pesticides.

A second political development that affects the economics of organic farming is the authorization in the 1990 Food, Agriculture, Conservation and Trade Act (the five-year domestic farm bill) of national standards for organic food and the implementation of a federal organic certification programme. Federal certification standards will almost certainly streamline the organic marketing process, making it simpler for organic producers to sell their products across state boundaries. However, in the short term, this act may slow down the expansion of organic production because compliance could increase costs of production, wholesaling and retailing (Jolly and Norris, 1991).

This chapter draws on experimental studies, economic models, and case studies of certified organic farms, as well as case studies and surveys of 'near-organic' farms that use no synthetic fertilizers, pesticides, or growth regulators but do not label their products 'organic'. Presumably many of the 'near-organic' producers would be certified if there were standard certification processes and costs across states, and a federal label to identify foods produced with organic methods. When no synthetic fertilizers or pesticides were used, but the study cited did not label the farming systems organic, this review uses the label 'low-input' for consistency. In discussions on the effects of premium prices on the demand for or economic viability of organic production, only certified organic is included since there is no consistent labelling for food produced in transitional or 'near-organic' systems. This chapter reviews a selection of US studies published since 1988, emphasizing work that can be widely accessed.

The Characteristics of Organic Farms and Farmers

Very few surveys have been conducted since 1988 that assess the characteristics of organic farms and farmers in the USA. A major publication by the National Research Council (NRC, 1989) included case studies of several farms that do not use chemical fertilizers and pesticides on at least some of the land in production. Other case studies of organic producers have been completed in Kansas and the Northwest. Such case studies are not appropriate for assessing characteristics of the population of organic producers and their farms. However, they document the existence of successful organic farms; describe organic practices used on a diversity of cereal, legume, fruit, nut, vegetable and livestock enterprises; and often provide useful economic information.

There was a brief heyday of studies of organic farming in the USA in the late 1970s and early 1980s. The literature of this period was dominated by comprehensive socioeconomic studies conducted among mixed-crop and livestock farmers in the Midwest, but studies were also performed in California, Maine, Michigan and New Jersey. Midwestern studies emphasized the similarities between organic and conventional farmers in practices, markets, and sociological characteristics; but these generalizations did not necessarily hold true in other regions with better-developed organic markets, higher land prices, and greater urbanization. In the same way that results from the Midwest should not be extrapolated to other regions, studies from the 1970s and early 1980s cannot necessarily be extrapolated to the present.

Studies prior to 1988 showed that organic farms, compared with conventional farms, tended to have the following characteristics: less land, smaller gross farm incomes, higher proportions of owned land, higher labour requirements, and more enterprise diversity, with vegetables a more frequent commodity choice. Organic producers often were younger than conventional producers and less likely to have grown up on a farm. They frequently had higher average educational levels and less farm experience. They tended to use different information sources than other producers, relying more on farming magazines, books and informal sources of advice such as neighbours.

A few farm surveys since the early 1980s included organic producers in their sample, but none reported results for organic producers as a separate group. Anderson (1990) surveyed reduced-chemical-input producers in North Carolina who were garnered from several sources, including members of two organizations promoting organic farming methods. She also surveyed a random sample of farm operators, distinguishing three groups of producers based on their actual practices. About a quarter of the low-input producers used synthetic fertilizers, and only 4% used synthetic herbicides and insecticides. Almost all of the full-time conventional producers used synthetic fertilizers, and about three-quarters used synthetic pesticides. Low-input producers, compared with other producers, had significantly smaller farms, a higher ratio of owned to rented land in operation, smaller gross farm incomes, greater reliance on off-farm income, less farming experience, and more formal education. They were less likely to have grown up on farms or to be operating the farms where they grew up. Low-input farmers also used very different information sources: compared to full-time conventional farmers, they relied less on extension agents' recommendations about farm chemicals and they more frequently read books and magazines promoting reduced-chemical farming.

A study of Northeastern organic farmers showed that the typical organic farm is much smaller than its conventional counterpart: the 1990 median size of 80 certified organic vegetable farms in Maine was less than 2 ha (Green, 1994). A survey of 738 organic producers in California found an average farm size for respondents of 23 ha, compared to a state-wide average of 158 ha in 1982. When the six largest organic growers were excluded, the average size was 11 ha. The largest 5% of farms was responsible for 80% of the total volume of sales, a somewhat greater concentration than in conventional agriculture in the state (Cook, 1988).

Consumer Demand for Organic Produce

Consumer demand for organic food has been studied more extensively in the last few years than organic farm and farmer characteristics. The organic food market has been analysed in California, Washington, Georgia, New Jersey and Kansas; and several national polls have assessed public opinions about organic foods and willingness to purchase organic produce. A large majority of the public is willing to buy organic products, and at present demand far exceeds supply. The organic foods market is in disequilibrium, with volatile prices and rapid expansion (Hall *et al.*, 1989). In California, total wholesale organic sales rose from US$ 1 million in 1977 to approximately US$ 50 million in 1987. The largest distributor of organic foods on the East Coast had sales of between US$ 20 million and US$ 22 million in 1988 (ERS, 1989).

Media publicity about food safety in the spring of 1989, fuelled by reports of alar in apples and cyanide in Chilean grapes, created a surge of public interest in organic foods. However, demand for organic produce fell in the months thereafter. Chain-store produce managers surveyed in 1989 and 1990 thought that organic produce had worse appearance and shelf life than conventional produce, but rated organic foods highly on environmental impacts, chemical residues, flavour and nutrition (Jolly and Norris, 1991).

Forty per cent of consumers surveyed in California believed that organic foods are better than their conventional counterparts, but only 23% bought organic foods regularly (Jolly *et al.*, 1989). Similar discrepancies between positive opinions about organic foods and low frequency of purchase have been found in other states.

Physical Productivity of Organic Farming Systems

Information about the productivity and financial performance of organic farming systems comes from four sources: private or public research plots, models constructed with data from organic farms, case studies of actual farms, and farm surveys. Comparisons between organic and conventional systems have been performed for each type of study, and results are separated by type in the following three sections. Two caveats on reported yields in farm-level studies should be heeded. First, results cannot be extrapolated beyond the region of the country in which they were performed, because they are highly dependent on climatic variables, pest problems, labour availability and other regional characteristics. Second, the short term of many studies limits their applicability because it does not allow results to be averaged over weather variability or phases of the transition from conventional to organic production (Lee, 1992).

Experimental studies

The Rodale Farming Systems Experiment is the most comprehensive and well-documented recent US experimental study of organic farming, but it is not included in this chapter because it is discussed in detail in Chapter 15.

Long-term experimental trials in Nebraska begun in 1975 have compared an organic rotation with three cropping systems having conventional chemical applications. The organic rotation is maize–soyabeans–maize–oats/clover with feedlot manure as a fertilizer. Two of the conventional cropping systems use the same four-year rotation with synthetic fertilizer, one with and one without synthetic herbicides. The third conventional system is continuous maize with synthetic fertilizers, herbicides and insecticides. After one four-year rotation cycle, average maize yields in the rotations were similar: 4.96, 5.09 and 5.28 $t\,ha^{-1}$ in the organic, fertilizer-only and fertilizer-and-herbicide management systems respectively (Sahs *et al.*, 1992). Yields of maize in the organic treatments were higher than all other treatments in 1990 and 1991; this was attributed to the 'buffering effect' of organic systems during drought stress. Organic maize yields usually were lower following oats/clover because poor and uneven maize stands made cultivation difficult, more weeds escaped, and soil water was used by weeds rather than maize. However, maize grown after oats/clover usually gave significantly lower yields than maize grown after soyabeans in all management systems (Sahs and Lesoing, 1991).

Experiment station field studies in South Dakota (Mends *et al.*, 1989) analysed two different low-input crop rotations that did not use chemical fertilizers, herbicides or a mouldboard plough. They were compared with systems using standard chemical applications. The first rotation compared a low-input oats/alfalfa–alfalfa–soyabeans–maize rotation with maize–soyabeans–spring wheat rotations using conventional and ridge tillage. The second low-input rotation was oats/clover–clover–soyabeans–spring wheat, compared to soyabeans–spring wheat–barley in conventional and minimum tillage systems. Average maize yields between 1985 and 1988 were slightly lower in the first low-input rotation than in the other two systems (Table 10.1). However, during the drought year of 1988, maize yields were significantly lower (49%) in the conventional system than in the low-input

Table 10.1. Comparison of crop yields, United States.

Crop	State	Period	Organic ($t\ ha^{-1}$)	Low input[a] ($t\ ha^{-1}$)	Conventional ($t\ ha^{-1}$)	Relative (conv = 100)	Source
Experimental studies							
Maize	Nebraska	1979–1991	5.0	n/a	5.3	92	Sahs *et al.* (1992)
Maize	S. Dakota	1985–1988	n/a	4.6	5.3–5.6	82–87	Mends *et al.* (1989)
Soyabeans	S. Dakota	1985–1988	n/a	1.5	1.5–1.8	85–105	Mends *et al.* (1989)
Spring wheat	S. Dakota	1985–1988	n/a	2.8	2.7–2.8	100–105	Mends *et al.* (1989)
Modelling studies							
Rice[b]	California	1992	5.6	n/a	8.4	n/a[c]	Williams *et al.* (1992a, b)
Almonds	California	1992[e]	1.7	n/a	2.2	77	Asai *et al.* (1992a,c)
Wine grapes	California	1992	10.6	n/a	n/a	–[d]	Klonsky *et al.* (1992a, b)
Maize	S. Dakota	1987[e]	n/a	4.7	5.1–5.3	89–91	Dobbs *et al.* (1988)
Soyabeans	S. Dakota	1987[e]	n/a	1.8–1.9	2.0–2.1	90–93	Dobbs *et al.* (1988)
Spring wheat	S. Dakota	1987[e]	n/a	2.7	2.8	95	Dobbs *et al.* (1988)

[a] No synthetic fertilizers or pesticides used, but not labelled 'organic'.
[b] Dryweight (13% moisture).
[c] Yields not comparable because organic rice is grown in a two-year rotation with a cover crop.
[d] Similar yields for the same varieties, locations, and growing conditions.
[e] Normalized yields derived from 1985–1986 data.
n/a, Not available.

system. In the second rotation comparison, low-input soyabean yields averaged slightly less than in the other two management systems. Low-input spring wheat yields were equal to or slightly greater than yields in the other systems.

A long-term experiment at Cornell University, begun in 1988, has compared weed control in maize silage production with conventional and low-input cropping systems. Alternative treatments included interseeding with red clover or ryegrass, cultivating, and ridge tilling. Plots that received no herbicides in 1989 had higher weed levels than other plots, but yield reductions were less than 10% (Janke *et al.*, 1991).

Modelling studies

Several models detailing yields and returns from organic production of various commodities have been prepared by the Co-operative Extension Service at the University of California. In most cases, companion budgets for conventionally produced commodities are also available. These models were constructed using practices and costs common to particular regions of California, with information gathered through interviews with organic growers.

Models of organic rice production in California using water-seeded and no-till drill-seeded methods were for a representative 243 ha farm in the Sacramento Valley with 121 ha in rice production in any given year. Organic rice yields can vary substantially from year to year and field to field, ranging from 1.3 to 9.0 t ha^{-1} on a dry weight basis (Williams *et al.*, 1992b). Reported yields for organic almonds grown on 38 ha in the Northern San Joaquin Valley range from 0.8 to 2.6 t ha^{-1} (shelled), depending on the year of production and growing conditions (Asai *et al.*, 1992c). Yields of organic wine grapes in the North Coast region range from approximately 6.7 to 20.1 t ha^{-1}, similar to conventionally produced wine grapes. Yields vary depending on factors such as vineyard design, grape variety, vineyard age, county of production and growing conditions (Klonsky *et al.*, 1992a). Table 10.1 includes representative organic and conventional yields used in companion budgets for rice, almonds, and wine grapes. The budget for conventional wine grape production was not comparable because it dealt with only one county and one variety of grape, rather than the combination of counties and varieties included in the organic model. Table 10.2 shows yields and returns of selected organic vegetables grown intensively on 47 ha in the Central California Coast region. High and low production figures depend on growing conditions and pest or disease infestation (Klonsky *et al.*, 1994). Actual yields and returns depend on factors such as crop damage and market competitiveness in any given year. Cooling and marketing fees have been subtracted from returns.

Table 10.2. Yield ranges for selected organic vegetable crops, Central California Coast, 1992/93.

		Yield (boxes per hectare)			Returns (US$ per box)	
Crop	Variety	Low	High	Box content	Low	High
Cabbage (green)		500	2250	22.7 kg	4	18
Cauliflower		1500	2500	12 count	4	14
Cucumber		1500	2500	10.9 kg	4.5	10
Garlic		750	1250	13.6 kg	36	68
Lettuce	Leaf	875	2125	24 count	5	10
	Romaine	750	1875	24 count	5	11
Onions	Red	1250	2000	18.1 kg	5	16
	Yellow	1500	2500	22.7 kg	5	16
Peas (non-staked)	Snap	1250	2500	4.5 kg	5	18
	Snow	625	1250	4.5 kg	5	18
Peppers	Green bell	1500	3000	12.7 kg	4	14
	Red bell	750	1750	12.7 kg	5	27
Sweetcorn		500	1000	48 count	5	12
Winter squash	Large	1250	2125	15.9 kg	5	12
	Small	500	2125	9.1 kg	6	13

Source: Klonsky *et al.* (1994).

In models constructed from South Dakota data reported above, 'normalized' organic yields were calculated based on a combination of field data, literature reviews, historical data and scientific judgement (Table 10.1). Attempts were made to account for yield interactions caused by rotations (Dobbs *et al.*, 1988).

Case studies

The National Research Council included yield information in several of its case studies of farms that did not use synthetic chemical fertilizers or pesticides on at least some land, or sub-therapeutic antibiotics or growth stimulators on livestock sold as 'natural' (Table 10.3). On the Spray Brothers' diversified grain and livestock farm in Ohio, average production for the years 1981–1985 was greater than the county average for maize, soyabeans, wheat and oats, ranging from 105% for wheat to 140% for soyabeans. On the Kutztown farm, a mixed crop and livestock farm in Pennsylvania, crop yields for 1978–1982 ranged from

Table 10.3. Comparison of crop yields from case study farms, United States.

Crop	Organic (t ha^{-1})	Conventional[a] (t ha^{-1})	Relative (conv=100)
Spray Brothers, Ohio, 1981–1985			
Maize	9.1–9.4	7.0	130–135
Soyabeans	3.2	2.3	140
Wheat	3.0	2.9	105
Oats	2.9	2.3	122
Kutztown farm, Pennsylvania, 1978–1982[b]			
Maize	6.8	5.4	127
Alfalfa hay	7.4	6.7	110
Corn silage	32.0	32.0	100
Rye	1.8	2.0[c]	92
Soyabeans	2.6	2.0[c]	126
Wheat	2.5	2.5	97
Pavich & Sons, California/Arizona, 1986			
Grapes	16.8	13.4	125
Lundberg Family Farms, California, 1985[b]			
Rice	4.9	8.3[d], 12.3[e]	59, 40

[a] County average unless otherwise indicated.
[b] Not certified organic.
[c] State average.
[d] Lundberg's conventional yields.
[e] Highest yielding farms in county.
Source: NRC (1989).

92% of the state average for rye to 127% of the state average for maize. Maize yields averaged 147% of the county average yield in a drought year (NRC, 1989).

At the Stephen Pavich & Sons grape farm, the production (total yield minus culls) was 125% of the estimated normal yields for conventionally raised grapes. Culls were only 1%, compared to 15–30% for conventional grapes. At the Coleman Natural Beef operation, feeder cattle were estimated to be 11–23 kg smaller than conventionally raised cattle, and daily rates of gain were slightly less. However, the death loss from calving and disease was only 2%, compared with an industry average of 6%. Lundberg Family Farms reported yields from 40 ha of

Table 10.4. Comparison of crop yields from case study farms, Dryland Northern Plains and Intermountain Northwest United States.

Crop	Organic ($t\ ha^{-1}$)	Conventional[a] ($t\ ha^{-1}$)	Relative (conv = 100)
Gould & Partners			
Winter wheat	1.9	2.3	82
Barley	1.7	n/a	n/a
Buckwheat	1.6	n/a	n/a
Quinn Farm			
Winter wheat	2.4	2.4	103
Durum	1.7	n/a	n/a
Kamut	1.7	n/a	n/a
Alfalfa	2.8	n/a	n/a
Marschner Farm			
Spring wheat	1.7–2.0	2.0–2.7	63–100
Sunflowers	0.9–1.1	0.9–1.6	57–122
Thomas Farm			
Wheat	2.0	1.6	125
Sunflowers	1.4	1.2	114

[a] County average.
n/a, Not available.
Source: Matheson *et al.* (1991).

organic rice and 769 ha of rice raised with synthetic chemical fertilizers and pesticides. Their organic rice yields were 59% of their conventional yields and 40% of the average yield on the most productive rice farms in the county (NRC, 1989).

Several of the case studies of certified organic farms in the Northern Plains and Pacific Northwest reported in Matheson *et al.* (1991) included average area yields for some crops for comparison with yields on organic farms (Table 10.4). On two farms, yields of spring wheat on clover fallow compared favourably with county averages; however, spring wheat following sunflowers on one of these farms yielded only half as much as the spring wheat on clover fallow. Winter wheat and sunflower yields were similar between organic and conventional systems.

Case studies on three certified organic farms in different parts of Kansas that raised cereals, legumes and livestock compared yields with state district averages (Table 10.5). On the Tangeman farm, average yields of dryland maize, soyabeans, wheat and oats between 1984 and

Table 10.5. Comparison of crop yields from case study farms, Kansas.

Crop	Organic ($t\ ha^{-1}$)	Conventional[a] ($t\ ha^{-1}$)	Relative (conv = 100)
Vogelsberg-Busch Farm			
Wheat	2.7	n/a	n/a
Oats	1.8	n/a	n/a
Soyabeans	2.0	n/a	n/a
Collman Farm			
Wheat	1.9	n/a	–[b]
Tangeman Farm, 1984–1986			
Wheat	1.7	2.4	74
Soyabeans	2.2	1.9	114
Maize	5.1	5.3	96
Oats	2.3–2.5	1.6	141–152
Bennett Farm			
Wheat	1.7	2.2	82
Soyabeans	3.0	2.4	122
Vogelsberg Farm, 1983–1986			
Maize	6.0, 5.7[c]	4.3[a], 2.9[c]	141, 198[c]
Soyabeans	2.0	1.7	115
Oats	2.6	2.1	126
Alfalfa/red clover	9.0	6.9[d]	129
Wheat	1.8	2.4	75

[a] County average unless otherwise indicated.
[b] Similar to county average.
[c] Drought conditions.
[d] County average for alfalfa only.
n/a, Not available.
Source: Cavigelli and Kois (1988).

1986 ranged from 74% of the district average for wheat to about 150% for oats. The Bennett farm reported yields similar to district averages during years when records were kept. The Vogelsberg farm had yields consistently above county averages for maize, soyabeans, oats and alfalfa/red clover hay; but wheat yields averaged 75% of conventional county yields. Several farmers in the Kansas case studies raised livestock, but livestock output was not reported (Cavigelli and Kois, 1988).

Table 10.6. Comparative economic analysis of dairy farms with different levels of chemical application, Wisconsin, 1989–1991.

	No chemical[a]	Low chemical[a]	Conventional[a]
Crop enterprise comparisons			
Cash costs (US$ ha^{-1})	163	222	262
Net cash return (US$ ha^{-1})	388	326	326
Net return (US$ ha^{-1})	259	247	165
Dairy enterprise comparisons			
Cash costs (US$ per cow)	1,081	1,064	1,349
Total costs (US$ per cow)	1,287	1,237	1,484
Milk yield (kg per cow)	7,219	7,219	8,444
Net cash return (US$ per cow)	1,377	1,317	1,356
Net return (US$ per cow)	431	417	112
Whole-farm comparisons			
Labour (h $year^{-1}$)	3,848	5,004	5,882
Net return to labour (US$ h^{-1})	9.11	8.03	10.42
Total cash costs (US$ $year^{-1}$)	51,490	56,958	101,194
Gross income (US$ $year^{-1}$)	102,717	114,843	184,594
Net cash return (US$ $year^{-1}$)	51,227	57,885	83,400
Net return (US$ $year^{-1}$)	15,801	15,172	31,854

[a] See text for explanation of division into groups.
Source: Kamp (1993).

Farm surveys

The Wisconsin Rural Development Center coordinated a comparative economic analysis of 34 dairy farms having 35–60 cows and a mix of feed grains, forage crops, and small grains (Kamp, 1993). Farmers were divided into three groups distinguished by pesticide and fertilizer application rates. The 'no-chemical' group had 'average chemical costs' of US$ 2.77 ha^{-1}, compared with US$ 79.02 ha^{-1} for the 'conventional-chemical' group and US$ 23.94 ha^{-1} for the 'low-chemical' group. Annual average milk production per cow in the 'no-chemical' dairies was 85% of the production in 'conventional-chemical' dairies (Table 10.6). It is important to note that 'no-chemical' is not organic.

Crop and Livestock Enterprise Financial Performance

Each of the studies reviewed above that reported crop yields, except the experiment at Cornell University, also analysed the economic performance of organic enterprises, or of the whole farm in whole-farm models and case studies. In comparing data on costs and returns in the next two sections with data from other countries, the reader must keep in mind that these figures are highly dependent on growing conditions, markets and government programmes.

Not all US certified organic producers sell their products at premium prices. Some growers lack access to organic markets, but some farmers intentionally avoid selling at premium prices. Their reasons include a commitment to making organic foods available to low-income consumers and a wish to demonstrate to other farmers that economic viability for organic farms does not depend on premiums (Green, 1994). In the Northeast, price premiums historically have been important to organic growers' survival. Produce premiums in this region usually are 10–50% above conventional prices, but they are highly variable and dependent on the particular market, the season, the commodity, and the growers' wishes (Green, 1994). Price premiums well over 100% have been reported for some commodities in California (ERS, 1989).

Experimental studies

Nebraska rotation comparisons between organic and conventional systems showed highest variable costs for continuous maize, and lowest costs for the organic rotation. The total production costs of the four-year organic rotation were 89% of costs for the same rotation treated with synthetic herbicides and fertilizer. However, since yields were also lower, the net returns (crop sales minus total production costs) were lower for the organic rotation: a total of US$ 825 versus US$ 865 ha^{-1} over four years. Crediting the organic rotation with a soil appreciation allowance did not make it more profitable than the conventional rotation. However, the comparison did not include resource costs for ground water contamination, because these cannot be estimated reliably with present data (Faeth *et al.*, 1991). Reduced ground water pollution may be an important advantage of the organic system.

In the rotation comparisons performed in South Dakota (Mends *et al.*, 1989), the two rotations using no fertilizer or pesticides had lower gross incomes than the corresponding conventional rotations during the first three years of the study, but surpassed the other systems during a drought year. The favourable drought-year income was due to higher yields of maize and spring wheat in the low-input systems and higher drought-induced alfalfa prices. The low-input systems had significantly lower direct costs other than labour.

Case studies

Most economic analyses in case studies published since 1987 were performed for the whole farm rather than by enterprise. However, Matheson *et al.* (1991) published partial budgets for each commodity grown on certified organic farms in the Northern Plains and Pacific Northwest (Table 10.7). Almost all the crops grown were sold at a premium; certified organic wheat from one farm received a premium between US$ 0.04 and US$ 0.05 per kilogram above local market prices. Income per hectare above variable costs ranged from a low of US$ 66 for spring wheat following sunflowers on one farm to a high of US$ 879

Table 10.7. Partial budgets for organic crops on case-study farms, Dryland Northern Plains and Intermountain Northwest United States, 1989/90 prices.

	Variable costs (US$ ha^{-1})[a]	Income above variable costs (US$ ha^{-1})
Gould and Partners		
Flax on fallow	99	879
Winter wheat after flax	104	262
Spring barley after winter wheat	71	422
Buckwheat after spring barley[b]	101	183
Spring wheat after buckwheat	76	121
Winter wheat after spring wheat	87	280
Quinn Farm		
Winter wheat on alfalfa fallow	73	304
Durum after winter wheat	92	242
Kamut after spring wheat	93	277
Winter wheat on clover fallow	66	312
Marschner Farm		
Spring wheat on clover fallow	135	254
Sunflowers after spring wheat	176	158
Oats/barley after sunflowers	80	68
Thomas Farm		
Spring wheat on clover fallow	177	238
Sunflowers after spring wheat	179	377
Oats after sunflowers	195	126
Spring wheat after sunflowers	142	66

[a] Seed, machinery, labour and transport.
[b] Production over 1000 lb. held for seed.
Source: Matheson *et al.* (1991).

for flax on another. The high net income for flax reflects a high premium sale price of US$ 0.87 kg^{-1}. Winter wheat income per hectare above variable costs averaged US$ 290 on three farms; spring wheat averaged US$ 170.

In the Kansas case studies, one farmer had collected sufficient data to allow enterprise analysis on maize, soyabeans, oats, alfalfa, wheat and hogs. Compared with district averages, his farm had performed very well: he kept production costs low, channelled almost all of his relatively high income gains back into his farm, and expanded his operation in difficult economic times. His variable cash costs (hired labour, seed, herbicides, fertilizer and lime, fuel and oil, machinery and equipment repairs, drying, machinery rental, and interest on variable costs) ranged from 83% of the district average for soyabeans to 16% for alfalfa (soyabeans: US$ 3.63 ha^{-1}; alfalfa: US$ 8.97 ha^{-1}). Total variable costs per litter of hogs averaged US$ 340 for an eight-pig litter, 84% of the district average for a seven-pig litter.

Farm surveys

On Wisconsin dairy farms, crop enterprise comparisons showed that 'no-chemical' farmers had lower cash costs per hectare and higher net cash returns and net returns per hectare compared to conventional-chemical farmers. In dairy enterprise comparisons, 'no-chemical' farmers had lower purchased feed, veterinarian and medicine expenses than conventional farmers. Total costs for 'no-chemical' farms were US$ 197 less per cow than for conventional dairies; however, because of higher production for conventional producers, total costs per unit of milk production equalled 'no-chemical' farmers' production expenses. Despite higher production per cow in conventional-chemical dairies, dairy enterprise profits and net returns were higher for 'no-chemical' producers because of special marketing of organic milk and higher butterfat premiums (Kamp, 1993; see Table 10.6).

Whole-farm Financial Performance

Modelling studies

Models created at the University of California detailed representative costs of producing several organic commodities. In some cases, companion budgets are available for the same crops produced with conventional methods; however, the wine grape budget for conventional production is not appropriate for economic comparison. For each budget, a ranging analysis showed net returns above operating costs,

Table 10.8. Model budgets for organic and conventional rice, California, 1992.

	Organic		
	No-till, drill-seeded	Water-seeded	Conventional
Per hectare costs (US$ ha^{-1})			
Cultural	299	343	701
Harvest	287	287	324
Post-harvest	–	–	20
Interest on operating capital	17	20	32
Total operating	603	647	1,074
Cash overheads	287	289	136
Non-cash overheads	724	739	348
Total operating and overheads	1,615	1,675	1,561
Annual costs (US$ per farm)			
Equipment	35,390	36,809	17,807
Investment	82,269	82,269	54,747
Business overheads	14,038	14,038	5,900
Break-even point			
Yield (t ha^{-1})	4.3	4.3	8.4
Price (US$ kg^{-1})	0.36	0.39	0.19

Source: Williams *et al.* (1992a, b, c).

cash costs and total costs for different yields.

For organic no-till drill-seeded rice (Table 10.8), the break-even point for positive net returns above total costs was a yield of 4.3 t ha^{-1} and a price of US$ 0.36 kg^{-1} (Williams *et al.*, 1992b). For water-seeded rice, net returns were positive with the same yield at a minimum price of US$ 0.39 kg^{-1} because total cultural costs and annual equipment costs were slightly higher (Williams *et al.*, 1992c). In conventional rice production on 162 ha, total costs per hectare were slightly lower than organic costs, mainly because cash and non-cash overhead costs were lower. However, total gross returns per hectare were only US$ 1361 in this model, giving a negative net return above total costs of –US$ 198 ha^{-1}. (These figures do not include income from government support programmes.) At yields of 8.4 t ha^{-1}, the break-even price to cover total costs is US$ 0.19 kg^{-1} (Williams *et al.*, 1992a).

For organic almond production (Table 10.9), operating costs, cash overhead costs, and non-cash overhead costs were higher than in

Table 10.9. Model budgets for organic and conventional almonds, California, 1992.

	Organic		Conventional	
	Flood-irrigated	Sprinkler-irrigated	Flood-irrigated	Sprinkler-irrigated
Per hectare costs (US\$ ha^{-1})				
Cultural	1,383	1,699	1,897	2,223
Harvest	2,579	2,579	827	827
Interest on operating capital	106	116	106	116
Total operating	4,068	4,394	2,831	3,167
Cash overheads	659	701	459	494
Non-cash overheads	1,667	1,971	1,289	1,289
Total operating and overheads	6,394	7,069	4,463	4,950
Annual costs (US$ per farm)				
Equipment	11,836	10,443	11,175	11,427
Investment	30,112	37,278	48,955	54,609
Business overheads	5,179	5,179	7,125	7,125
Break-even point				
Yield (meat t ha^{-1})	1.2	1.5	2.2	2.2
Price (US\$ meat kg^{-1})	5.18	5.18	1.98	2.20

Source: Asai *et al.* (1992a, b, c, d).

conventional production with both flood-irrigated and sprinkler-irrigated systems. Annual organic growers' investment costs and business overhead costs were lower than conventional costs, and their annual equipment costs were similar. The range of prices received by growers contributing data to the organic almond model was US$ 4.52 to US$ 5.95 per meat kg. At an average price of US$ 5.18 per meat kg, almonds produced with flood irrigation showed positive net returns above total costs when yields were at least 1.2 meat t ha^{-1}. For production with sprinkler irrigation and the same average price, returns above total costs were positive if yields were at least 1.5 meat t ha^{-1} (Asai *et al.*, 1992c,d). At yields of 2.2 t ha^{-1} in conventional production (typical for year 7 and on), the break-even price was US\$ 1.98 kg^{-1} to cover total costs under flood irrigation, or US\$ 2.20 kg^{-1} under sprinkler irrigation (Asai *et al.*, 1992a, b).

Organic wine grape production models assumed no price premium. For grapes grown with resident vegetation or an annually sown

Table 10.10. Model budgets for organic wine grapes, California, 1992.

	Resident vegetation	Annually sown vegetation
Per hectare costs (US$ ha^{-1})		
Cultural	3,512	3,653
Harvest	1,408	1,408
Assessment	86	94
Interest on operating capital	173	183
Total operating	5,187	5,340
Cash overheads	1,507	1,509
Non-cash overheads	4,115	4,137
Total	10,809	10,987
Annual costs (US$ per farm)		
Equipment	7,223	7,855
Investment	70,188	70,188
Business overheads	9,698	9,698
Break-even point		
Yield (t ha^{-1})	10.6	10.6
Price (US$ t^{-1})	1100	1100

Source: Klonsky *et al.* (1992a, b).

cover crop, net returns above total costs (Table 10.10) were positive at a grape price of at least US$ 1100 t^{-1} for yields of 10.6 t ha^{-1}. For prices at or below US$ 550 t^{-1} and for yields at or below 20 t ha^{-1}, it was not profitable to produce organic wine grapes (Klonsky *et al.*, 1992a,b).

The South Dakota rotations comparison study (Mends *et al.*, 1989) modelled whole-farm results for a hypothetical 219 ha farm. The low-input system in the first rotation had a positive net income every year, but the conventional and ridge-till systems suffered significant losses in the 1988 drought year. The low-input system for the second rotation had a slight loss in 1985 and comparable or higher net incomes than the other two systems in 1986–1987. It broke even in the drought year, while the other two systems studied showed profits in 1985–1987 and substantial losses in the drought year.

Case studies

Several organic farmers in the case studies compiled by the National Research Council reported indicators of their farms' economic performance. The Spray Brothers' mixed grain and livestock farm showed strong overall economic viability: per hectare production costs were below county averages, and yields were well above county averages (NRC, 1989).

The portion of the Kutztown Farm using organic production methods probably had higher labour requirements than a comparable conventional farm because it relied on cultivation for weed control, had a complex crop rotation plan, and had a very large number of small fields (98) in its contour strip-cropping system. Several economic analyses have been performed on this farm. In the most recent study reported (1986), a comparable profit maximizing conventional farm was 3% more profitable than the Kutztown Farm when soil erosion was not limited. However, the Kutztown Farm would have been more profitable than the conventional farm if soil erosion had been limited to a maximum of 11 t ha^{-1} (NRC, 1989).

Few economic data were available for the Ferrari Farm in California, most of which was certified organic. The Ferraris received a price premium of US$ 0.09 kg^{-1} on organic walnuts sold in the shell, and US$ 0.22 kg^{-1} (5% of the sale price) for shelled walnuts. Prices for organic Granny Smith apples ranged during the season from US$ 2.00 below to US$ 3.00 above the conventional price per 20 kg box (NRC, 1989).

The Stephen Pavich & Sons grape farm sold US$ 5.7 million in grapes in 1985, 1.7% of all US fresh grape sales. The Paviches incurred several costs that were higher than conventional costs. For example, they paid labourers a higher wage than the standard, they used three rather than two fallow years, and their disease and pest control materials were more expensive. The Paviches charged a 12–25% premium for grapes certified to comply with each buying state's organic foods regulations; the higher price covered costs of certification and special handling and storage. The Pavich pre-harvest cost per 10.4 kg box of grapes (US$ 2.20) was about the same as the University-estimated cost per box (US$ 2.14) for conventional production (NRC, 1989).

Coleman Natural Beef (CNB) in Colorado received a premium price of up to 25% for beef raised without synthetic chemical inputs. However, the owners reported that the ranch earned a return on labour and management that was below the wages they paid hired personnel. Their marketing operation appeared to support the ranch: 12,000 additional cattle were produced to CNB specifications by neighbours and sold by CNB. Given the low return on investment, the premium price was necessary to keep this operation functional (NRC, 1989).

Lundberg Family Farms reported that their organic rice production was substantially less profitable than conventional, despite premium prices. In 1986, the California rice price averaged US$ 0.17 kg^{-1}, and certified organic rice was US$ 0.26 kg^{-1}. Some production costs for the

Lundbergs' organic rice were lower than conventional costs; for example, no direct costs were incurred for pest control, and rice was planted without tillage into flail-chopped purple vetch, at a cost 30% of conventional seedbed preparation. The total direct cash costs per hectare for organic rice were approximately one-half the conventional costs (US$ 423 versus US$ 809). The net return over direct cash operating costs per hectare of rice harvested was US$ 854 for the organic rice and US$ 636 for conventional rice. However, the Lundbergs raised their organic rice in a rotation of one year of rice followed by one year of legume-fallow. The conventional rotation is two to five years of rice followed by one year of fallow for relevelling, and sometimes rice is grown as a continuous crop without fallow. Fallow costs were higher for organic than conventional rice, and the average annual net return over cash cost per hectare was higher for conventional rice: US$ 370 versus US$ 260 (NRC, 1989).

All five farmers in the Kansas case studies had lower variable costs, including hired labour, but most also had lower returns. Some of the farmers had net losses, especially during transition years, partially because their yields were lower than conventional yields and because some land was in green manures rather than cash crops. Premium prices for organic grains allowed some of these farmers to earn returns similar to or even higher than district averages. One farm (the same farm with enterprise budgets reported in the previous section) had very good economic performance compared with district averages. Cash operating expenses as a fraction of gross income were 47% (compared to 80% as the district average), total expenses to gross income were 51% (compared to 93%), and net farm income to gross income was 49% (compared to 7%). This farm showed a more than ninefold increase in net worth between 1983 and 1987, despite a 27% decrease in land value (Cavigelli and Kois, 1988).

Dobbs *et al.* (1991) calculated whole-farm economic performance indicators for a matched pair of low-input and conventional mixed crop and livestock farms. Direct costs were US$ 96 ha^{-1} higher on the conventional farm, but the higher gross income (US$ 534 ha^{-1} compared to US$ 410 ha^{-1}) more than offset the difference. The conventional farm had a net income excluding costs of land, labour and management of US$ 255 ha^{-1}, compared to US$ 217 ha^{-1} for the low-input farm. When labour was included in costs, the difference per hectare increased from US$ 38 to US$ 47 because the low-input farm required 58% more labour than the conventional farm. Both farms had positive net returns to management in all five years of the study, but the low-input farm had higher net returns in only one year. When farmer-estimated yields rather than hand-harvesting estimates were used in calculations, the low-input farm had negative net returns to management in two of the five years.

Farm surveys

On Wisconsin dairy farms, conventional-chemical users earned higher average gross incomes than 'no-chemical' farmers primarily through additional livestock and crop sales and government programme payments (see Table 10.6). Labour (milking, cleaning the barn, and feeding livestock) was greater for the conventional-chemical farmers but the net return per hour of labour was higher as well. Whole-farm net returns and net cash returns also were higher for conventional-chemical farmers (Kamp, 1993).

Summary and Conclusions

The general conclusion of US studies of organic farming has been that lower yields on organic farms contrasted with conventional farms are balanced by lower production costs, allowing organic farmers and growers to compete on the basis of net income (Knoblauch *et al.*, 1990; Green, 1994 for the Northeast). However, Lee (1992) summarized recent economic studies of reduced-chemical systems by claiming that lower variable costs may not offset lower yields and the necessity to include lower-value crops in rotations. This conclusion is supported by studies reviewed in this chapter. Under current government policies, organic farming enterprises generally are slightly less profitable than comparable conventional enterprises.

Some of the differences noted between economic performance of organic and other farms may be due to farm size rather than farming system, as several studies have shown that low-input farms are usually smaller than conventional farms. When high-income and low-income dry land wheat farms in Kansas were compared, total variable costs per hectare were higher for low-income farms (which correspond to small-scale farms), with the greatest differences in machinery hire and repair and agricultural chemicals. Successful low-input farms tend to have much lower fertilizer, pesticide and machinery expenses per unit area; but these savings may not offset the scale-of-production factors (NRC, 1989, pp. 203–4).

Comparisons of profitability between organic and conventional systems have limited applicability because of several intrinsic problems. Differences in management costs between organic and conventional production are difficult to assess (Lockeretz, 1989) and are not included in existing studies (Lee, 1992). Also, the financial performance of organic farming is complicated by difficulties measuring externalities and discounting benefits over time (Lockeretz, 1989). Farm profits depend on many conditions beyond the farm gate, such as farm subsidies, commodity price fluctuations, whether the farmer can get premium prices and how large they are, and whether cheap, effective synthetic pesticides for prevalent pests are available to farmers.

Federal programmes historically have induced farmers to deplete natural resources beyond their recharge rates, use marginal lands, and use excessive, inefficient amounts of chemical inputs (NRC, 1989, p. 207). Recent farm bills have begun to link payments to environmental performance, but US organic producers do not yet benefit financially from the environmental advantages of their farming systems except to the extent that consumers are willing to support these systems by paying a premium. If natural-resource depletion is included in the profitability calculation, organic or low-input farming in the USA can be competitive with or economically superior to conventional farming (Faeth *et al.*, 1991).

References

Anderson, M. (1990) Farming with reduced synthetic chemicals in North Carolina. *American Journal of Alternative Agriculture* 5(2), 60–68.

Asai, W., Hendricks, L., Klonsky, K. and Livingston, P. (1992a) *Sample Costs to Establish and Produce Almonds in the Northern San Joaquin Valley – Flood Irrigated and Mowed Centers – 1992.* Department of Agricultural Economics, Cooperative Extension, University of California, Davis, California.

Asai, W., Hendricks, L., Klonsky, K. and Livingston, P. (1992b) *Sample Costs to Establish and Produce Almonds in the Northern San Joaquin Valley – Sprinkler Irrigated and Mowed Centers – 1992.* Department of Agricultural Economics, Cooperative Extension, University of California, Davis, California.

Asai, W., Hendricks, L., Verdegaal, P., Klonsky, K., Ingels, C., Livingston, P. and Tourte, L. (1992c) *Sample Costs to Produce Organic Almonds in the Northern San Joaquin Valley – Flood Irrigation – 1992.* Department of Agricultural Economics, Cooperative Extension, University of California, Davis, California.

Asai, W., Hendricks, L., Verdegaal, P., Klonsky, K., Ingels, C. and Livingston, P. (1992d) *Sample Costs to Produce Organic Almonds in the Northern San Joaquin Valley – Sprinkler Irrigation – 1992.* Department of Agricultural Economics, Cooperative Extension, University of California, Davis, California.

Cavigelli, M. and Kois, J. (1988) *Sustainable Agriculture in Kansas: Case Studies of Five Organic Farms.* Kansas Rural Center, Whiting, Kansas.

Cook, R.L. (1988) *Marketing Organic Commodities in California: Structure and Obstacles to Expansion.* University of California Agricultural Extension Service, Davis, California.

Dobbs, T.L., Leddy, M.G. and Smolik, J.D. (1988) Factors influencing the economic potential for alternative farming systems: Case analyses in South Dakota. *American Journal of Alternative Agriculture* 3(1), 26–34.

Dobbs, T.L., Smolik, J.D. and Mends, C. (1991) On-farm research comparing conventional and low-input sustainable agricultural systems in the Northern Great Plains. In: National Research Council (ed.) *Sustainable Agriculture Research and Education in the Field: A Proceedings.* National Academy Press, Washington, DC, pp. 250–265.

ERS (1989) *Vegetables and Specialties.* Situation and Outlook Report TVS-248. Economic Research Service, United States Department of Agriculture, Washington, DC.

Faeth, P., Repetto, R., Kroll, K., Dai, Q. and Helmers, G. (1991) *Paying the Farm Bill: US Agricultural Policy and the Transition to Sustainable Agriculture.* World Resources Institute, Washington, DC.

Green, J. (1994) Economics and management issues. In: Smith, M. and members of the Northeast Organic Farming Association and Cooperative Extension (eds) *The Real Dirt. Farmers Tell About Organic and Low-input Practices in the Northeast.* Northeast Region Sustainable Agriculture Research and Education Program, Burlington, Vermont.

Hall, D.C., Baker, B.P., Franco, J. and Jolly, D.A. (1989) Organic food and sustainable agriculture. *Contemporary Policy Issues* 7(4), 47–72.

Janke, R.R., Mt. Pleasant, J., Peters, S.E. and Bohlke, M. (1991) Long-term, low-input cropping systems research. In: National Research Council (ed.) *Sustainable Agriculture Research and Education in the Field: A Proceedings.* National Academy Press, Washington, DC, pp. 291–317.

Jolly, D.A. and Norris, K. (1991) Marketing prospects for organic and pesticide-free produce. *American Journal of Alternative Agriculture* 6(4), 174–179.

Jolly, D.A., Schutz, H., Johal, J. and Knauf, K.D. (1989) *Marketing Organic Foods in California. Opportunities and Constraints.* University of California Sustainable Agriculture Research and Education Program, Davis, California.

Kamp, M. (1993) *Profits or Production: The Economics of Low-input and Conventional Dairy Management.* Wisconsin Rural Development Center, Mount Horeb, Wisconsin.

Klonsky, K., Tourte, L. and Ingels, C. (1992a) *Sample Costs to Produce Organic Wine Grapes in the North Coast with an Annually Sown Cover Crop – 1992.* Department of Agricultural Economics, Cooperative Extension, University of California, Davis, California.

Klonsky, K., Tourte, L. and Ingels, C. (1992b) *Sample Costs to Produce Organic Wine Grapes in the North Coast with Resident Vegetation – 1992.* Department of Agricultural Economics, Cooperative Extension, University of California, Davis, California.

Klonsky, K., Tourte, L., Chaney, D., Livingston, P. and Smith, R. (1994) *Cultural Practices and Sample Costs for Organic Vegetable Production on the Central Coast of California.* Information Series No. 94–2. Gianniní Foundation of Agricultural Economics, University of California, Davis, California.

Knoblauch, W.A., Brown, R. and Braster, M. (1990) *Organic Field Crop Production. A Review of the Economic Literature.* Department of Agricultural Economics Research 90–10. Cornell University, Ithaca, New York.

Lee, L.K. (1992) A perspective on the economic impacts of reducing agricultural

chemical use. *American Journal of Alternative Agriculture* 7(1/2), 82–88.
Lockeretz, W. (1989) Problems in evaluating the economics of ecological agriculture. *Agriculture, Ecosystems and Environment* 27, 67–75.
Matheson, N., Rusmore, B., Sims, J.R., Spengler, M. and Michalson, E.L. (1991) *Cereal-legume Cropping Systems: Nine Farm Case Studies in the Dryland Northern Plains, Canadian Prairies, and Intermountain Northwest.* Alternative Energy Resources Organization, Helena, Montana.
Mends, C., Dobbs, T.L. and Smolik, J.D. (1989) *Economic Results of Alternative Farming Systems Trials at South Dakota State University's Northeast Research Station: 1985–1988.* Research Report 89–3. Economics Department, South Dakota State University, Brookings, South Dakota.
NRC (1989) *Alternative Agriculture.* National Academy Press, Washington, DC.
Sahs, W.W. and Lesoing, G. (1991) *Progress Report. Research Project 12–58.* Agricultural Research and Development Center, University of Nebraska, Mead, Nebraska.
Sahs, W.W., Lesoing, G.W. and Francis, C.A. (1992) Rotation and manure effects on crop yields and soil characteristics in eastern Nebraska. *Agronomy Abstracts* 84, 155.
Williams, J., Klonsky, K. and Livingston, P. (1992a) *Sample Costs to Produce Rice in Sutter, Yuba, Placer and Sacramento Counties – 1992.* Department of Agricultural Economics, Cooperative Extension, University of California, Davis, California.
Williams, J., Wick, C., Scardaci, S., Klonsky, K., Chaney, D., Livingston, P. and Tourte, L. (1992b) *Sample Costs to Produce Organic Rice in the Sacramento Valley – No-till Drill Seeded – 1992.* Department of Agricultural Economics, Cooperative Extension, University of California, Davis, California.
Williams, J., Wick, C., Scardaci, S., Klonsky, K., Chaney, D., Livingston, P. and Tourte, L. (1992c) *Sample Costs to Produce Organic Rice in the Sacramento Valley – Water Seeded – 1992.* Department of Agricultural Economics, Cooperative Extension, University of California, Davis, California.

11 Economics of Organic Farming in Australia

ELS WYNEN

Introduction

As in many other countries, organic agriculture in Australia is practised by less than 1% of all producers. Conacher and Conacher (1991) estimated the total number of commercial organic farmers in 1990 to be between 950 and 1200, which represented between 0.75% and 0.95% of the total of 126,000 producers. This figure is up from their estimate of less than 500 in 1982, although, due to changes in definitions, those figures are not strictly comparable.

The only detailed study on the economics of organic farming in Australia published to date was carried out in the cereal-livestock sector (including mixed cropping where cereals and other crops are grown) (Wynen, 1989a). This sector constitutes 29% of the total organic farm population, judging on the response to a call for participants for a survey on organic farming between 1984 and 1986. Combining this percentage with Conacher and Conacher's (1991) estimates would suggest that the number of commercial cereal-livestock farms in Australia in 1990 was between 275 and 350. However, in early 1994 two of the three main national organic/bio-dynamic licensing organizations reported a combined number of licensed cereal-livestock farmers of 115. Assuming that the third, bio-dynamic, licensing organization is of similar size as the two others, 180 licensed cereal-livestock farmers should be considered the maximum. In addition, some farmers are licensed by more than one organization, so that the total number is likely to be between 100 and 150.

Characteristics of the Australian cereal-livestock industry

The cereal-livestock sector was the largest single broadacre industry in Australia in 1990/91. In that year 23,800 out of a total of 82,200 broadacre operators were cereal-livestock producers, which compares with 21,700 farmers in the sheep-only and 20,400 in the beef-only sectors. The 'sheep-beef' and 'wheat and other crops' sectors are considerably smaller, with 10,100 and 7100 producers, respectively. These figures exclude the dairy and horticulture industries, with 14,100 and 7700 enterprises, respectively, in 1989/90 (Australian Bureau of Agricultural and Resource Economics (ABARE), 1992, pp. 3–5).

In eastern Australia the cereal-livestock industry is situated in those areas where the land is not too steep for cropping, and where the amount of rainfall allows a cereal crop to be grown. This means that, in general, cereals are grown on the plains west of the Great Dividing Range (a mountain range stretching from the north to the south near the eastern Australian coast). Rainfall rate decreases, and the area becomes less suitable for cropping, mainly according to the distance from the coast.

Crop production (wheat or barley) has traditionally been considered the highest income-producing enterprise in the Australian cereal-livestock industry. Livestock are kept as a means of diversification, and to make use of resources on the farm not used for cropping. On most farms, some non-arable land can be grazed; and, due to rotational requirements on the arable land, some arable land can be available for livestock. Sheep complement cropping occurs in those areas where rainfall is less reliable, whereas cattle are preferred in higher, or more reliable, rainfall areas. Over the years, farmers change their mix of grain or livestock depending on expected relative output prices.

More recently the trend on these cereal-livestock farms has been towards more and more cropping. When world wheat prices are low, many farmers look towards including other crops such as lupins, peas and beans in their rotation. However, this is not always possible in all areas.

In general, Australian farmers use superphosphate on their farms, applied at the time of planting of the crop. In some areas nitrogen is also applied, and trace elements are sometimes considered essential. Application of fertilizer is at lower rates than in some European countries, and rates of between 50 and 100 kg ha^{-1} fertilizer with 11.5% nitrogen and 21% phosphorus on wheat are common.

Synthetic pesticides used on farms are mainly herbicides (pre- and postemergent); a fungicide as seed dressing; and an insecticide to protect stored grain. Sometimes an insecticide is applied in the soil against pests such as the red-legged earth mite (*Halotydeus destructor*). In the growing crop no pesticides are used as a rule, except in very wet years when fungicides may be applied against various forms of rust. However, such years are exceptional, as the Australian farmers' enemy is mainly dry weather in the latter part of the growing season.

Livestock are treated against lice (dips), internal parasites (drenches), diseases (vaccines), and fly-strike (crutching to prevent fly-strike, and a pesticide when sheep are fly-struck, that is, the fly has laid eggs in the sheep's flesh, and the resulting maggots are alive).

Methods

This study on the economics of organic farming was carried out in 1986 and 1987, and pertained to the financial year (July to June) 1985/86. It covered farmers in southeastern Australia, which includes southern Queensland, New South Wales, Victoria and South Australia.

In total 13 organic farmers were interviewed, of whom eight were 'fully sustainable',[1] and five 'semi-sustainable'. All had farmed that way for at least five years. At the time of the interviews no organic licensing scheme was in operation in Australia. However, it is likely that those classified as 'fully sustainable' would have been able to qualify as level A ('organic') in one of the national organic certification schemes (The National Association for Sustainable Agriculture, Australia), while those in the 'semi-sustainable' category would most probably have received level B ('organic-in-conversion') or a similar qualification. The results of the study are mainly those of the eight 'fully sustainable' farmers. The exception is the data on demographic characteristics and on farmers' opinions. In the rest of the chapter these farmers are referred to as 'organic'.

The organic farmers were each compared with one neighbour who was a broadacre cereal grower. To find an appropriate comparison, local officers of the Department of Agriculture were asked to nominate a conventional farmer who, in their opinion, was at least as good a manager as the organic farmer, while similar in other factors such as soil type, local climate and farm size. Where possible, the organic farmers were asked for their opinion, which was taken into account when making the final decision.

All 13 pairs of farmers were asked to grade themselves and their

---

[1] At the time of the study the word 'sustainable' was used in Australia. For the purposes of this study it was defined according to the definition of 'organic' used by the US Department of Agriculture (1980).

neighbour as managers. Of the group of organic farmers two of the ten respondents thought that they were better managers than their neighbour, six that they were similar, and two that they were worse managers. Of the 11 conventional farmers, six considered that they were better managers than their neighbour, and five that they were similar to their neighbours. For most pairs of farmers the grading was similar (only one point different on a scale from 1 to 9) except for two pairs, where both farmers thought that the conventional farmer was a considerably better manager (three or four points different).

Characteristics of Organic Farms and Farmers

For the comparison between the organic and conventional cereal-livestock farmers, all producers were asked about demographic factors as well as about the physical and financial measurements on their farms.

Of the 13 pairs of farmers analysed, the average age, years farmed and education level of the two groups were similar. The organic farmers were, on average, 48 years old and their conventionally farming neighbours 51. The organic farmers had farmed 29 years on average (20 of which as organic farmers), and the conventional farmers 35 years.

Compared to the conventional farms, the eight fully organic farms in the survey were of similar size, with similar improved capital value, which includes land and improvements (Table 11.1). The average

Table 11.1. Some characteristics of organic and conventional farms in south-eastern Australia, 1985/86.

	Organic	Conventional	Relative (conv=100)	Statistical significance[a]
Area operated (ha)	755	928	81	n.s.
Improved capital value (A$ ha^{-1})	1011	1038	97	n.s.
Arable area (% area operated)	74	82	90	n.s.
Area cropped (% arable area)	47	77	61	***
Stocking rate (DSE[b] ha^{-1})	1.9	1.3	68	n.s.

[a] Calculated with the paired Wilcoxon test: n.s. = not significant at 90%, * = significant at 90%, ** at 95% and *** at 99% confidence levels, respectively.
[b] Dry sheep equivalent.

figures given for the different components of farming are non-weighted averages.

Arable land as a percentage of the total area operated was not statistically different between the two groups of farms, while conventional farmers cropped over 50% more than their organically farming neighbours (77% as compared to 47% of arable land). No statistical difference could be shown in the stocking rate of the two types of farms.

On organic cereal-livestock farms in Australia lack of phosphate is usually solved by using rock phosphate, while the nitrogen requirements are often provided by nitrogen-fixing plants in the pasture phase.

In crop production, weeds are the main problem for organic farmers. A large part of the farming enterprise is directed to coping with them, for example by way of:

- design of rotation: the lower cropping rate in the rotation on organic farms facilitates management of weeds, apart from the soil nutrient management;
- choice of enterprise: the livestock enterprise is often considered to be essential on an organic farm;
- manipulation of planting dates, where later planting allows for better weed control;
- mowing the crop before harvesting: when a crop has too many weeds, an organic farmer might use the crop for hay, and mow it before the weed can set seeds;
- choice of cultivation technique: organic farmers tend towards
  (a) not burning stubble after a cereal crop;
  (b) using a ripping implement instead of a disc or mouldboard plough;
- hand pulling of weeds;
- adaptation of seeding rate, with some farmers using considerably heavier rates than customary to smother out weeds.

With one exception, none of the organic farmers was using a seed protectant (fungicide). They considered that this did not result in problems.

Some organic farmers had expected, in the early years of conversion, damage by the red-legged earth mite (*Halotydeus destructor*), but that soil pest problem seemed to resolve itself without intervention from the farmer.

To prevent insect damage in storage, farmers needed a special kind of storage (airtight containers) which does not require a synthetic insecticide (although some other substance such as carbon dioxide is still needed to keep the product insect-free). Several farmers had increased their storage space in order to accommodate produce which they could not sell at harvesting time.

Insect pests in stock are generally more of a problem. For external parasites, all farmers used dips to some degree, although not necessarily in each year and not necessarily of a kind which is not permitted under organic management. The same picture emerges for the internal parasites, where some farmers drench only some of the flock (such as the lambs at weaning), or when they are bought on to the farm. Against fly-strike many organic farmers had the long-term strategy of culling. However, for the short-term strategy many used conventional methods on those sheep which were struck by flies. Vaccines are not used on a regular basis. Some organic farmers employ them on part of the flock, for example on lambs at marking time.

On average, organic farmers spent about one quarter of the amount spent by conventional farmers on veterinary chemicals. The expenses on organic farms included materials which are allowed by organic licensing schemes at present. In general, farmers tried to avoid having to use veterinary chemicals by keeping their sheep healthy by:

- feeding organically grown feed, possibly supplemented with minerals;
- breeding their own replacement stock so that less problems occurred with internal and external parasites;
- buying stock from specific areas which are known to have no or few pest problems;
- having relatively small paddocks, so that stock can be moved regularly, to minimize internal pest problems;
- having good boundary fences, so that stock do not get lice from neighbours' sheep.

Most farmers who planted into stubble (from a grain in the previous year) reported that, in the early years of organic management, they experienced problems with seeding (for example, blockage of the machinery which resulted in poor germination). They mentioned that they had to adjust their existing equipment or buy machines which could cope with the extra stubble.

None of the farmers mentioned availability of organic feed as a limiting factor in their operations. In another study on cereal-livestock farming (Wynen, 1992) where this question was specifically asked, none out of seven farmers bought feed in 1990/91. It is reasonable to assume that feed availability is generally not a problem.

For more details about practices of Australian organic cereal-livestock farming and comparisons with conventional farmers see Wynen (1990, 1992).

Productivity of Organic Farming Systems

Physical performance

Wheat yields are shown in Table 11.2. For livestock, enterprise gross returns are displayed. As a combination of sheep and cattle can be kept, and sheep can be kept for different purposes (such as wool, meat or a combination of the two), comparison of the physical output of the livestock enterprise does not make much sense.

The average wheat yields on organic farms were similar to those on conventional farms. The gross returns from livestock (meat and wool) were not statistically different between the two systems. At the time of the survey, no farmer received price premiums for livestock products. Although, at present (1993), there are some outlets for organic meat, price premiums are still not received by most organic farmers.

The variability of wheat yields was analysed to determine whether organic farmers achieve better results in dry years. Dry conditions in the maturing stage of the grains is an important determinant of cereal yield levels on Australian farms. Many of the organic farmers mentioned that, in dry years, their crops seemed to suffer later from lack of water than their neighbours' crops. This characteristic of crops surviving dry conditions longer could well lead to higher yields in dry years on organic farms relative to those on conventional farms. In wet years this trend could well be reversed due to the fact that nutrient availability may be the limiting factor.

The conditions described above would not be valid in extremely dry or wet years. For example, in years with a total drought (as in 1982/83) many farmers did not have a crop to harvest. In 1992/93 an extremely wet growing season was experienced. One organic farmer who was estimated to have close to 3 t ha^{-1} (a high yield for that area)

Table 11.2. Yields and returns on organic and conventional farms in south-eastern Australia, 1985/86.

	Organic	Conventional	Relative (conv=100)	Statistical significance[a]
Wheat (t ha^{-1})	2.4	2.5	96	n.s.
Livestock (A\$ ha^{-1} operated)	45.2	35.4	128	n.s.

[a] Calculated with the paired Wilcoxon test: n.s. = not significant at 90%, * = significant at 90%, ** at 95% and *** at 99% confidence levels, respectively.

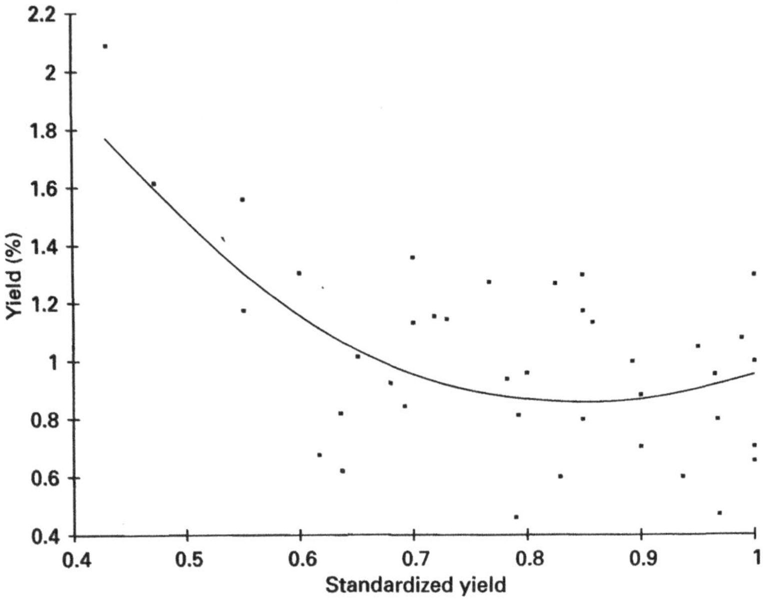

Fig. 11.1. Relative organic wheat yields compared with standardized conventional wheat yields, south-eastern Australia, 1985/86.

reported that his neighbour probably harvested twice as much. However, this situation was very exceptional, as the highest average yield recorded in the relevant local government area between 1945 and 1980 was 2.4 t ha^{-1}.

For those pairs of farms for which enough data were available, the standardized yield on the conventional farm (the yield expressed as a proportion of the highest yield achieved on that farm) was plotted on the horizontal axis (Fig. 11.1). The yield on the organic farm relative to that on the conventional counterpart farm for a given year was plotted on the vertical axis. Identical yields in the two management systems would result in a horizontal line at 1.0. A downward sloping curve would be consistent with higher yields on the organic farm relative to those on the conventional farm in years of low yields on conventional farms. As the limiting factor in most years is dry conditions, such a situation would be consistent with less variability on the organic farms. If the climatic conditions become too dry, crops are not viable in either system. A cut-off point in yields, below which level differences are not present, should therefore be expected.

Sufficient data were available from four pairs of farmers. Data were taken from the sixth year after transition, to exclude possible negative effects during that period. The year 1982/83 was excluded, since this was a year of major drought for most farmers.

The data for the four organic farms were combined in one series, and those of the conventional counterparts in another, to arrive at an aggregate level. This gave 43 observations, of which three were outliers. In those three cases the standardized yield was low (less than 0.4) compared with the other observations. This point could be considered the cut-off point, below which it was too dry to differentiate yields on organic or conventional farms.

Regression analysis was used to determine a trend in the data. The best fit of the curve is the quadratic function $y = 4.65 - 8.59x + 4.84x^2$ (see Fig. 11.1), with an adjusted R^2 of 0.410. This means that 41% of the total variation between the two yields can be explained by the seasonal conditions. It indicates that the yields on organic farms as a proportion of conventional yields increase as conventional yields decrease. This means that the yields on organic farms decrease less than on conventional farms in years with dry weather conditions.

In summary, although the data are not such that firm conclusions can be drawn, indications are that wheat yields are less variable on organic farms than those obtained under conventional management.

Financial performance

In general, organic farmers would, after the pasture phase, cultivate their land and plant a cereal crop, often wheat. In the winter-dominant rainfall areas most farmers in the survey grew another grain for one or two more years after the initial crop. This grain might be wheat, barley, oats or rye, depending on the state of the paddock (in connection with moisture content and weeds) and expected state of the market. In the summer-dominant rainfall areas (in Queensland and northern New South Wales) summer crops can be grown, making the rotation system considerably more flexible and complicated.

Input costs

Input costs on organic and conventional farms on a per hectare cropped basis are displayed in Table 11.3. The inputs listed are used mainly, but not exclusively, in the cropping phase. Fertilizers (including soil nutrients allowed under organic management such as rock phosphate) are mainly applied in the cropping phase, but might benefit growth of the pasture in later years. Pesticides on organic farms (also encompassing materials allowed under organic standards) are used mainly, but not exclusively, on stock, while on conventional farms they are used mainly in connection with cropping (weed control). Fuel is used for

Table 11.3. Input costs on organic and conventional farms in south-eastern Australia (A$ per hectare cropped), 1985/86.

	Organic	Conventional	Relative (conv=100)	Statistical significance[a]
Fertilizers	9	27	33	**
Pesticides	1	18	6	***
Fuel	35	33	106	n.s.
Machinery	89	101	88	n.s.

[a] Calculated with the paired Wilcoxon test: n.s. = not significant at 90%, * = significant at 90%, ** at 95% and *** at 99% confidence levels, respectively.

Table 11.4. Input use on organic and conventional farms in south-eastern Australia (A$ per hectare operated), 1985/86.

	Organic	Conventional	Relative (conv=100)	Statistical significance[a]
Fertilizers	3	19	16	**
Pesticides	0	14	0	***
Interest	5	16	31	n.s.
Fuel	12	21	57	**
Machinery and equipment	31	74	42	*
Labour	35	41	85	n.s.

[a] Calculated with the paired Wilcoxon test: n.s. = not significant at 90%, * = significant at 90%, ** at 95% and *** at 99% confidence levels, respectively.

cropping activities such as ploughing, planting and harvesting and for checking stock in the paddocks and marketing animals. Most of the machinery and equipment is employed in cropping.

On the organic farms, expenditure on fertilizers was relatively low, and on pesticides negligible. Contrary to expectation, there was no significant difference in use of fuel on organic and conventional farms. This may reflect the greater role of livestock in the rotation on organic farms, influencing weed control, and so number of cultivations.

Costs per hectare operated are shown in Table 11.4. Differences between Tables 11.3 and 11.4 reflect the longer rotations of organic farmers. With the exception of interest and labour, all inputs are used less in organic than in conventional agriculture.

Returns

Returns to farming are calculated according to the measures used by the Australian Bureau of Agricultural and Resource Economics (ABARE). Those measures include the total cash costs (TCC), total cash receipts (TCR) and the farm cash operating surplus (FCOS = TCR − TCC). Table 11.5 shows that the organic farmers had lower TCC than their conventionally farming neighbours, while no statistical difference could be shown for the TCR and the FCOS.

The measures discussed so far include only the cash costs and returns, and do not take into account the change in stock (both crop and livestock), the decrease in value of capital items such as machinery (depreciation), and payments for family labour. If values for these factors are deducted, the resulting income (which could be considered the return to capital and management), is also similar for the organic (A\$ 37 ha^{-1}) and the conventional farmers (A\$ 21 ha^{-1}). If the same measure is taken as a percentage of capital invested (in land, improvements, machinery and livestock), the organic farmers had, on average,

Table 11.5. Financial aspects of organic and conventional farms in south-eastern Australia (A\$ per hectare operated), 1985/86.

	Organic	Conventional	Relative (conv=100)	Statistical significance[a]
Cash returns and costs				
Total cash receipts	181	262	69	n.s.
Total cash costs	76	128	59	**
Farm cash operating surplus	105	134	78	n.s.
Returns to capital and management				
A\$ per hectare operated	37	21	176	n.s.
% of capital invested	3.0	1.2	250	n.s.
Adjusted returns to capital and management				
A\$ per hectare operated	42	37	114	n.s.
% of capital invested	3.4	2.2	155	n.s.

[a] Calculated with the paired Wilcoxon test: n.s. = not significant at 90%, * = significant at 90%, ** at 95% and *** at 99% confidence levels, respectively.

a return of 3.0% in 1985/86, while the conventional farmers had a return of 1.2%, which statistically was not different.

The returns to capital and management include interest costs and rent for the farm. In order to compare the financial aspects of farming itself, the cost of interest and rent were deducted from the total cost of farming. These 'adjusted' figures showed no statistically significant difference between the groups of organic and conventional farmers (A$ 42 and A$ 37 per hectare operated, and 3.4% and 2.2% per unit capital invested, for organic and conventional farms, respectively). Interest deductions for conventional farmers are greater than for organic farmers, which results in adjusted returns closer to those of organic farmers.

Influence of product prices

It is sometimes suggested that the viability of organic farms depends on the higher prices received for produce relative to conventional prices. An attempt was made to obtain some indication of the influence of prices on the viability of the surveyed farms. For that purpose financial returns were calculated assuming average conventional prices for the major crop, wheat. That is, premiums received for wheat produced under organic farming conditions were disregarded.

In Table 11.6 the same financial measures are shown as in Table 11.5 for the adjusted returns to capital and management. The difference is that these are calculated with conventional wheat prices. As can be expected, figures show a decrease in financial benefits for the organic farmers from A$ 42 to A$ 31 per hectare operated. This results in a decrease in the gap between the two groups of farmers for these variables. Again, no statistical difference can be shown.

An issue in comparing financial performance of organic and

Table 11.6. Adjusted returns to capital and management assuming conventional wheat prices in south-eastern Australia.

	Organic	Conventional	Relative (conv=100)	Statistical significance[a]
A$ per hectare operated	31	35	89	n.s.
Percentage of capital invested	2.5	2.2	114	n.s.

[a] Calculated with the paired Wilcoxon test: n.s. = not significant at 90%, * = significant at 90%, ** at 95% and *** at 99% confidence levels, respectively.

Table 11.7. Indices of prices received by farmers in Australia, 1982/83–1987/88.

	1982/83	1983/84	1984/85	1985/86	1986/87	1987/88
Wheat	114	106	111	109	95	106
Livestock sector	103	111	119	119	134	165
Wheat/livestock	1.11	0.95	0.93	0.92	0.77	0.64

Sources: BAE (1985) and ABARE (1989).

conventional farmers is the prices of crops relative to livestock and livestock products for 1985/86, as compared to other years. Since organic farmers are likely to have more livestock and less crop than conventional farmers, relative movements of the prices of the different enterprises can be important in determining relative profitability. In Table 11.7 the indices of prices received by farmers in Australia are shown. From 1982/83 to 1986/87 the ratio of prices for wheat, the main crop on most of the farms surveyed, to prices for livestock and livestock products decreased considerably. In 1985/86 the ratio was similar to that of the two previous years, but higher than that of the two following years. This means that, if the analysis had been carried out on data for the two years previous to 1985/86, relative financial benefits for the two groups of farmers would not have differed from relative benefits in 1985/86 on account of output prices. However, in 1986/87 and 1987/88 prices for livestock increased absolutely and relative to wheat prices. This would benefit the organic farmers in general more than the conventional farmers, increasing the profitability of organic farming absolutely and relative to that of conventional farming in comparison with 1985/86. In 1982/83 this situation would have been reversed.

Marketing of Organic Produce

In 1985/86 organic grains were mainly sold in speciality shops, and directly from farmer to customer. Farmers who wanted to sell their grains as organic had to develop their own market. In addition, they encountered institutional impediments up until 1989.

In summary, these institutional problems were as follows. All wheat in Australia had to be sold to the Australian Wheat Board (AWB). With the exception of 2–3 million tonnes (Mt) which is used for domestic consumption, most of Australia's wheat (between 8 and 16 Mt) is exported. The AWB did not have storage, neither had it developed markets, for organic wheat. This meant that the only way in which organic farmers were able to receive a premium for their produce was by selling the wheat to the AWB, buying it back, and then, with a permit, selling it in their own market. The farmer had to pay for this 'transaction' with the AWB, during which the wheat did not leave the farm. The exact cost depended on the state in which the farmer lived and on what value was put on certain services. It ranged between A$ 18.80 and A$ 38.80 t^{-1} in 1986/87 (for a more detailed explanation of the issue, see Wynen, 1989b). This was as compared to the net pool price (the price received for wheat sold via the AWB) of A$ 96 t^{-1} in that year.

In 1985/86, three of the eight organic farmers sold all their wheat to the AWB as conventional wheat, without receiving premium prices. Two farmers sold all their wheat for a premium, and three sold some wheat for a premium and some in the conventional market.

Those organic farmers who sold their produce as conventional wheat received an average price of A$ 132 t^{-1}, and their conventional farmer neighbours A$ 129 t^{-1}. This indicates that, on average, the quality of the wheat, as defined by the AWB, on the two types of farms was similar.

The average wheat price for the two organic farmers who sold all their wheat for a premium was A$ 197 t^{-1}. Their neighbours received an average of A$ 125 t^{-1}. The organic farmers looked after their own marketing.

In 1989 the Wheat Marketing Act 1984, which caused these institutional impediments to the marketing of organic wheat (similar acts have been in operation since 1948), was changed. The new Act allowed domestic trading of wheat without interference from the AWB, while export licences could be supplied by the AWB to potential exporters of wheat without a financial disincentive such as that present in the 1984 Act.

In the early 1990s two large companies became interested in the market, dealing mainly with breakfast cereals, cake flour and feed. Although these markets are still rather small at present (less than 8000 t wheat and 4000 t oats), it is likely that they will grow in the future.

Conclusions

An analysis of data obtained from a comparison of eight organic cereal-livestock farmers in Australia with one of their conventionally farming neighbours for the year 1985/86 indicates statistically significant differences for some variables, while for others, statistical significance cannot be established.

Although wheat yields were similar in that year, total production on the two farming systems was different. This was due to the fact that a considerably higher percentage of the arable area was cropped on conventional farms than on organic farms. There was also an indication that the variability of wheat yields on organic farms was lower than on conventional farms.

The financial results of the two groups of farmers per hectare operated were remarkably similar. No differences could be established for the total cash receipts (also when cash costs were deducted), and for the (adjusted) returns to capital and management. This was found to be the case even when the receipts for wheat on organic farms were adjusted to reflect conventional prices.

References

ABARE (1989) *Indexes of Prices Received and Paid by Farmers.* Australian Government Publishing Service, Canberra.

ABARE (1992) *Farm Surveys – Financial Performance of Australian Farms 1989–90 to 1991–92.* Australian Government Publishing Service, Canberra.

BAE (1985) *Indexes of Prices Received and Paid by Farmers.* Australian Government Publishing Service, Canberra.

Conacher, J. and Conacher, A. (1991) An update on organic farming and the development of the organic industry in Australia. *Biological Agriculture and Horticulture* 8, 1–16.

US Department of Agriculture (1980) *Report and Recommendations on Organic Farming.* US Government Printing Office, Washington, DC.

Wynen, E. (1989a) Sustainable and conventional agriculture: an economic analysis of Australian cereal-livestock farming. PhD thesis, School of Economics, La Trobe University, Bundoora.

Wynen, E. (1989b) *Wheat Marketing and Sustainable Producers in Australia.* Discussion Paper No. 2. National Association for Sustainable Agriculture Australia, Sydney.

Wynen, E. (1990) *Sustainable and Conventional Agriculture in South-eastern Australia: A Comparison.* Economics Research Report No. 90.1. School of Economics and Commerce, La Trobe University, Bundoora.

Wynen, E. (1992) *Conversion to Organic Agriculture in Australia: Problems and Possibilities in the Cereal-Livestock Industry.* National Association for Sustainable Agriculture Australia, Sydney.

12 Farm-level Performance of Organic Farming Systems: An Overview

SUSANNE PADEL AND NICOLAS H. LAMPKIN

Introduction

In the preceding seven chapters, results of empirical research into the physical and financial performance of established organic farming systems in different countries have been presented. The quantity and quality of data available reflect the emphasis given to organic farming in each country, with the availability of time-series data from the early 1980s in Germany and Switzerland contrasting markedly with the limited information available from individual studies in other countries. Most of these studies focus purely on the financial performance of the systems, leaving out the wider economic benefits for the environment and society (Chapter 4). A complete economic assessment would require these benefits to be included, but the lack of studies quantifying the environmental benefits of organic farming makes this difficult. Nevertheless, information about financial performance is essential for individual as well as for policy decision-making. It is widely seen as an important factor determining the acceptance of organic farming by conventional farmers, although until now financial motives have not been the only or even the main reasons for converting to organic farming (Chapter 17).

The aim of this chapter is to bring together some of the common themes identified in the preceding chapters, to establish whether there is a theoretical basis for some of the findings, and to consider the implications for future research. The social and other characteristics of organic producers presented in the preceding chapters, such as motivation, farming background, education and farm size, are discussed in the context of studies dealing with the process of conversion to organic farming in Chapter 17.

Physical Performance

Crop yields

From the evidence presented in the preceding chapters, there are two main conclusions which can be reached relating to crop yield levels in organic farming systems.

First, *absolute* yield levels under organic management are increasing over time, but at a slower rate than for comparable conventional systems (Chapters 6 and 8, this volume; Lampkin, 1993), and are significantly higher than conventional yields achieved in pre-1950s agriculture, to which organic farming is sometimes mistakenly likened. Absolute yields are, however, subject to considerable variability due to a number of factors, including variety selection and plant breeding (Chapter 5, this volume; Stöppler, 1988), soil type, rotation design and manuring (Chapters 6 and 16), length of time under organic management (Chapters 14–17), as well as management ability and developments in scientific knowledge and technology.

Second, yields *relative* to comparable conventional systems are directly related to the intensity of the prevailing conventional system. This is the case not only for comparisons between regions, but also between crops within a region, and for individual crops over time. In most of the studies reported, crop yields under organic management are somewhat lower than for comparable conventional systems. However, the studies from Canada, the United States (see also Lockeretz *et al.*, 1984; NRC, 1989) and Australia report yield reductions of 10–20% in some cases and similar or higher yields in others, while reductions of up to 40% are reported in Britain, Germany, Denmark and the Netherlands (Kleijn *et al.*, 1990). The relative yield differences are greatest for crops such as wheat where produced intensively, as in northern Europe, and least for crops such as oats and field beans, or in North America and Australia, where conventional production systems are typically less intensive.

Within individual countries, relative yield differences for specific crops such as winter wheat and potatoes have increased over time as the intensity of conventional production has increased (Chapter 6, this volume; Lockeretz *et al.*, 1984; Stanhill, 1990). Similar conclusions were reached by Lampkin (1993) on the basis of an analysis of cereal yield comparisons in northern Europe between 1970 and 1990, where the typical yield gap for winter wheat widened from 10–20% in the 1970s to 30–40% in the late 1980s. The Swiss data indicate a much smaller widening of the yield gap for wheat and barley, which might be related to the lower intensity of the conventional comparison farms.

Grazing livestock output

The performance of livestock under organic management has received much less attention in research, which is surprising since the organic philosophy emphasizes the integration of cropping and livestock on the farm (Lampkin, 1990; Neuerburg and Padel, 1992). The majority of published studies of organic livestock enterprises relate to milk production (Chapters 5–9 and 25, this volume; Lampkin, 1993; Schulze Pals, 1993). Typically, milk yields per cow are around 10% lower than conventional, while stocking rates are 20–30% lower, resulting in milk yields per hectare some 30–40% lower. Individual studies have recorded up to 30% lower milk yields per cow (Chapters 5 and 7). Milk yield variability appears to be smaller than for crops, but breed is one important factor. The preference for Channel Island and other specialist breeds among organic farmers in Britain and Denmark can partly explain the lower performance figures in those countries.

Stocking rates are not a reliable guide to forage output from organic systems as they also reflect the reduced reliance on 'borrowed' land in the form of bought-in concentrates. Dry matter yields are also not reliable, due to the higher quality and efficiency of utilization of forage containing clover (Thomas *et al.*, 1991). Milk yields from forage, or utilizable energy output from grassland, may therefore be more appropriate measures of the potential of the organic system. On these measures, output per forage hectare on organic farms may be similar to or only slightly lower than conventional (Chapters 6 and 13).

Discussion

The physical productivity of organic farming systems has often been one of the most contentious issues with regard to their viability. From the evidence available, it seems that the likely yield penalties have frequently been overstated. In part, this is due to misconceptions and unrealistic assumptions like 'organic farming is a return to pre-1950s farming'. More importantly perhaps, the traditional analysis of yield response to input use has relied on comparative-static production response curves, where the use of a single input is varied while all other factors of production are held constant.

In the case of organic farming, there is a need to consider not just *modifications to agricultural practices*, such as the level of input use, or the combination of individual inputs, but also the *restructuring of the*

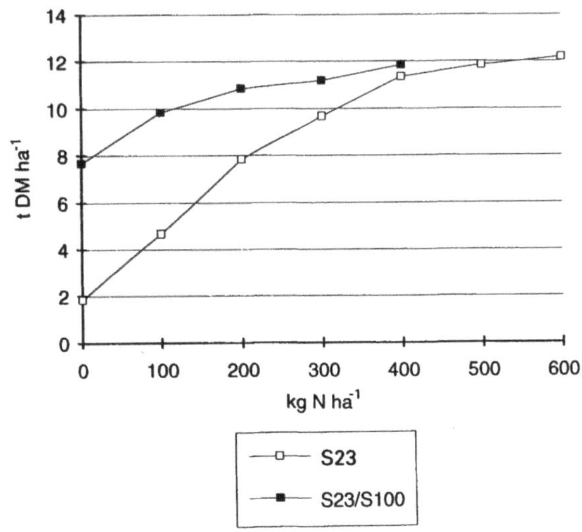

Fig. 12.1. Nitrogen fertilizer production response curves for perennial ryegrass only (S23) and perennial ryegrass–white clover mixtures (S23/S100). DM, dry matter. Source: Thomas *et al.* (1991).

farming systems, a strategic rather than tactical change involving a large number of different variables. The organic system relies on both management (redesign of the system) and alternative inputs (labour, machinery, organic manures, etc.) to replace those inputs which have been withdrawn. In the context of the traditional production response curve, changes in factors other than the variable input will lead to a shift to a different production function. The prohibition of certain inputs such as synthetic nitrogen fertilizers in organic farming, and the emphasis instead on biological nitrogen fixation, represents just such a shift to a different production response curve, *not* a shift along an existing conventional one to the zero nitrogen input level. This is well illustrated by data on yield responses to nitrogen use in a grass monoculture compared with a clover–grass mixture (Fig. 12.1). Not only are the yields under zero nitrogen use conditions substantially higher in the grass–clover case, but it may actually be more profitable, depending on input costs, to choose this zero nitrogen option rather than the optimal nitrogen use level for a grass-only sward.

The comparative-static analysis also fails to reflect fully the biological and therefore dynamic nature of agricultural systems. For example, dynamic responses to increased nitrogen use may include increased pest and disease incidence or a decline in soil pH or organic matter levels. In such situations, the *ceteris paribus* condition that all other factors remain fixed no longer holds. In the short term, some of these dynamic feedback effects may be reflected in diminishing marginal returns in a particular production cycle. In the long term, the negative residual impacts of these changes will result in a lower level of output in response to the same level of input use in subsequent production periods, and will require an increase in the use of other inputs to compensate. It is therefore possible that economically rational, short-term decisions relating to levels of input use may prove irrational in the long term, but intensification – the technological treadmill – continues because short-term decisions take precedence.

Although these factors may have led to the potential productivity of organic systems being underestimated, it is also true that improvements in the level of crop yields, and reductions in the high degree of yield variability, have been limited by the lack of extension, research and technological development directed specifically at the needs of organic farmers.

In any case, the prevailing view that lower than maximum yields *per se* are undesirable needs to be challenged. This is not just because of the environmental and social problems of subsidizing overproduction in developed countries or the damaging consequences of subsidized exports on the economies of traditional agricultural exporters and developing countries. Ever increasing yields, in the face of static demand, are not in the long-term financial interests of farmers themselves. Due to the low price elasticities for food in developed countries, the decisions made by thousands of individual farmers to intensify production lead, at an industry level, to declining real output prices and increased real input costs. The residual element, farm income, declines as a consequence – the classic cost–price squeeze illustrated by Cochrane's Treadmill theory (Cochrane, 1958). Reduced yields and input use, at an *industry* level, could reverse this trend, leading to increased farm incomes.

Policy measures aimed at maintaining farm incomes by supporting output prices or subsidizing input costs cannot be successful in the long term, as they simply reinforce the underlying economic processes, leading to further intensification, specialization and overproduction, with the now familiar environmental consequences. An individual farmer seeking change is powerless to break this cycle, since the actions of the majority in the industry will continue to determine the underlying economic trends. In the short term, there is therefore no incentive for an individual farmer to change, even though this may be economically rational and preferable in the long term. This is a clear case of market failure, which can only be addressed if some consumers are prepared to place a higher value on an individual farmer's actions, as in the case of organic food, or if policy-makers act to bring about coordinated change through voluntary agreements, regulation or appropriate financial signals.

Crop and Livestock Enterprise Financial Performance

Prices

Premium prices for organic crops are widely available in northern European countries such as Britain, Germany, Denmark and Switzerland, but the size of the premium varies from crop to crop and country to country. For example, premiums for milling wheat range from up to 300% in Germany, over 100% in Britain, to 40% in Switzerland, reflecting both the level of demand for organic food in each country and the level at which conventional prices are supported by agricultural policy measures (the conventional price for wheat in Switzerland is more than three times as high as in the European Union). Recently, financial aid schemes for organic farming have been influencing the supply of organic crop products, leading to a reduction in premiums (Chapters 23 and 25). Premium prices for livestock products are less widely available, due to a lack of market outlets and problems of supply coordination. Where available, premiums for milk and meat range from 10% to 25% (Chapters 5–8). Frequently, milk and livestock are sold at conventional prices. The situation is somewhat different in Canada, the United States and Australia where premiums for non-horticultural organic crops are less widely available and many farmers consider that their organic systems should be able to function profitably without premium prices (Chapter 10).

Although premium prices are available, their level and availability are closely related to the choice of market outlets, and the commitment of time and resources to market development and added-value activities such as on-farm processing (Schulze Pals, 1993). The costs of these activities, in particular the increased labour and capital requirements, are not usually specifically identified, so that it is difficult to judge whether the emphasis on premium markets is really worthwhile. In the absence of higher prices to compensate for reduced yields, good financial performance of the organic system depends entirely on maintaining output and cost reductions.

Variable input costs

The replacement of external inputs by farm-derived resources normally leads to reduced variable input costs under organic management.

For example, total variable costs are typically 50–60% lower for organic cereals and grain legumes, and 10–20% lower for potatoes and horticultural crops. Expenditure on fertilizers and sprays is substantially lower than in conventional systems in almost all cases (Chapters 5–11, this volume; Kleijn *et al.*, 1990). In a few, higher input costs due to the purchase of composts and other organic fertilizers have been reported (e.g. Sellen *et al.*, 1993). This may reflect the contrast between production with organic *inputs* rather than as part of an organic farming *system*, where rotation design should help to keep fertilizer costs low, although the latter might be difficult to implement on horticultural units with limited land resources. With vegetable and row crops, higher input costs may result from the use of organically raised transplants and casual labour for hand weeding and harvesting, or the use of special equipment such as flame weeders, but few of the European studies examine horticultural crops in detail. Additional variable costs may also be due to a higher level of packaging or processing on the organic farms, which usually would not be present on conventional holdings.

Total variable costs for dairy cows are typically 20–25% lower, mainly due to the reduced use of concentrates (Chapters 5, 6, 8 and 9). Premium prices for purchased concentrates, or the opportunity cost of home-grown concentrates, may cancel out such benefits, particularly in pig and poultry production (Chapters 5 and 6). Lower veterinary costs (Chapters 5, 8 and 9) and higher costs for organically reared stock (Chapter 5) are mentioned, but their overall effect on the total costs of livestock production appears to be insignificant. Reduced expenditure on fertilizers and sprays for forage production further lowers the production costs of grazing livestock (Chapters 5 and 6), but is not always included in livestock enterprise comparisons.

Enterprise gross margins

Most of the studies from Europe and Canada report similar or higher gross margins for organic crops (Chapters 5, 6, 8 and 9), but with wide variations in organic gross margins between farms and regions. These examples clearly illustrate that higher prices are required, in addition to lower costs, in order to compensate for reduced yields. The achievement of similar gross margins simply on the basis of lower costs would require a percentage reduction in costs much greater than the percentage reduction in yield (Fig. 12.2). For example, if, in the high-intensity conventional case, variable costs account for 40% of output, and output falls by 40%, then costs would have to fall to zero to compensate, which is clearly not feasible. If variable costs are only reduced by 50%, as is typical for organic cereal crops in northern Europe, a 33% premium would be needed to maintain the same gross

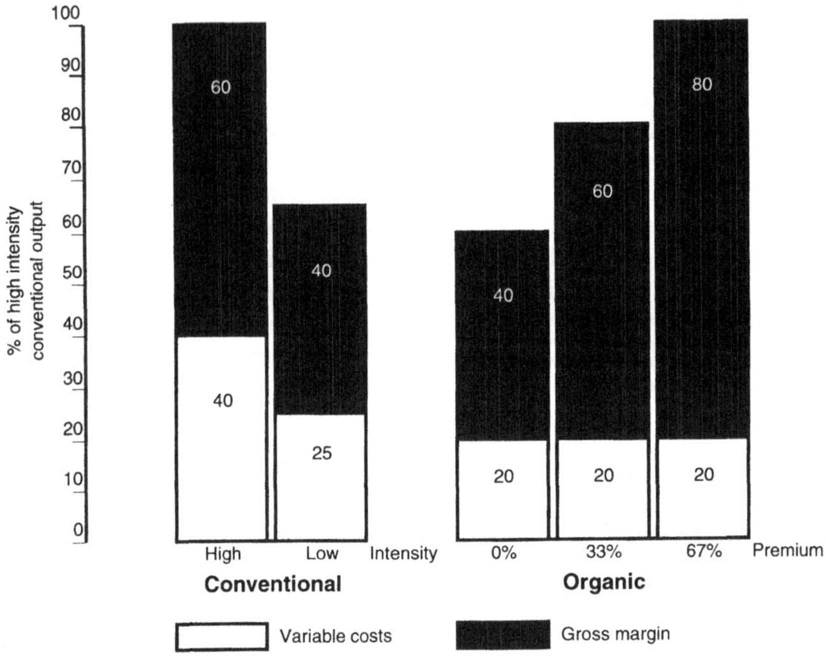

Fig. 12.2. Schematic presentation of the relationship between output, variable costs, premium prices and gross margins in organic and conventional agriculture. Source: Based on Vine and Bateman (1981).

margin, while the 70% plus currently achievable for many crops will result in higher gross margins and output than on conventional farms. On the other hand, where conventional production is already less intensive, as in the United States and Australia (Chapters 10 and 11), and output falls by only 10%, variable cost reductions may be sufficient on their own to compensate for the lower output.

Studies of organic milk production show similar (Chapters 5, 6 and 8) or even higher (Chapter 9) gross margins per *cow*, with only a small or without any premium for the milk. Because of lower stocking rates, gross margins per *hectare* are lower. However, in situations where quotas are limiting, and not land, the margin per litre produced (or more accurately margin per litre quota or kilogram fat quota) is a more relevant performance measure. On this basis, the organic systems perform as well as or better than conventional in the studies presented in Chapters 5 and 6. Equal performance to conventional systems can be achieved with little or no premium in the case of livestock enterprises due to the benefits of biological nitrogen fixation and the higher nutritional quality of clover (see above).

Although gross margins are useful for comparative purposes, whether between enterprises of the same type on organic and conventional holdings, between different organic holdings, or between different enterprises on the same farm, there are some important limitations to their use. First, high individual enterprise gross margins for cash crops do not reflect the potentially very different enterprise mix on the organic farms and the need for fertility-building crops in the rotation. Taken out of the whole-farm context, they can therefore be very misleading (see below). Second, gross margins represent the difference between enterprise output and variable costs, to the exclusion of fixed costs. Gross margin comparisons between enterprises with different fixed cost structures can be misleading, particularly where conventional variable costs, such as fertilizer and crop protection inputs, have been substituted by fixed costs (machinery and labour) in the organic context.

Whole-farm Performance and Farm Incomes

Enterprise diversity

Organic farming systems in most European countries are more diverse in terms of enterprise mix than comparable conventional systems (Chapters 5–7 and 25). Cereals, forage maize, sugarbeet and oilseed rape tend to account for a much lower proportion of organically managed land, while grain legumes, temporary grass, fodder crops, potatoes and vegetables are relatively more important. Pigs are a much less important form of livestock in all countries except Switzerland, but are only partly replaced by grazing livestock. The greater similarity between organic and conventional farms in the Swiss studies may be due to the less specialized and less intensive nature of conventional systems there, or it may be a result of the method of pairing farms with similar characteristics.

Whole-farm output, variable costs and gross margins

The relationship between output levels and inputs is such that gross output cannot be considered to be a useful comparative performance measure for comparing farming systems with different enterprise structures, levels of intensity and reliance on external inputs. A larger proportion of high-value crops, such as vegetables, in the rotation might contribute to higher output levels (Chapter 7), even in the absence of premium prices, while fewer intensive livestock enterprises, such as pigs, may lead to lower output levels (Chapter 6). Gross margins and *net* output, on the other hand, reflect the value added to external inputs by the farming operation, and are therefore better suited to comparisons between systems (Vine and Bateman, 1981).

The combination of the different enterprise structures and performance of individual enterprises leads to mixed results in the different countries. In Britain, Germany (see Fig. 6.3) and Switzerland, lower whole-farm output (combining lower yields and higher prices) and lower variable costs on organic farms result in gross margins similar to comparable conventional systems. Where premium prices are relatively high, or where high-value crops such as vegetables are grown, whole-farm output and gross margins may both be higher than conventional, as in Denmark and in Germany for the more closely matched comparison group (see Table 6.1). Where premium prices are relatively low, whole-farm gross margins or net returns may be lower, as in Australia, or higher than conventional, as in Canada, due to the yield–cost relationships discussed above (Fig. 12.2).

The interactions between enterprises in an organic farming system make an important contribution to the overall financial performance of the system, particularly where external inputs are no longer available to fulfil a particular function. Grazing livestock on organic farms provides a means of utilizing the fertility building phase of a rotation which is usually based on forage legumes. The forage enterprises can therefore be seen as part of the 'costs' of achieving high returns for individual crops like wheat and potatoes. In a similar way, although beef cattle may give lower returns per hectare than sheep, the parasite control benefits from interactions between the two species are such that the beef enterprise can, in certain situations, be seen as a justifiable 'cost' of sheep production.

These interactions are illustrated using hypothetical data in Fig. 12.3, assuming that cereals and forage each account for 50% of the land area, and that cattle and sheep each account for 50% of the forage utilization. The total gross margin for cereals, including the opportunity cost of cereals and straw utilized by livestock, comes to 65 units, while the total gross margin for all forage-using enterprises amounts to only 30 units, clearly indicating the relative financial advantage of the cereals. The gross margin for sheep at 30 units is also higher than that for cattle at 10 units. However, if values are attached to the other benefits resulting from interactions between the enterprises, such as fertility gains (carbon and nitrogen fixation), weed, pest and parasite control, the *net* benefit of the cereals falls to 30 units, while the total net benefit from forage enterprises increases to 65 units, with beef at 20 units contributing more to overall performance than sheep at 10 units.

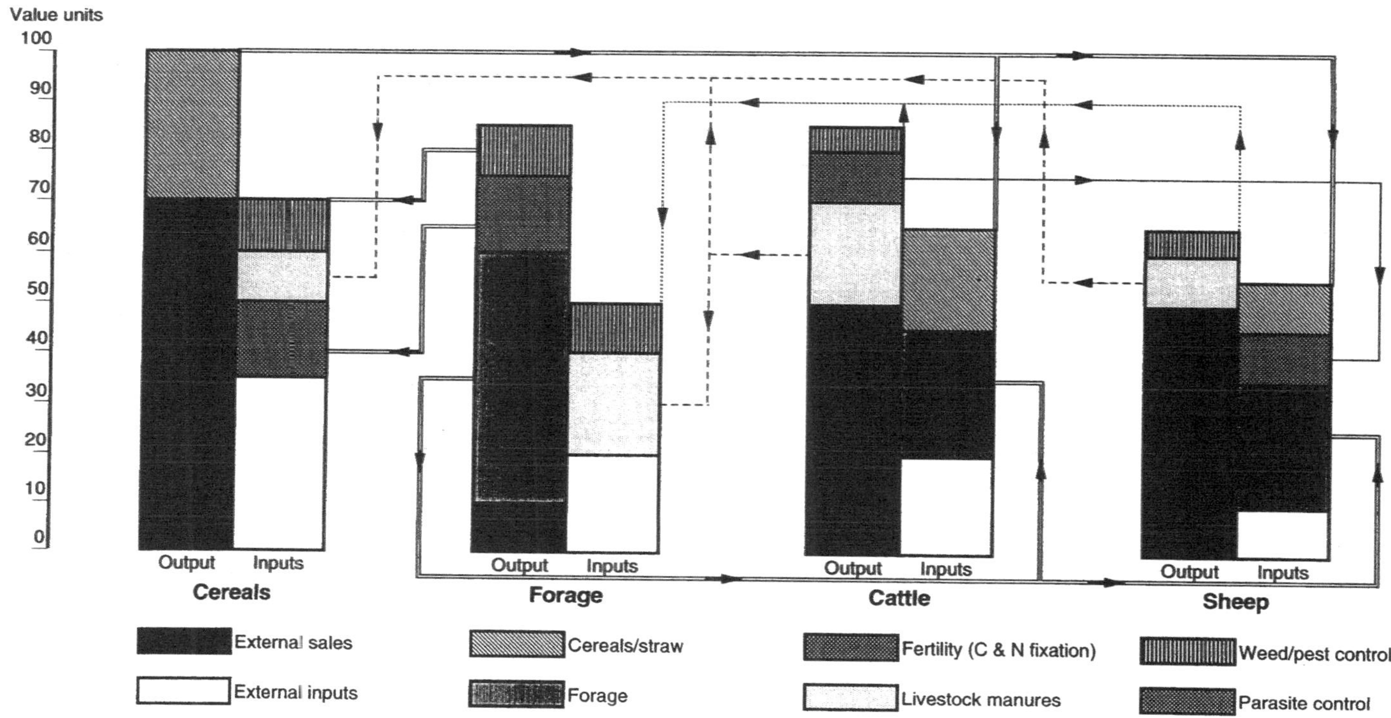

Fig. 12.3. Schematic presentation of the flow of benefits between enterprises in an organic farming system. Source: Based on Heißenhuber (1993).

The total farm net benefit at 95 units is the same as the total farm gross margin, but the contributions of the individual enterprises to this amount are very different.

The use of other hypothetical values could have produced a different result, but the example illustrates that it is often inappropriate to discuss the economics of a single enterprise, such as organic wheat, outside the context of the whole-farm system. Optimizing individual elements does not necessarily result in the optimization of the whole system. The complementary interactions between enterprises lie at the heart of the holistic systems approach to agriculture to which organic farming aspires (Chapter 3, this volume; Scofield, 1986; Dabbert, 1990). However, while increased diversity and complexity allow for complementarity to be exploited, for labour requirements to be more evenly distributed over time, and for the stability of the system to be enhanced reducing risk and uncertainty, a balance has to be achieved with the benefits which can be obtained from economies of size and specialization.

Labour use and labour costs

Labour use on organic farms, whether in terms of hours or full-time labour units, is generally higher on organic farms than on comparable conventional farms, at least in the northern European context. The figures reported vary between 20% more labour in Germany and Switzerland and twice as much labour on organic holdings in Denmark. The reasons for this include the greater diversity of enterprises, marketing and processing activities and the higher proportion of vegetables and root crops in the rotations of organic farms. The analysis of labour requirements for individual enterprises on Swiss organic farms in Chapter 8 found that the additional labour requirement for crops is between 49% and 72%, whereas no additional labour is required in orchards and for animal production. Overall, the organic farms in the Swiss study use 12.4% more labour than their conventional counterparts, if work for processing and marketing is excluded. Several studies show the influence of farm size on labour requirements, with small farms having higher labour use and requirements per hectare than larger organic holdings (Chapters 5 and 6, this volume; Kleijn *et al.*, 1990; Mühlebach and Näf, 1990).

Differences in expenditure on wages were greater than the differences in labour use reported in Chapters 5–9. In Canada, Burgoyne (1992) reported 50% higher labour costs, although total labour use was similar. In Britain, wage costs were on average 40% higher on a per hectare basis and twice as high per farm (excluding horticulture, where costs were only one third of the conventional level per hectare and less than 10% per farm), reflecting differences in farm size. In Switzerland and Germany, where farm sizes were more comparable, wage costs were two to four times as high as conventional, although total labour use was only around 20% higher. In all three cases, family labour use was similar; the higher wage costs reflecting the need to meet additional labour requirements through hired rather than family labour. Differences in family labour use per hectare in Britain were again primarily due to differences in farm size. Only in Denmark, where wage costs were 2.5 times as high as conventional, was family labour use also higher at 40% more than comparable conventional farms.

Other fixed costs

Other fixed costs include items such as power, repairs and depreciation of machinery, property charges and depreciation, and capital costs, although definitions vary between countries. These costs are typically similar between organic and conventional farms in most countries, although higher levels have been reported for machinery and rents on organic holdings in Germany. In contrast, substantially lower machinery costs have been reported from Australia and individual studies in the United States (Chapter 15).

Farm income measures

Farm profits

Farm profits represent the return to the farm family's own labour, land and capital. Costs for wages and interest payments for loans are usually accounted for in farm profits, as they represent real cash costs to the farmer. For comparative purposes, however, this approach can be misleading, due to differences in land tenure, availability of unpaid family labour and reliance on borrowed capital. In most countries, standard approaches have been developed to obtain comparative income measures. These either include some notional charges, such as the value or opportunity cost of family labour and the rental value of family-owned land, or exclude certain payments, such as interest and, in some countries, rent. The effect is to put all farms on the same basis, treating them as if all labour is paid, all capital is family-owned, and all farms have similar land tenure. However, some differences exist in standard practice between countries, so that any international comparisons need to be interpreted with care.

Standardized farm profits were found to be comparable with or greater than the average for similar conventional farms in Germany,

Denmark, Wales (Chapter 5) and Canada (Burgoyne, 1992). Profits were slightly lower for organic farms in Switzerland, but were still seen as providing a sufficient income for the farming family. In Britain, profits (net farm incomes) were slightly lower than conventional overall, although there were large variations between farm types. In the Australian and some US studies, the *net* returns for the organic farms were lower than conventional.

Return to labour

Traditionally, wages for farm workers are treated as costs to the farm business, while returns to family labour are derived from profit. For farm management purposes, this may be appropriate. However, the return to total labour (labour income) is a more useful measure in assessing the contribution that organic farming can make to rural employment and incomes. It is calculated by adding the income of the farm family to expenditure on hired labour after any necessary adjustments for real and notional returns to land and capital. Higher labour incomes per hectare were reported from Germany, Denmark and Switzerland, although in some cases labour incomes per hour or per day were lower than conventional. In Britain, labour incomes per hectare were similar, although farm size differences influenced per farm values. In Canada, Burgoyne (1992) reported higher returns to labour per cow and per litre, but lower returns per hectare. In some cases, the results indicate that organic farming may help to maintain the full-time operation of smaller farm units.

Return to capital and changes in net worth

Information on returns to capital invested is more limited. There are some indications that organic farms in Switzerland and Australia were less capital-intensive, but the other northern European studies indicate similar levels of capital intensity. Burgoyne (1992) reported slightly higher returns to capital for organic milk production in Canada, while returns to capital were much higher in Australia, and would still have been similar to conventional in the absence of premium prices. The return to capital invested was negative for the organic and two of the conventional samples in Denmark. The size of the deficit was highest for the organic group, because higher profits were cancelled out by the higher notional value placed on the opportunity cost of family labour. However, the poorer returns to capital are counterbalanced by better returns to labour.

Change in net worth provides an indicator of business growth, and in particular growth of the farm family's capital stake in the business.

In Germany, the average net worth of the organic sample increased by DM 263 ha^{-1} while the net worth of comparable conventional farms increased by DM 44 ha^{-1} (see Table 6.1). In Switzerland, however, the net worth increase per farm on the organic farms was slightly lower at 92% of the conventional level.

Factors influencing whole-farm performance measures

Whole-farm performance is influenced by yields, prices, variable costs, enterprise structure and labour requirements under organic management. The performance *relative* to conventional farms will also be influenced by the intensity of the conventional system. This is clearly illustrated by studies from the USA, where early studies found similar income levels (Lockeretz *et al.*, 1984), but recent studies (Chapter 10) present a picture more similar to that in northern Europe; i.e. lower variable costs alone may not offset lower yields and the necessity to include lower-value crops in rotations.

In several chapters, the selection of the comparison group also has a strong influence on the results. Farm size, farm type, location, land value and business size (including full-time or part-time) are all important criteria in selecting suitable comparison groups (Chapter 3), but these factors have not always been taken into account. The German data, for example, show poorer relative performance for the organic sample when compared with the average for all conventional main-income holdings, but the organic sample differs in size, type, location and land value (Chapter 6). When compared with the more carefully selected group, the organic farms perform as well or better. In Denmark, three different sets of comparison data are presented, giving very different results, particularly if the organic sample (with its high proportion of full-time holdings) is compared with the sample of conventional full-time holdings (see Tables 7.4 and 7.5). The importance of the selection of the conventional comparison group according to farm size and business size was stressed in Chapter 5 (see Tables 5.8 and 5.9).

Risk and Uncertainty

Very few studies have tackled the issue of risk and uncertainty in organic farming. While some indicate a high degree of variability in performance between organic farms, they do not contrast this with variability on conventional farms. Most do not include basic statistical measures of dispersion, let alone significance testing of differences between means. It is thus difficult on the basis of the available evidence to come to a conclusion on the issue of risk in organic farming.

There is some evidence for greater variability in crop yields and prices under organic management due to climatic factors. Under drought conditions, some studies indicate even lower yields (Chapter 5) while others report cases of better yields (Chapter 10 and 11, this volume; Lockeretz *et al.*, 1984). However, the time-series data which are available (Chapters 6 and 8) and long-term experimental trials (Helmers *et al.*, 1986) indicate a close correlation between mean organic and conventional yields and other parameters over time, suggesting that exogenous factors including climate may be a more important cause of yield and financial variability over time than the management system itself – a hypothesis which would bear further study.

Organic farming is certainly *perceived* as more risky than conventional farming by many non-organic farmers. This may be a false perception due to lack of knowledge, because information which does exist may not be readily available to those not actively seeking it (Chapter 17). It is also true that in many cases the necessary information is simply not available or sufficiently reliable, particularly with respect to *actual* levels of risk and uncertainty in organic farming systems. Research is therefore needed to characterize actual levels of risk and uncertainty, and the factors contributing to them, in established organic farming systems as well as during the conversion process (Lampkin, 1993). These questions need answers, so that perceived and actual risk and uncertainty can be reduced to levels considered acceptable by the majority of farmers. In particular, future research should examine the extent of:

- increased yield and quality variability, and risk of crop failure, due to the lack of opportunities to intervene with fertilizer or biocide applications should preventive management strategies fail;
- reduced yield and income variability due to increased diversity and complexity in organic systems;
- price instability due to the small-scale, immature nature of the organic market and the lack of government intervention to stabilise prices;
- lack of opportunity to undertake actions which would alleviate the impact of uncertainty, such as flexibility of action, hedging or insurance (including prophylactic treatments);
- farmers' objectives and attitudes to risk, as the objectives of existing organic producers may reflect their status as innovators in the adoption process, implying a greater willingness to accept higher risk levels than the majority of farmers;
- the contribution which improved extension and conversion planning procedures can play in improving the communication of

existing information and the effectiveness of its application to farm management decision-making (Chapter 17).

Conclusions

From the evidence presented in this chapter, it can be concluded that on organic farms:

- absolute yields are generally lower, but yield differences relative to conventional systems vary depending on the enterprise type and the intensity of the conventional system;
- variable input costs, particularly agrochemicals, are lower;
- gross margins may be similar or higher if premium prices can be obtained, but are otherwise likely to be lower, and individual enterprise gross margins need to be considered in the context of the whole-farm system;
- labour use is higher, but this is generally related to an increase in the number and range of enterprises on the farm, and to the development of new marketing and processing activities, rather than to increases in labour use for specific crop or livestock enterprises;
- other fixed costs are generally similar;
- premium prices are needed to compensate for reduced output *and* increased labour use for farm incomes to be similar (this situation exists in Germany, Denmark and Switzerland, and on some farm types in Britain);
- there is little evidence available concerning levels of risk, with some studies suggesting that the increase in enterprise diversity reduces risk, but others suggesting that risk increases because yields for a given enterprise are more variable.

Where no premiums are available for crops, lower incomes may result unless the conventional systems are already relatively extensive. For livestock farmers, comparable incomes might be achieved without access to premium prices due to the benefits of forage legumes, but the relative results will also depend on the intensity of the comparable conventional system.

In Chapters 4 and 26, the case has been made for the environmental and other benefits of organic farming and the contribution which the market can make towards compensating producers for internalizing their external costs. Consumers, by being willing to pay a premium for organic food, are in some cases expressing their preference for a different, more environmentally friendly method of agricultural production. But it also has been argued by Bateman (Chapter 4) that the protection of the environment cannot be left only to market forces. Premium levels may well decline in the face of widespread conversion to organic farming. The need for a premium to maintain similar income levels on organic farms could be decreased if financial support were to be given to organic farms and farmers in conversion (Chapters 23–26).

References

Burgoyne, D. (1992) Analyse économique comparée de l'impact du niveau d'intensification, de l'utilisation d'intrants et de la production biologique sur la rentabilité des entreprises laitières au Québec. MSc Thesis, FSAA, Université Laval, Quebec, Canada.

Cochrane, W.W. (1958) *Farm Prices – Myth and Reality.* University of Minnesota Press, Minneapolis.

Dabbert, S. (1990) *Zur optimalen Organisation alternativer landwirtschaftlicher Betriebe. Untersucht am Beispiel organisch-biologischer Haupterwerbsbetriebe in Baden-Württemberg.* Agrarwirtschaft Sonderheft 124. Verlag Alfred Strothe, Frankfurt am Main.

Heißenhuber, A. (1993) Produktionskosten im ökologischen Landbau und Einfluß der EG-Agrarreform auf Rentabilitätskriterien. In: Zerger, U. (ed.) *Forschung im ökologischen Landbau – Beiträge zur zweiten Wissenschaftstagung im ökologischen Landbau.* Stiftung Ökologie und Landbau, Bad Dürkheim, pp. 336–340.

Helmers, G., Langemeier, M.R. and Atwood, J. (1986) An economic analysis of alternative cropping systems for east-central Nebraska. *American Journal of Alternative Agriculture* 1, 153–158.

Kleijn, E.H.J.M.d., Groenwold, J.G., Hack, M.D., Jager, A.d. and Wijngaarden, G. (1990) *Produktie en afzet van BD- en EKO-Produkten.* Mededeling 425. Landbouw-Economisch Instituut, The Hague.

Lampkin, N.H. (1990) *Organic Farming.* Farming Press, Ipswich.

Lampkin, N.H. (1993) The economic implications of conversion from conventional to organic farming systems. PhD thesis, Department of Agricultural Sciences, University of Wales, Aberystwyth.

Lockeretz, W., Shearer, G., Kohl, D.H. and Klepper, R.W. (1984) Comparison of organic and conventional farming in the Corn Belt. In: Bezdicek, D.F. and Power, J.F. (eds) *Organic Farming: Current Technology and its Role in a Sustainable Agriculture.* ASA Special Publication 46. American Society of Agronomy, Madison, Wisconsin, pp. 37–48.

Mühlebach, J. and Näf, E. (1990) *Die Wettbewerbsfähigkeit des biologischen Landbaus.* Schriftenreihe der Eidg. Forschungsanstalt für Betriebswirtschaft und Landtechnik, 33. Tänikon, Switzerland.

Neuerburg, W. and Padel, S. (1992) *Organisch-biologischer Landbau in der Praxis.* BLV-Verlag, Munich.

NRC (1989) *Alternative Agriculture.* National Academy Press, Washington, DC.

Schulze Pals, L. (1993) Ökonomische Wirkungen des Extensivierungsprogrammes auf die Umstellung landwirtschaftlicher Betriebe. In: Zerger, U.

(ed.) *Forschung im ökologischen Landbau – Beiträge zur zweiten wissenschaftstagung im ökologischen Landbau.* Stiftung Ökologie und Landbau, Bad Dürkheim, pp. 328–335.

Scofield, A.M. (1986) Organic farming – the origin of the name (Editorial). *Biological Agriculture and Horticulture* 4(1), 1–5.

Sellen, D., Tollman, J.H., McLeod, D., Weersink, A. and Yiridoe, E. (1993) *The Economics of Organic and Conventional Horticulture Production.* Working Paper, Department of Agricultural Economics and Business, University of Guelph, Canada.

Stanhill, G. (1990) The comparative productivity of organic agriculture. *Agricultural Ecosystems and Environment* 30, 1–26.

Stöppler, H. (1988). Zur Eignung von Winterweizensorten hinsichtlich des Anbaues und der Qualität der Produkte in einem System mit geringer Betriebsmittelzufuhr von aussen. PhD thesis, Fachbereich Landwirtschaft der Gesamthochschule Kassel, Witzenhausen.

Thomas, C., Reeve, A. and Fisher, G.E.J. (eds) (1991) *Milk from Grass* (2nd edn). British Grassland Society, Reading.

Vine, A. and Bateman, D.I. (1981) *Organic Farming Systems in England and Wales: Practice, Performance and Implications.* Department of Agricultural Economics, University College of Wales, Aberystwyth.

Part 3

Converting to Organic Farming: The Economics of the Transition Process

13 Changes in Physical and Financial Performance during Conversion to Organic Farming: Case Studies of Two English Dairy Farms

NICOLAS H. LAMPKIN

Introduction

The conversion or transition from conventional to organic farming systems is subject to a number of physical, financial and social influences which differ from those associated with established organic farming systems as discussed in the preceding chapters. The process of conversion is a complex one involving a significant number of innovations and changes in the farm system as well as in production methods. Key factors which determine the success of conversion on a particular farm include the extent of changes to farming practices and the introduction of new enterprises and marketing activities, as well as the possibility of additional yield and financial penalties related specifically to the conversion process.

There is some evidence of a decline in yield during the conversion period greater than that which would be expected in an established organic system, as biological processes such as nitrogen fixation and rotational effects on weeds, pests and diseases become established. Conversion-specific yield reductions may also be associated with mistakes or inappropriate practices during the conversion period. Reductions in forage production for livestock may be of particular significance, due to the need to take land out of production during the establishment of new leys, or due to the time taken for clover to regenerate following the cessation of nitrogen fertilizer use.

Conversion-specific financial changes relate not only to the possible conversion-specific yield reductions discussed above, but also to:

- the loss of income relative to established organic systems due to the lack of access to premium prices during the official two-year conversion period;
- the start-up costs for establishing new enterprises, including the purchase of livestock and costs of investment in buildings, fences, water supplies, manure handling systems and other facilities, in particular when converting from very specialized arable or livestock systems;
- the write-off costs for discontinued enterprises (e.g. battery cages);
- the costs of advice, training and other types of information gathering (including farm visits, study tours and conferences) as well as the costs of additional labour associated with absences at training events;
- loss of income and additional costs due to experience gathering, trialling new crops and techniques, and mistakes resulting in higher than necessary yield reductions.

The purpose of this and the following chapters is to examine the extent and influence of these issues using different methodologies including case studies, experimental trials and cross-sectional analysis of survey data. In this chapter, the process of conversion on two dairy farms over periods of eight to ten years is examined. The results, together with case studies of two arable farms, are reported in more detail in Lampkin (1993).

Farm Characteristics and the Process of Conversion

Two contrasting dairy farms were selected for this study. The first, Farm A, is a small (46 ha), tenanted, specialist dairy farm. The second, Farm B, is a medium-sized (102 ha), owner-occupied, mixed dairy-arable farm. Both have Friesian/Holstein dairy herds, the predominant type of cattle used for milk production in the United Kingdom.

Farm A

Farm A is typical of the smaller specialist dairy farms in England and Wales. The farm was taken on by the farmer from his parents in 1976. Prior to the conversion, the farmer carried out significant improvements including the installation of a milking parlour, dairy, cubicle housing and slurry store, leaving the farm reasonably well equipped in terms of buildings and machinery.

The conversion started with a trial field in the late 1970s, but did not start in earnest until 1984. A staged approach was followed, in which fields were progressively changed to organic management following reseeding with clover-based pastures. The last conventional inputs were applied in 1989. Two important external factors in 1984 influenced the process and the rate of conversion. First, a gas pipe was installed across the farm. The consequent reseeding requirements meant that the conversion could proceed faster than initially planned. Second, the introduction of milk quotas forced a reduction in the number of cows anyway. The cropping pattern on Farm A became more diverse as the conversion proceeded, with an increase in the area down to arable crops such as cereals, fodder beet and short-term red clover/Italian ryegrass leys, which together accounted for over one third of the total land area by 1990. Total numbers of livestock units on Farm A remained relatively constant throughout the conversion period, apart from a marked fall in the number of dairy cows in 1984 due to the combined effects of the introduction of milk quotas and the installation of the gas pipe. However, the total land area increased by just over 5 ha in 1983/84. Dairy replacements increased significantly in importance: at the start of the conversion, all replacements were bought in; by the end of the conversion, most replacements were reared on the farm.

Farm B

Farm B is in many respects typical of mixed dairy–arable farms in the South-West of England, slightly smaller than the average in terms of farm size, but with a similar enterprise structure (two thirds livestock, one third arable). The predominantly east-facing, shallow, flinty soils over chalk tend to restrict production. The farm was taken on as part of a generation change in 1969, initially as a tenancy, but later additional land was bought and the tenanted land was inherited so that the farm is now owner-occupied. Milking parlour, cow kennels and other facilities were installed during the 1970s so that the farm was reasonably well equipped at the start of conversion.

The conversion started in 1983, with the last conventional inputs used in 1986. The conversion proceeded more rapidly than originally planned, although in this case much of the land entered the transition phase through cessation of fertilizer use rather than by reseeding as in the case of Farm A.

Farm B maintained a relatively constant proportion (about 50% of total land area) of medium- to long-term leys and permanent grassland throughout the conversion period. The proportion of cereals declined (from 33% in 1983 to 20% in 1990) and was replaced, to a large extent, by lucerne and short-term red clover/Italian ryegrass leys. The absolute decrease in cereals area was not so great (from 30 to 21 ha) due to an increase in overall farm size from 91 to 101 ha over the study period. The total number of livestock units on Farm B remained relatively constant, but the relative importance of different types of livestock changed markedly over the conversion period. The number of sheep and pigs increased and the number of dairy replacements fell. Although all replacements were home-reared throughout the conversion period, the replacement rate fell from 29% in 1983 to 14% in 1990.

Conventional comparison data

Three different sources, detailed in Lampkin (1993), have been used for comparative data on conventional milk production. The most closely matched samples, in terms of region, farm type and size, come from the regional Farm Business Survey (FBS) reports. However, these cover mainly whole-farm statistics and do not provide detailed information on the physical and financial performance of the dairy enterprise itself. For this reason, two further sources of data have been used for the enterprise comparisons: Milkminder, which differentiates results by region but not by farm size and type (i.e. specialist or mainly dairy), and Milk Marketing Board Farm Management Services (MMB-FMS), which differentiates by farm type but not by region. The different conclusions which can be drawn from the comparisons using these different groups are a good illustration of the methodological problems involved in comparing farming systems (see Chapter 3). None of the sources uses identical samples throughout the study period. Changes in the conventional comparison group may therefore be associated with changes in the composition of the group, which means that long-term trends based on these samples should be treated with caution.

Dairy Enterprise Physical Performance

Stocking rates

The number of grazing livestock units (LU) per forage hectare on Farm A declined by 20% during the conversion period (from 2.0 LU ha^{-1} in 1982/83 to 1.6 LU ha^{-1} in 1984/85), but subsequently increased to levels approaching the original conventional system (1.9 LU ha^{-1} in 1990/91). Although the conversion involved taking some land out of production for reseeding, it is not possible to attribute all the stocking-rate decline to the conversion alone, as other factors including the introduction of milk quotas and the installation of the gas pipeline played a role.

The combined effect of static livestock numbers, the increase in forage area and the increase in farm size on Farm B was a decrease in stocking rates per hectare. The number of grazing livestock units per forage hectare fell from 1.9 to 1.4 LU ha^{-1} over the conversion period. The fall in the number of livestock units per farm hectare was less dramatic, from 1.28 to 1.12 LU ha^{-1}, due to the increase in the proportion of total land area utilized by livestock.

Milk yields and concentrate feed use

Milk yields per cow on Farm A fluctuated from year to year due to changing policies on concentrate feed use. Over the conversion period, yields per cow declined by 10% from about 5500 l in 1982/83 to around 5000 l, due to a 35–40% reduction in concentrate use (from 1500 to 900 kg per cow), consistent with the findings of Houghton and Poole (1990) and Redman (1991) reported in Chapter 5.

Yields per cow on Farm B fell dramatically in the early stages of the conversion period (from 4878 l in 1982/83 to 4225 l in 1984/85), which coincided with the introduction of milk quotas and dry weather conditions in 1984, but increased subsequently to over 5000 l, more than under previous conventional management. Unlike Farm A, Farm B already relied heavily on home-grown cereals for feed before conversion, and concentrate use only declined by about 20%, from 1500 to 1200 kg per cow. Yields were also generally lower than on the comparable conventional farms, reflecting a lower level of intensity which was present from before the start of the conversion. Towards the end of the conversion period, the differences between Farm B and the comparison groups declined.

Reductions in concentrate feed use also occurred in 1984/85 on the conventional farms, indicating that the introduction of milk quotas may have been a more important influence on the level of concentrate use on the case study farms than any consideration of organic livestock management, at least in the early stages of the conversion process.

Milk yields per forage hectare declined steeply at the start of the conversion period on both Farm A (from 11,000 to 9000 l ha^{-1}) and Farm B (from 9000 to 7000 l ha^{-1}), and remained at levels significantly below those of the conventional comparison groups. A number of factors complicate the picture, in particular the introduction of milk quotas, so that it is not possible to attribute this decline entirely to the change to an organic system.

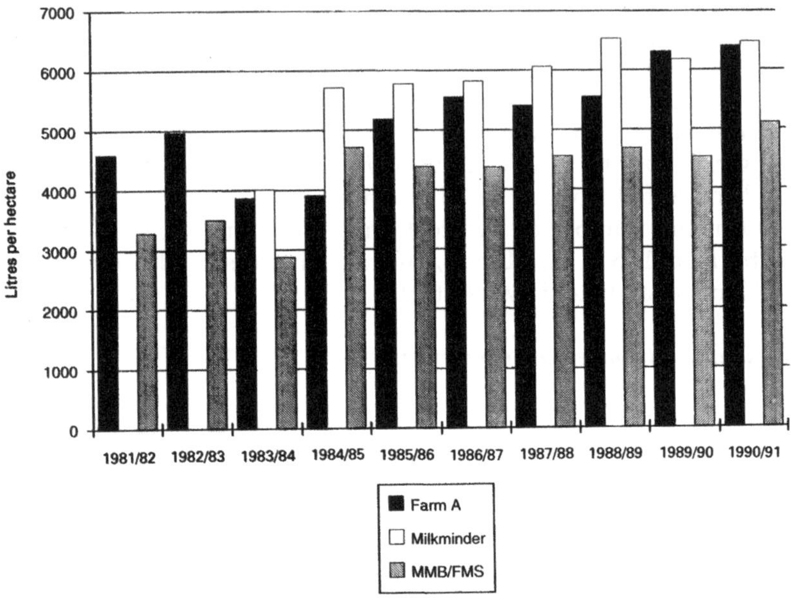

Fig. 13.1. Milk yields from forage and purchased bulk feeds per hectare on Farm A, other dairy farms in the region (Milkminder), and other specialist dairy farms (MMB-FMS), North-West England, 1981/82–1990/91. Source: Lampkin (1993).

Milk production from forage

Stocking rates and milk yields are related to the inputs of nutrients and energy from outside the farm. A decline in these variables may therefore reflect the reduction in external inputs, rather than a reduction in the productive capacity of the farm itself. An alternative approach to measuring the productivity of a dairy farm or enterprise is to calculate the milk produced from forage, excluding concentrates and other purchased bulk feeds (Genus, 1991). It was not possible to exclude purchased bulk feeds in this study because information on the quantity of bulk feeds fed was not available for the comparison groups. As a result, the values calculated for milk produced from forage may overstate the actual case by including purchased bulk feeds, and should therefore be treated with caution.

Farm A consistently produced higher levels of milk from forage per cow than the comparison groups, with levels increasing from 2480 to 3362 l per cow compared with an increase from 1659 to 2527 l per cow for the MMB-FMS sample over the study period. Milk yields from forage per hectare also increased by 35% relative to pre-conversion

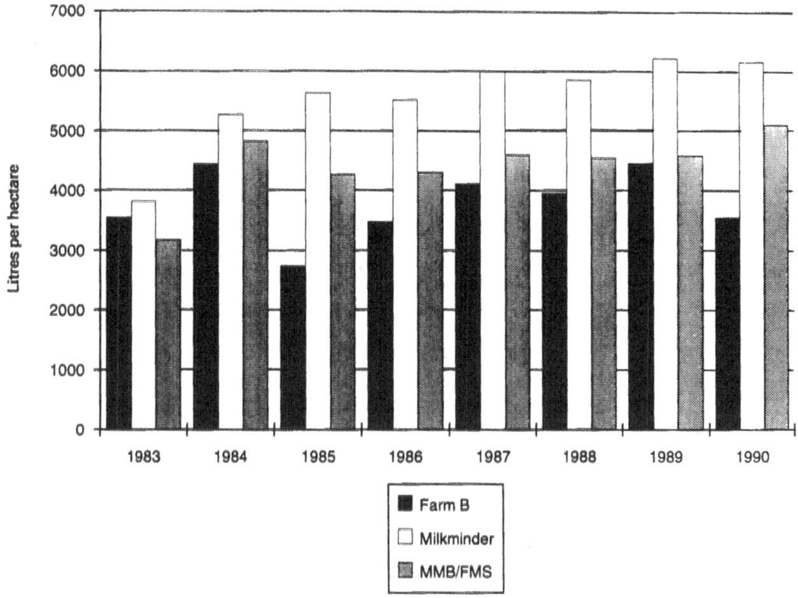

Fig. 13.2. Milk yields from forage and purchased bulk feeds per hectare on Farm B, other dairy farms in the region (Milkminder), and other mainly dairy farms (MMB-FMS), South-West England, 1983–1990. Source: Lampkin (1993).

levels (Fig. 13.1). Farm A outperformed the MMB-FMS comparison group in terms of milk from forage per hectare throughout the conversion period, and achieved levels comparable to the Milkminder group, despite lower stocking rates and lower overall milk production per hectare. Part of this improvement in relative performance on Farm A can be attributed to greater reliance on purchased bulk feeds, which in 1990/91 amounted to £43 per cow (£82 ha^{-1}) compared with £14 per cow (£33 ha^{-1}) for the Milkminder comparison group. This difference is approximately equivalent to 300 l milk from forage per cow, or 700 l ha^{-1}.

Farm B achieved yields of milk from forage per cow which were similar to the average for the Milkminder group and slightly higher than the MMB-FMS group, increasing from around 2000 to nearly 3000 l per cow by 1989, but falling back to 2500 l per cow in 1990 due to the influence of drought conditions the previous year on the availability of conserved forage. Milk produced from forage per hectare was lower than the average for both comparison groups, although by the end of the conversion period, levels on Farm B had recovered and exceeded the levels achieved under previous conventional management (Fig. 13.2). Expenditure on purchased bulk feeds was slightly lower on Farm B than on the comparison farms in 1990/91, so that their inclusion in these estimates is less likely to distort the comparison, and if anything exaggerates the difference between Farm B and the conventional groups.

Dairy Enterprise Financial Performance

Milk price, sales and enterprise output

The milk price per litre obtained on both Farm A and Farm B and the conventional comparison groups was similar throughout most of the study period, ranging from 17 to 18 pence per litre (ppl) at constant 1988 prices, except for the last year of the study when, on both farms, a premium was obtained which averaged about 3 ppl (18%). Farm A also obtained a small premium which averaged 0.8 ppl in the previous year.

On Farm A, the value of milk sales per cow was very similar to the comparison values, except in the last two years of the study. In 1989/90, the fall in milk yields per cow due to a drought management strategy was not compensated for by the premium price. In 1990/91, once yields had recovered, the premium price led to the value of milk sales exceeding the conventional averages by about £100 per cow. Without the premium the result would have been £50 per cow lower. Output per forage hectare allocated to the dairy enterprise fell by 17% at constant 1988 prices, from over £1900 to around £1600 ha^{-1} following conversion, but recovered to nearly £1800 ha^{-1} in 1989 and exceeded £2000 ha^{-1} in 1990/91 with the benefit of premium prices. The average for the MMB-FMS comparison group fluctuated between £1900 and £2000 throughout the study period.

On Farm B, the premium received in the final year of the study had a marked influence on the value of milk sales per cow, which increased from levels significantly below the average for the conventional farms to a level slightly above the average despite lower milk yields. Output (at constant 1988 prices) per forage hectare allocated to the dairy enterprise fell by nearly 30%, from £1759 to £1244 ha^{-1} immediately following the start of conversion, but only recovered to levels just over £1400 ha^{-1} even with the benefit of premium prices.

Variable costs and gross margins

Variable costs per cow on Farm A were similar to previous conventional management in the first two years of conversion, with feed and forage costs similar to the conventional comparison groups at around

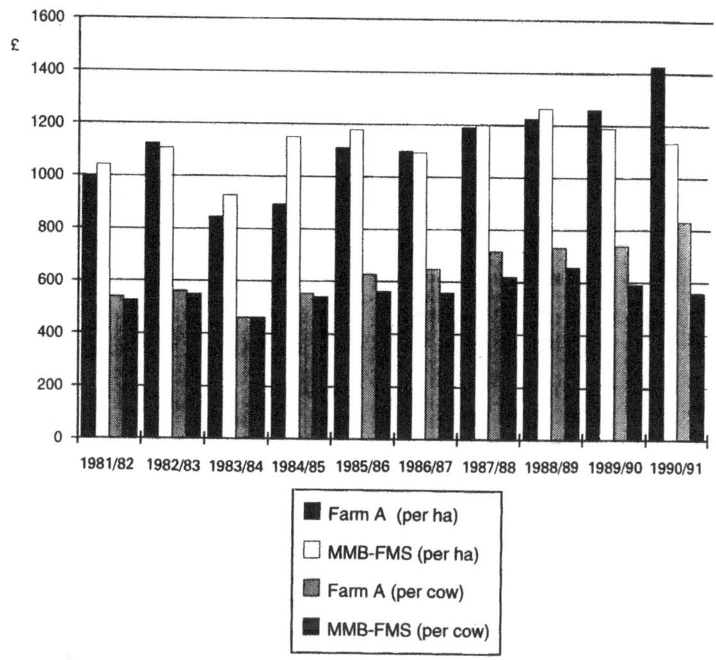

Fig. 13.3. Dairy enterprise gross margins per hectare and per cow on Farm A and other specialist dairy farms (MMB-FMS), England, 1981/82–1990/91, at 1988 prices. Source: Lampkin (1993).

£350 per cow. Feed and forage costs then fell rapidly, due to reduced purchases of feeds and fertilizers, to around £225 per cow, compared with a fall to around £300 per cow on the conventional farms in response to the introduction of milk quotas. In the final year of the study, these costs increased to just over £250 per cow because of greater reliance on contractors and home-grown cereals valued at organic premium prices.

The cost of all feed including forage on Farm B remained below £250 per cow throughout the study period and had declined to £160 per cow by 1990. On the conventional farms, the cost of all feeds including forage remained over £300 per cow throughout the period, and by 1990 was twice as high as on Farm B. Casual labour and contracting costs increased on both case-study farms in line with their diversification into new enterprises and marketing activities.

The one source of increased variable costs related specifically to conversion which can be identified is the cost of reseeding grassland (Houghton and Poole, 1990). On Farm A, these costs averaged £1700 per year above reseeding costs under established organic management over the three-year period 1984–1987, but were to a large extent

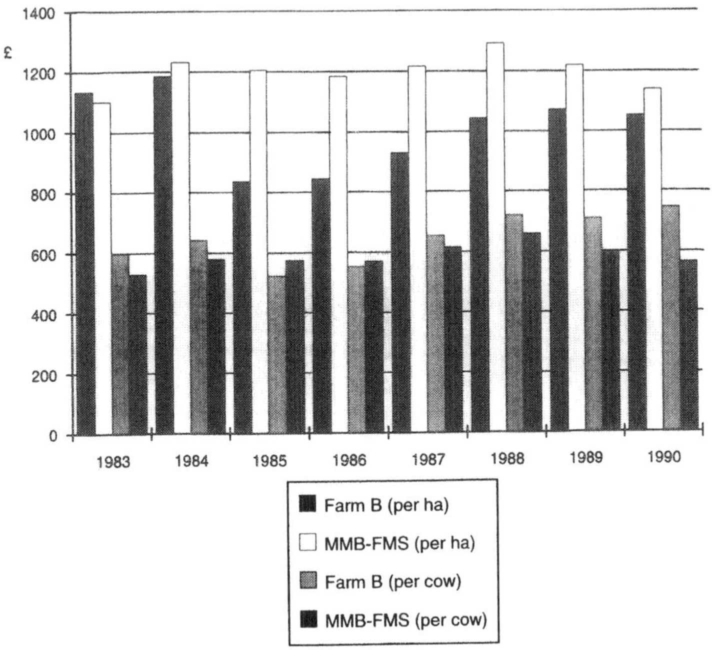

Fig. 13.4. Dairy enterprise gross margins per hectare and per cow on Farm B and other mainly dairy farms (MMB-FMS), England, 1983–1990, at 1988 prices. Source: Lampkin (1993).

associated with the gas pipe installation, and financed by compensation for this disruption.

If all costs, including the value of home-grown feed and other sundry costs, are taken into account, the gross margins per cow and per hectare allocated to the dairy enterprise on Farm A, at constant 1988 prices, are very close to the MMB-FMS average throughout almost all of the study period (Fig. 13.3). Higher per cow performance compensated for slightly lower stocking rates to give similar per hectare results, although the lower intensity commonly associated with larger farm size (a characteristic of the MMB-FMS comparison group) may mean that this comparison is unduly favourable for Farm A. Premium prices in the last two years helped compensate for lower yields due to the drought in 1989/90 and led to above average performance in 1990/91. If no premium price had been received in 1990/91, the gross margin per hectare on Farm A would have been £300 lower at £1123 ha^{-1} (at 1988 prices), just under the level for the MMB-FMS comparison group.

Farm B outperformed the MMB-FMS comparison group on a per

cow basis, but this did not make up for the significantly lower stocking rates, resulting in lower per hectare performance (Fig. 13.4). Premium prices for organically produced milk in the last year of the study period allowed gross margins per hectare to approach those on conventional farms, despite also valuing home-grown feed at premium prices. The weakness of MMB-FMS comparison data in this case is that the farm-type definitions are not consistent throughout the period. The farm-type definitions used in the period 1985–1988 most closely compare with Farm B. The comparison for this farm provides some indication of a reduction in income due to the conversion, if the differences of up to £370 ha^{-1} after 1985 are compared with 1983 and 1984 where similar results were achieved (at constant 1988 prices). By 1990, this difference had fallen to £86 ha^{-1} at 1988 prices.

Since the introduction of milk quotas, quota limitations have in some cases become a more important constraint than land. In these circumstances, costs and margins per litre of milk produced may provide a better basis for comparing performance. On this basis, Farm A outperformed the conventional farms from early in the conversion period, even without the benefit of premium prices. Farm B's low-intensity approach and reliance on home-grown feeds resulted in better per litre performance under conventional management prior to conversion. This continued during the conversion period, and by the end, when premium prices were obtained, was nearly 50% higher than the conventional comparison group.

On Farm A, the proportion of enterprise output retained as gross margin increased from 50% in 1981/82 to over 70% in 1990/91. The comparison group followed a similar trend, but by the end of the study period, the amount retained as gross margin was less than 60%. On Farm B, the proportion of enterprise output retained as gross margin increased from 62% at the start of the conversion period to about 75% at the end. The proportion of enterprise output retained as gross margin on the conventional farms also increased, from 49% to 61%, but remained at a much lower level than on Farm B throughout the study period.

Whole-farm Financial Performance

Farm A

Total farm output per hectare on Farm A declined by 20% in real terms (1988 price levels), from a peak of £1916 ha^{-1} in 1981/82 to around £1500 ha^{-1} throughout the conversion period, recovering slightly in the final year of the study to £1636 ha^{-1} due to the premium price obtained

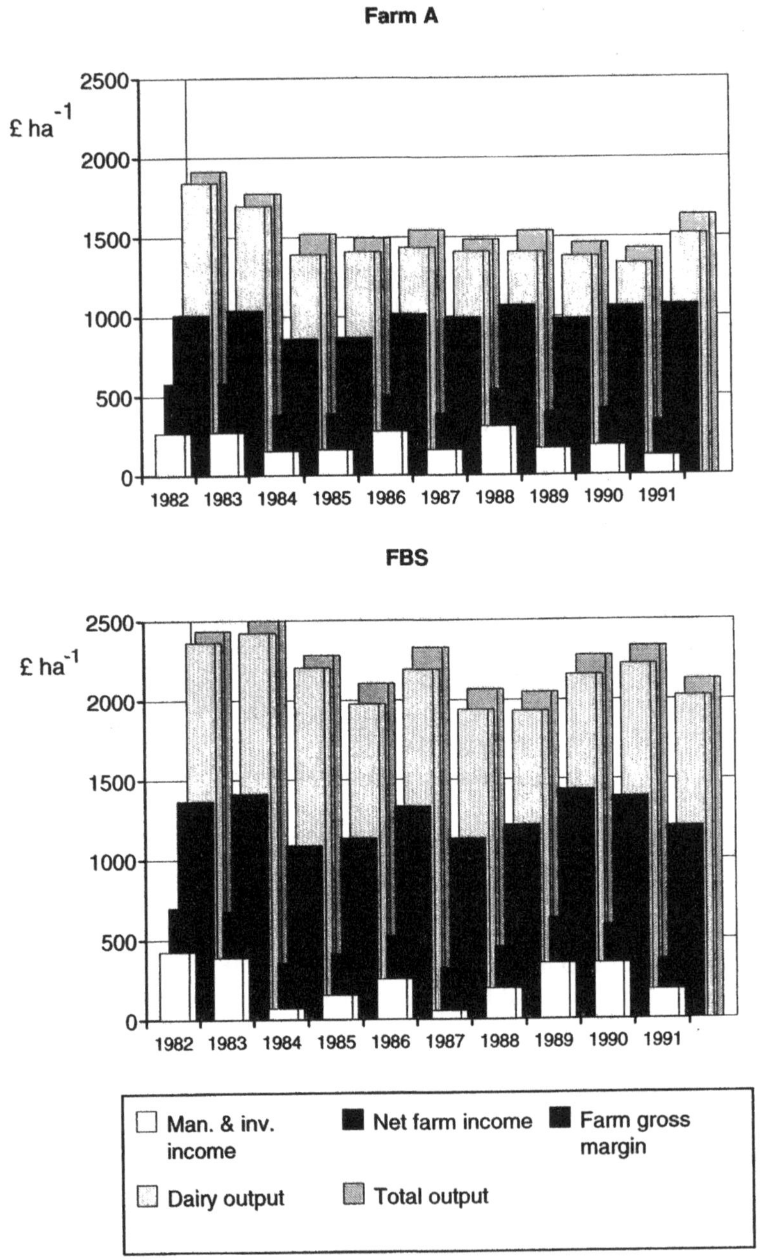

Fig. 13.5. Output, margin and income values per farm hectare on Farm A and similar specialist dairy farms (FBS), North-West England, 1982–1991, at 1988 prices. Source: Lampkin (1993).

for the milk which boosted output by nearly £200 ha^{-1} (Fig. 13.5). Output levels for the Farm Business Survey (FBS) comparison group of conventional farms were much higher in absolute terms, but they also fell, from a peak of £2525 ha^{-1} in 1982/83 to a low of £2048 ha^{-1} in 1987/88 indicating the impact of the introduction of milk quotas in 1984. In contrast to Farm A, output from the conventional group fell in 1990/91, a decline reflected throughout the dairy industry.

The reduction in variable costs on Farm A following conversion meant that total farm gross margin could be maintained despite the fall in output, at around £1000 ha^{-1} in real terms throughout the study period. The reduction to £860 ha^{-1} in 1983/84 and 1984/85 may be due to the conversion, but can also be attributed to the introduction of milk quotas. Although total farm gross margins were still higher on the conventional farms than on Farm A, the difference was smaller than for output because of the lower variable cost levels on Farm A.

The gap was further closed by the lower fixed cost levels on farm A, which increased from around £750 ha^{-1} under conventional management to over £900 ha^{-1} in real terms at the end of the study period. A major factor in this was the increase in paid labour costs from £43 ha^{-1} in 1981/82 to £140 ha^{-1} in 1990/91, although farmer and spouse labour declined by a smaller amount over the same period. Fixed cost levels on the comparable conventional farms remained around £1000 ha^{-1} throughout the study period, while expenditure on paid labour declined.

The combination of lower (and falling) variable costs and lower fixed costs on Farm A was sufficient to compensate for the lower level of output, to the extent that management and investment income (the return to management and all capital invested) and net farm income (which also includes the return to the farmer and spouse's manual labour) on Farm A were similar to the conventional comparison group for much of the study period. In 1981/82 and 1982/83, when Farm A was under conventional management, income levels on Farm A were around £100 ha^{-1} lower than for the comparison group. Incomes fell to a greater extent on the FBS farms during the period of adjustment to quotas and only recovered to a higher level than Farm A in 1988/89 and 1989/90. Farm A failed to benefit from a sharp improvement in the fortunes of the conventional dairy farms, in part because of the increasing levels of fixed costs, with the result that income levels were around £200 ha^{-1} lower on Farm A, but the difference was reduced considerably in the final year of the study due to the availability of premium prices for organic milk.

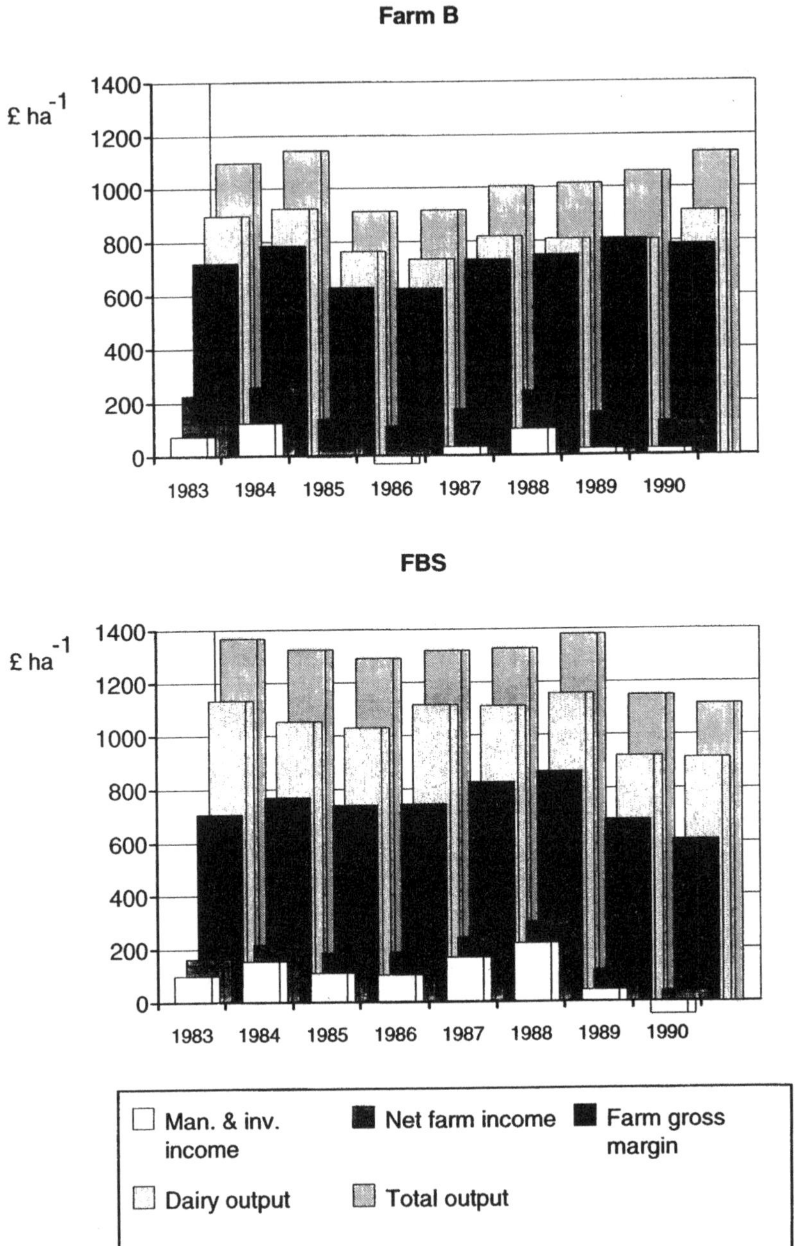

Fig. 13.6. Output, margin and income values per farm hectare on Farm B and similar mainly dairy farms (FBS), South-West England, 1983–1990, at 1988 prices. Source: Lampkin (1993).

Farm B

Total enterprise output per hectare at 1988 prices on Farm B fell by about £200 ha^{-1} in 1985, to levels nearly £400 ha^{-1} less than the average for the FBS sample, and only recovered to pre-conversion levels in 1990 with the availability of premium prices for organic milk (Fig. 13.6). Premium prices for crops helped to compensate for reduced cereal yields and boosted arable output and gross margins above levels obtained under previous conventional management and above levels achieved by conventional farms. However, new high-value enterprises such as thatching straw also contributed to this increase, and some cereals were fed to livestock, reducing the overall benefit to the farm. In 1989 and 1990, the years when organic cereal prices were at their highest, cereal premiums contributed about £100 ha^{-1} to farm output. In 1990, this was supplemented by a further £100 ha^{-1} due to the organic milk premium of 2.9 ppl. Enterprise output for the FBS sample of similar mainly dairy farms remained relatively constant and at higher levels than on Farm B until 1989, when output levels fell by over £200 ha^{-1} to levels similar to Farm B.

The level of variable costs on Farm B fell from over £350 ha^{-1} under conventional management to a low of £252 ha^{-1} in 1989 (at 1988 prices), although there was a marked increase in 1990. However, variable costs on Farm B were much lower than on the comparable conventional farms, where average variable costs fluctuated typically between £500 and £550 ha^{-1}. As a result, total farm gross margins on Farm B were similar to the FBS group. Although Farm B suffered a decline in farm gross margin in the early stages of conversion, levels soon recovered to the pre-conversion situation and exceeded the FBS average by £180 ha^{-1} in 1990.

Fixed costs on Farm B increased from around £650 ha^{-1} at the start of the study to over £750 ha^{-1} at the end. These levels were higher than for the FBS comparison group, which fluctuated between £600 and £650 ha^{-1} at constant 1988 prices. As a consequence, net farm income and management and investment income levels on Farm B were lower than for the FBS group, except in the last two years of the study when income levels for the FBS group fell markedly.

Discussion and Conclusions

Although stocking rates and arable crop yields declined on both farms due to the reduction in purchased feed and fertilizer use, milk yields from forage per hectare increased on both farms relative to previous conventional management. Changes in physical output levels and input use at the start of the conversion were also influenced by exogenous factors such as the introduction of milk quotas in 1984. Although Farm A experienced an additional reduction in output in the early stages of conversion, the extent to which this can be related specifically to conversion is debatable, as similar reductions also occurred to the conventional comparison groups. On Farm B, stocking rates and yields remained at the lower levels over a long period of time so that there is no real evidence of a conversion-specific yield decline.

The lower variable costs on Farm A compensated for output changes to give dairy gross margins similar in most years to the average for comparable dairy farms, even on a per hectare basis and without the benefit of premium prices. This performance was better than recorded by other studies, including Houghton and Poole (1990) and Murphy (1992). The reduced stocking rates on Farm B resulted in a lower gross margin per hectare despite similar per cow performance, a result more in line with the findings of other studies of organic milk production.

Labour use increased by 20–25% on both farms. This was primarily associated with diversification into new high-value and labour-intensive enterprises, such as fodder beet and thatching straw, as well as new marketing initiatives on Farm B. There was no evidence to suggest that the labour requirements for existing enterprises had increased significantly with organic management.

Apart from the increases in labour use resulting from the introduction of new enterprises, there was no evidence for an increase in fixed costs directly associated with conversion to organic production. Conversion-related expenditure was incurred on both farms, including modifications to farm buildings and advice, training and certification costs, but the overall impact on farm financial performance was small. In some cases, the capital investment costs could be defrayed by capital released due to reductions in breeding livestock numbers. Houghton and Poole (1990) have suggested that on some farms, the sale or leasing of milk quota might also be used in this way.

Net farm income on Farm A was comparable to the average for the conventional farms for much of the conversion period, but declined relatively and absolutely towards the end. In part, this was due to the changing characteristics of the FBS comparison group and to the general decline in income experienced by dairy farmers in 1989/90 and 1990/91. Premium prices helped to cushion Farm A against the worst effects of these. It is not possible to identify a conversion-related income reduction on Farm A which can not also be related to economic circumstances affecting all dairy farmers. Net farm income on Farm B also followed similar trends to the conventional farms, reflecting wider economic pressures, while premium prices helped cushion the decline in incomes experienced by dairy farmers in 1989 and 1990. However, there is more evidence of a conversion-related reduction in income in this case, linked to changes in stocking rates, output and gross margins without the benefit of premium prices. At constant 1988 prices, this reduction in income is estimated to be of the order of £60 to £120 ha^{-1} over a four-year period, which contrasts with price premiums worth £100 to £200 ha^{-1} in 1989 and 1990, respectively.

The results from the case-study farms indicate the considerable potential for cost savings from the use of legumes and low-concentrate feed regimes in milk production. Other studies confirm the potential for biological nitrogen fixation in association with clovers and other legumes to be a financially attractive alternative to purchased nitrogen fertilizers in milk and livestock production systems, particularly on farms using a moderate level of nitrogen fertilizers (Holliday, 1989).

The conclusions relating to these case-study farms are only generalizable to a limited extent. Both Farm A and Farm B (to a lesser extent) are above average in performance when compared with other organic farms. As such, they are not representative of other existing organic farms, but the better performance may reflect the greater level of experience of farmers now adopting organic farming. Both farms were lower than average intensity farms (relative to their respective FBS comparison groups) under previous conventional management, so that these conclusions might not hold for high-intensity dairy farms considering conversion. The results confirm the need to relate performance assessments to the previous performance of the farms under conventional management, as well as to the performance of other conventional and organic farms. The results also indicate the need for studies of the conversion process to be conducted over a sufficiently long time period for trends to become apparent.

Case studies of two arable farms converting to organic farming, conducted as part of the same study (Lampkin, 1993), differed from those of the dairy farms to the extent that both arable farms converted in a single step, instead of gradually over a period of years, and one of the farms was established as a new business. Comparisons with previous conventional management were difficult in both cases. The need for premium prices was much more significant on the arable farms than on the dairy farms, and even with currently available premium prices, intensive arable farms may not be able to achieve comparable incomes to their conventional counterparts.

These case studies do not on their own provide an adequate basis for generalization. However, if the case-study results are contrasted with models of the conversion process (Lampkin *et al.*, 1987; Lampkin and Midmore, 1988; Midmore and Lampkin, 1989; Ramsay and O'Mara, 1991) and other published literature, some general conclusions relating to the costs of conversion on different farm types, and the role of premium prices, can be obtained. Conversion costs over a five-year period are likely to be of the order of:

- £150 and £200 ha^{-1} $year^{-1}$ for specialist cereal and mainly arable farms, respectively;
- £100 ha^{-1} $year^{-1}$ for specialist dairy farms, but possibly much lower depending on the initial level of intensity;
- £50 to £100 ha^{-1} $year^{-1}$ for other livestock and mixed livestock/arable farms;
- less than £50 ha^{-1} $year^{-1}$ for hill and upland holdings.

Across the range of lowland farm types, premium prices may be expected to contribute £150 to £200 ha^{-1} $year^{-1}$ toward net farm income. Improved information, conversion planning and marketing can help to reduce the costs of conversion and to improve returns during the conversion period, but the constraints on marketing produce as organic during the official two-year conversion period represent the most significant cost of conversion. A change to organic farming legislation and standards which would make organic status and premiums available after one year instead of two could contribute significantly to reducing the costs of conversion.

These results represent the situation in the economic, policy and market context of the period from the introduction of milk quotas in 1984 to the 1992 CAP reform (Chapter 26). It is likely that the shift from commodity price support to arable area payments, and the introduction of quotas on eligibility for some livestock premiums, will have a significant (and not necessarily beneficial) impact on the financial performance of farms converting to organic production. This will be further influenced by the introduction of payments to support conversion to organic farming under the agri-environment programme (Chapter 26). However, dairy farms are less likely than some other farm types to be influenced significantly by these changes.

References

Genus (1991) *Genus Milkminder Annual Report 1990–91.* Report 72. Genus Management, Crewe.

Holliday, R.J. (1989) *White Clover Use on the Dairy Farm.* FMS Information Unit Report 65. Milk Marketing Board, Thames Ditton.

Houghton, M. and Poole, A.H. (1990) *Organic Milk Production.* Genus Information Unit Report 70. Genus Management, Wrexham.

Lampkin, N.H. (1993) The economic implications of conversion from conventional to organic farming systems. PhD thesis, Department of Agricultural Sciences, University of Wales, Aberystwyth.

Lampkin, N.H. and Midmore, P. (1988) *Organic Production as an Alternative to Set-aside – A Response to the Agricultural Departments' Consultative Document on Extensification.* Department of Economics and Agricultural Economics, University of Wales, Aberystwyth.

Lampkin, N., Jones, W., Mann, M. and Midmore, P. (1987) *Proposal for an Integrated Rural Development Programme Focusing on Organic Agricultural Production in the Teifi Valley, Dyfed.* I. Department of Economics and Agricultural Economics, University College of Wales, Aberystwyth.

Midmore, P. and Lampkin, N.H. (1989) Organic farming as an alternative to set-aside and an option for extensification. In: Dubgaard, A. and Nielsen, H.A. (eds) *Economic Aspects of Environmental Regulations in Agriculture.* Wissenschaftsverlag Vauk, Kiel, pp. 267–278.

Murphy, M. (1992) *Organic Farming as a Business in Great Britain.* Agricultural Economics Unit, University of Cambridge, Cambridge.

Ramsay, D.A. and O'Mara, M.M. (1991) *Economics.* Final Report, III. Organic Farming Centre, University of Edinburgh, Edinburgh.

Redman, M. (1991) Organic dairy costings: update. *New Farmer and Grower* 33, 20–21.

14 Case Studies of Farms Converting to Organic Agriculture in Germany

BERNHARD FREYER, RUDOLF RANTZAU AND HARTMUT VOGTMANN

Introduction

Since the beginnings of organic agriculture in Germany, the conditions for conversion from conventional to organic farming have changed, with premium prices, subsidy payments, the changing attitudes of farmers and the world-wide economic situation all having an impact on the number and type of farms converting and the financial consequences of conversion (Fischer, 1982; Hotzelt, 1985; Rantzau *et al.*, 1990). The aim of this chapter is to characterize the main types of farmers who have started farming organically in the last few decades, the agropolitical and social conditions facing farmers during conversion, and the technical and financial changes associated with the conversion process.

On the basis of case studies of farms and farming families during the conversion period, the main social, production and financial patterns of recent years (1986–1990) are presented. The data are based mainly on the related studies of Freyer (1991) and Rantzau *et al.* (1990), as well as on a preliminary survey of a larger group of farms which had converted to organic production prior to 1986 (Vogtmann *et al.*, 1993). German farmers' descriptions of the conversion process and their experiences are also reported by Benecke *et al.* (1988).

The Socioeconomic Context of Conversion

Since the 1940s there have been three main periods in the development of organic farming, each producing a particular type of farmer converting to organic agriculture. Until the 1970s, the pioneers of the organic farming movement consisted of a small group of farmers working on the basis of traditional farming systems and using low levels of external inputs.

In the late 1970s and early 1980s, the organic movement gained new impulses, primarily from people with a strong political background and an increasing concern for the natural environment, but with professional backgrounds other than farming. These newcomers to farming simultaneously faced the problems of conversion and the problems of learning about farming. At the same time, another group of farming families was motivated to convert from conventional to organic farming as they observed declining soil fertility and a link between human or animal diseases and the use of pesticides. Social conflicts, between generations within families over the future development of the farm, and with other farmers practising intensive, conventional farming, made the process of conversion more difficult for these people.

Since the mid-1980s, two main types of farming families have been converting to organic agriculture. In both groups, ecological consciousness is a key factor. The first group consists of farmers who are well organized and innovative in farm development. Business and marketing strategies are of central importance. The second group consists of farmers whose farms were in financial difficulties because the area of cropping land available, or the condition of farm buildings and machinery, did not allow the continuation of the kind of management they had previously practised. In most cases, cost–benefit analysis showed that a completely new farming strategy was needed for the survival of the farm, and that the costs of conversion to organic agriculture were often lower than the costs for replacement investments and renovations under conventional management (Freyer, 1991).

The conversion period is now defined by organic production standards. It usually lasts for at least two years. In the first year, the so-called zero year, products have to be sold with no specific certification of the conversion status and mostly at conventional prices. This renders conversion more difficult. In the following two years, the products can currently be labelled as 'in conversion' and might achieve a premium in some cases. This is currently only possible on a temporary basis under European Community legislation (EC, 1991), although this exemption looks likely to be extended from 1994 for an indefinite period.

Conversion to organic farming requires considerable efforts to organize the marketing of one's own products in order to benefit from the premium prices which are available. Although marketing is often seen as central to reducing the risks of conversion, there are other ways of increasing or maintaining incomes such as avoiding losses in crop or livestock production (Freyer, 1991). But these are often ignored. Whole-farm conversion planning, involving both physical and financial analyses, is seldom conducted even today.

Major Processes and Changes during Conversion

Motivation and aims for conversion

Negative experiences with conventional farming are often the reason for taking up organic farming. In a survey of 63 farmers in Germany (Vogtmann *et al.*, 1993), the following difficulties were mentioned: decline in farm income (32%); soil exhaustion and soil erosion (14%); increase in animal diseases (32%); increase in pesticide costs and aversion to the use of pesticides (19%). These sorts of difficulties preceding conversion meant that the farmers had specific reasons to change their farm management.

Financial motives were often at the forefront for the case-study farms (Rantzau *et al.*, 1990; Freyer, 1991). However, a decline in farm income cannot be prevented by conversion to organic agriculture alone, because organic agriculture is not intended to be a solution for the structural problems of farming in general.

All the farmers surveyed also mentioned nature conservation and environmental protection. Bodenstedt (1988) reports that organic farmers do not seek to use domination of nature as a means to increase yields and profit, but they aim primarily to maintain and protect nature. Five of the case-study converters raised agricultural policy issues as a reason to convert. They were looking for a way to sustain a traditional agricultural structure based on small family farms. Other motives, such as making a contribution to enhancing the relationships between producers and consumers, or supporting the Third World, show that converting farmers deal with social and political questions in a comprehensive manner. In other words, the conversion is focused not only on the situation of the farm itself, but also on wider issues. The interaction with the social and political environment makes clear that conversion is experienced as an extremely complex and holistic process.

Conversion to organic farming is, as a rule, linked to the decision to continue in agriculture in the medium to long term. Farm investment decisions, which may have been postponed for several years,

Table 14.1. Preconditions for conversion to organic farming on 14 case-study farms, western Germany, 1986.

	Case number													
	Cropping farms							Dairy farms						
	1	2	5	8	9	12	14	3	4	6	7	10	11	13
Goal setting	–	+	–	–	+	+	+	+	+	+	–	–	0	+
Social situation														
Family	+	+	–	–	–	+	0	–	+	+	0	+	–	+
Qualifications	+	+	+	+	+	+	+	+	+	0	–	+	–	+
Preparation of the conversion	+	+	+	+	+	+	+	+	+	0	–	+	0	–
Contacts outside the village	0	+	0	+	+	+	+	+	0	+	0	+	+	+
Familiarity with the surroundings	+	–	+	–	+	+	–	+	+	+	+	+	–	+
Farm condition														
Buildings	–	+	+	–	+	+	–	–	+	+	–	0	–	+
Machinery	+	–	+	–	+	+	–	–	+	+	–	0	–	+
Soil	–	–	0	–	+	–	–	0	0	+	+	+	+	+
History														
Enterprise mix	–	0	–	0	–	0	0	+	+	+	+	+	+	–
Intensity before conversion	0	–	–	+	–	–	+	+	0	+	+	+	+	0
Need for change	–	–	–	–	–	+	–	+	+	+	+	+	–	–

Flexibility														
Financial reserves	–	–	+	–	+	+	–	–	+	+	–	–	–	+
Labour reserves	+	+	+	–	+	+	–	–	+	+	–	–	–	+
Availability of land to rent	0	0	+	–	0	0	+	+	+	–	+	–	+	+
Flexible use of buildings	–	+	+	+	+	+	+	0	0	0	+	+	+	+
Site														
Climate	–	0	0	–	+	+	0	–	0	+	+	–	+	0
Conditions for direct marketing	–	–	+	–	+	+	+	0	+	–	+	0	–	–

– = Difficult; + = favourable; 0 = neutral for the conversion.
Source: Adapted from Rantzau *et al.* (1990).

need to be considered as part of the future organization of the farm. That some of those fundamental investments (e.g. the building of new housing) cannot always be made during the conversion phase is clear from the 14 conversion case-study farms (Rantzau *et al.*, 1990).

Potential production and marketing goals for the future development of the farm generally only become more concrete as the conversion progresses, in line with the development of the farmer's own competence and experience. The farm adviser's task is to help plan the conversion process and to indicate potential development strategies along with their respective risks and opportunities.

Preconditions for conversion

In principle, every farm can be converted to organic farming. This statement is valid irrespective of the production system (Freyer, 1991). Occasionally, however, there are farm types and site-specific factors or social situations which may make conversion more difficult or may favour it (Table 14.1). The extent and the condition of the farm's resources (land, machinery, buildings and capital reserves), and the flexibility with which the family takes up innovations and integrates into the farming process, are significant factors which influence the conversion process.

Farms which are difficult to convert include:

- specialized, one-sided pig finishing or breeding farms, which, as a result of heavy investments in buildings, have to achieve high capital returns (Rantzau *et al.*, 1990);
- dairy farms with high on-farm labour requirements (e.g. case 4 in Freyer, 1991);
- small farms on poor soils (e.g. cases 1 and 7 in Freyer, 1991).

In contrast, the following may help to reduce the costs of conversion:

- changes in farm enterprise structure (e.g. the introduction of vegetable production and direct marketing);
- dairy farms with the capacity to extend the forage area (e.g. renting) and the possibility of obtaining a premium on milk.

In some cases, the conversion process does not always end with full organic certification (e.g. cases 6 and 7 in Rantzau *et al.*, 1990).

Social situation

It is not just the farm's structural and financial situation which can hinder or assist the conversion. Arguments within the family (e.g.

between generations over the future direction of the farm), the level of training, and the willingness to carry the risks of conversion, together have a decisive influence on the success of the conversion (Freyer, 1991).

Even though the acceptance of organic farming in the community has increased substantially (Rantzau *et al.*, 1990), the impact of local (village) social structures continues to represent a barrier to conversion for many farmers. Equally, the emphasis of agricultural training, with few exceptions, is still almost exclusively on conventional or integrated agriculture.

Crop production

Intensity before conversion

As a consequence of crop production intensity in the years preceding conversion, the initiation of conversion is very specific to the individual farm. The application of mineral fertilizers, herbicides and plant protection chemicals may either be reduced gradually two to three years before receipt of the conversion contract (so-called preparatory measures), or stopped in stages during the conversion period itself. The latter method was practised by two intensively managed farms studied by Rantzau *et al.* (1990).

Soil cultivation

No significant changes in soil cultivations were undertaken. A reduction in the ploughing depth was found generally. Exclusive reliance on rotary cultivators (an approach recommended by some organic farming advocates) on two farms was abandoned by the end of the conversion.

Weed control, plant protection and nutrition

With the exception of sugarbeet, all farms stopped using crop protection chemicals at the start of the conversion. In contrast to the problems with weed control often cited by producers, only on a few plots could weed levels be described as high. Typical weed problems were silky bent (*Apera spica venti*), couch (*Elymus repens*), cleavers (*Galium aparine*), and creeping thistle (*Cirsium arvense*). Recent research on farms in conversion in Switzerland (Freyer, 1994) shows that docks (*Rumex* spp.) and creeping thistle are most widely distributed, and considerable hand labour is needed for their control. However, increased weed pressure is seldom found on more than 10% of the farm area (see below).

Yield reductions due to diseases and pests were attributable to the occurrence of brown rust (*Puccinia recondita*) in wheat and rye, potato blight (*Phytophtora infestans*) and Colorado beetles (*Leptinotarsa decemlineata*) in potatoes, and chocolate spot (*Botrytis cinerea/fabae*) in field beans. Copper oxychloride was used against potato blight, and Spruzit (a *Pyrethrum* preparation) against the potato beetle larvae.

The use of purchased fertilizers was restricted to phosphorus and potassium on individual fields. The potassium applications were primarily on potatoes or on sandy soils. The phosphorus applications were based more on familiarity with previous practice or out of consideration for the views of the sceptical older generation, rather than on plant production or site-oriented criteria. The minimum requirements for the length of storage of on-farm manures were only fulfilled by one third of the farms during the conversion period. Spreading of the limited manure resources at the wrong time can lead to environmental pollution and yield losses. Advice is needed early and repeatedly in order to support the necessary changes.

Crop rotation

Three different approaches to the design of crop rotations were found (Fig. 14.1). The first strategy is that the farmer invests in soil fertility with a high proportion of legumes (red clover, lucerne). At the same time, the proportion of cash crops is reduced. The second method is to start conversion with a target proportion of legumes. The third way to design the rotation is to raise the proportion of cash crops at the beginning of conversion, mostly to the disadvantage of soil fertility. During conversion, the proportion of legumes is increased.

In a three-year survey of seven conversion farms (Freyer, 1991; see also Rantzau *et al.*, 1990) short-term changes in the actual crop rotation compared to the planned crop rotation were mainly observed on farms with a high proportion of cash crops including field vegetables (32% of all cases). This is primarily a reflection of the market situation. On dairy farms the forage supply is a factor which does not allow short-term changes in the crop rotation, and therefore only in 7% of all cases was the rotation adjusted. Other studies show that during conversion the yields tend to be lower than after (Dabbert, 1990).

The extreme weather conditions in both conversion years studied had a negative impact on crop yields. Yields of potatoes and field beans were further reduced due to the pest and disease problems referred to above. Other crops showed little or no difference in yields during the two conversion years and there was no consistent pattern between the farms which showed considerable yield variability. Permanent grassland and fodder crop yields were at a relatively high level in both years.

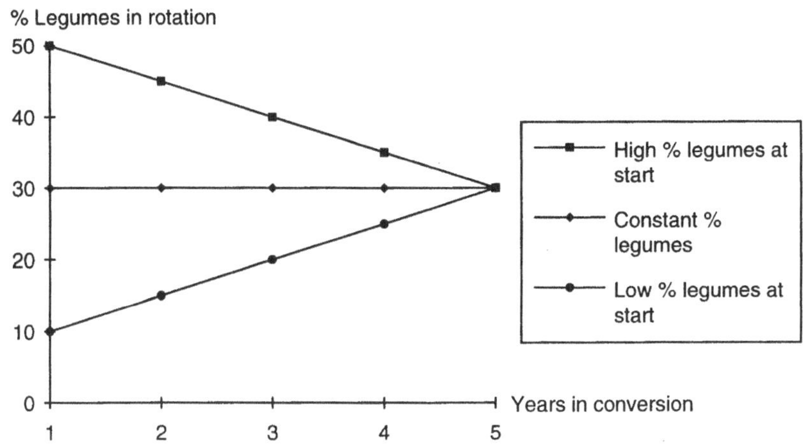

Fig. 14.1. Schematic analysis of differing approaches to the role of legumes in crop rotations during conversion to organic farming.

Yield reductions may be due to damage to soil structure or low soil fertility (Maire *et al.*, 1990). Soil fertility can only be improved by crop rotations with high proportions of legumes over several years, and through well-prepared organic compost. Another reason for lower yields during conversion may be problems with cultivation techniques (e.g. weed control). Weed problems can cause major difficulties during conversion. Weed control is best managed by observing rotational constraints and well-organized technical management. In this respect, an analysis of the rotation on individual fields in previous years also needs to be included.

Cropping patterns and total land area

While the proportion of cereals did not change significantly, there was a marked decrease in row crops and an increase in the area of legumes, undersowing and catch cropping. On forage-based livestock farms, this means that the area of cash crops is reduced in favour of grass/clover (lucerne) leys. On cropping farms, grain legumes were introduced. On 50% of the case-study farms, the land area increased following the start of the conversion. This is an indicator that conversion involves structural change, but it is also clear that the reduction in the cash crop area in favour of forage is compensated for by an increase in the size of the farm, to the extent that the availability of land for rent allows this.

Permanent grassland

On the majority of farms, grassland management and forage production were not given sufficient consideration. This is in part due to the subordinate role given to grassland management even before the conversion. The restricted application of organic manures as well as deficiencies in forage harvesting and conservation practices led to unsatisfactory forage quality and quantity as well as cattle performance. Appropriate advice in this area could help to improve performance during the conversion period.

Livestock production

Changes in livestock numbers

Because the production of feed grains such as barley, field beans or forage peas is less lucrative than the production of cash crops (winter wheat or rye), and because the purchase of concentrates is restricted by production standards, pig finishing or breeding sow numbers will be reduced during conversion. Also, only some farmers are able to realize a premium for meat. Income losses can be compensated for by the production of field vegetables or the intensification of direct marketing.

Generally, a reduction in the number of animals during conversion was observed, although not in all cases (Tables 14.2 and 14.3). If under conventional management income were obtained primarily from cash crops, and if animal numbers were small, the number of animals may increase during conversion, or a new species may be introduced.

Table 14.2. Stocking rates (livestock units per farm hectare) on seven cropping farms during conversion to organic farming, western Germany, 1986–1990.

	1	2	5	8	9	12	14
Initial farm size (ha)	32	22	31	39	32	56	25
Other main enterprises[a]	P, DM	DM		P, DM	FV, DM		FV, DM
Start (1986)	1.4	n/a	0.5	0.4	0.6	0.1	0.8
Year 1 (1987)	0.8	0.9	0.4	0.6	0.5	0.1	0.5
Year 2 (1988)	0.8	0.9	0.2	0.5	0.3	0.1	0.4
Target[b] (1990)	0.8	0.9	0.0	0.6	0.3	0.1	0.4

[a] In addition to cash crops: DM – direct marketing; FV – field vegetables; P – processing.
[b] Planned organic endpoint.
Source: Rantzau *et al.* (1990).

Table 14.3. Stocking rates (livestock units per farm hectare) on seven dairy farms during conversion to organic farming, western Germany, 1986–1990.

	3	4	6	7	10	11	13
Initial farm size (ha)	20	45	14	13	32	11	24
Other main enterprises[a]	DM	FV, DM	FV, DM	DM	DM	DM	
Start (1986)	0.8	1.2	2.4	1.5	1.1	1.0	2.0
Year 1 (1987)	0.8	1.1	2.0	1.1	1.1	0.8	1.0
Year 2 (1988)	0.8	1.0	2.1	0.9	1.0	0.7	1.0
Target[b] (1990)	0.8	0.9	1.6	0.7	0.9	1.1	1.2

[a] In addition to dairy cows: DM – direct marketing; FV – field vegetables.
[b] Planned organic endpoint.
Source: Rantzau *et al.* (1990).

Livestock nutrition

Changing the rotation and fertilizer management usually means less forage production with different quality. One typical fault in conversion is to underestimate the forage yields or the need for forage. In all cases except case 9, and in one year case 8, the area of forage production in the conversion period was lower than planned for the target organic system (Table 14.4). Purchased feed inputs were required on all dairy farms during the conversion. From a financial point of view the use of purchased conventional concentrates (within the limits allowed under the standards) is preferable to home production. However, the promotion of soil fertility requires that bulk forages (clover, lucerne) should

Table 14.4. Forage area (ha) per grazing livestock unit on seven dairy farms during conversion to organic farming, western Germany, 1986–1990.

	3	4	6	7	10	11	13
Initial farm size (ha)	20	45	14	13	32	11	24
Other main enterprises[a]	DM	FV, DM	FV, DM	DM	DM	DM	
Start (1986)	0.77	0.32	0.32	0.26	0.52	0.76	0.31
Year 1 (1987)	0.66	0.41	0.33	0.45	0.52	0.88	0.59
Year 2 (1988)	0.70	0.46	0.32	0.48	0.62	1.19	0.65
Target[b] (1990)	0.78	0.56	0.41	0.48	0.66	0.68	0.61

[a] In addition to dairy cows: DM – direct marketing; FV – field vegetables.
[b] Planned organic endpoint.
Source: Rantzau *et al.* (1989).

be produced completely on the farm. The central point of conversion planning should therefore be to coordinate available feed with feed requirements.

A large reduction in the proportion of concentrates fed can result in a decrease in milk production. The best feed quality and a high proportion of roughage can reduce this decrease. Generally, maintaining production at milk quota levels is desirable.

Labour use, financial performance and marketing

Labour use

The input of manpower will change during conversion as new production methods are introduced, as, for example, with mechanical or manual weed control, or compost preparation, which may require a higher input of labour. Additional issues include the integration of new crops such as potatoes and field vegetables or the local processing of milk and meat. The organization of market outlets is one of the time-consuming tasks. Public relations work is needed for the development of direct marketing during conversion. In summary, the increased labour required is a result of special production methods, but also of new production enterprises contributing to farm income.

On the 14 case-study farms, a variety of farm activities had an influence on the total labour force (Tables 14.5 and 14.6). Only in the case of one dairy farm (case 4) is the total labour force lower than on similar conventional farms. In this case, the crucial point is that the whole farm area is three times larger than, for example, in case 6. Usually, farmers

Table 14.5. Labour use (FTE[a] per 100 ha) on seven cropping farms during conversion to organic farming, western Germany, 1986–1990.

	1	2	5	8	9	12	14
Initial farm size (ha)	32	22	31	39	32	56	25
Other main enterprises[b]	P, DM	DM		P, DM	FV, DM		FV, DM
Start (1986)	4.7	n/a	3.2	3.7	8.7	2.1	8.8
Year 1 (1987)	4.7	6.7	2.7	4.0	8.7	1.9	7.7
Year 2 (1988)	5.1	5.8	2.3	4.2	8.4	1.9	9.2
Target[c] (1990)	5.1	5.7	2.0	5.3	8.1	2.0	6.8

[a] Full-time equivalent (2300 h $year^{-1}$)
[b] In addition to cash crops: DM – direct marketing; FV – field vegetables; P – processing.
[c] Planned organic endpoint.
Source: Adapted from Rantzau *et al.* (1989, 1990).

Table 14.6. Labour use (FTE[a] per 100 ha) on seven dairy farms during conversion to organic farming, western Germany, 1986–1990.

	3	4	6	7	10	11	13
Initial farm size (ha)	20	45	14	13	32	11	24
Other main enterprises[b]	DM	FV, DM	FV, DM	DM	DM	DM	
Start (1986)	6.5	4.0	12.7	10.9	6.5	8.7	6.1
Year 1 (1987)	5.8	4.1	11.2	8.9	7.1	6.0	5.5
Year 2 (1988)	4.8	4.0	9.9	7.4	7.1	5.7	4.8
Target[c] (1990)	6.4	4.0	0.7	6.7	7.0	8.5	5.3

[a] Full-time equivalent (2300 h year^{-1})
[b] In addition to dairy cows: DM – direct marketing; FV – field vegetables.
[c] Planned organic endpoint.
Source: Adapted from Rantzau *et al.* (1989, 1990).

employ trainees or seasonal labourers to compensate for the increased labour requirements. Paid regular labour was employed on only one farm. Unfortunately, the demand for trainees is currently much higher than the supply. The alternative, the employment of paid regular labour, means additional costs. Efficient work allocation, good work planning and technical equipment are therefore central features of successful farm management during conversion.

Financial performance

General statements about income trends during conversion are not possible because of the diversity of types of farming found on the case-study farms. Therefore, some examples of the main factors which determine changes in income during the process of conversion will be illustrated. In some farm enterprises there are typically opportunities to reduce costs. In other enterprises, an increase in costs at the start of the conversion process seems to be inherent in particular farming systems (Table 14.7). However, increased costs of processing and marketing need to be set against potential benefits in terms of increased prices.

Changes in fixed costs are especially difficult to analyse, because the decision to convert often also involves a decision to invest in parts of the farm. Typical investments included:

- machinery for weed control;
- equipment for dehumidification, cleaning and storage;
- machinery for new enterprises such as potatoes and field vegetables;

Table 14.7. Schematic analysis of changes in variable costs of crops and livestock enterprises as a consequence of conversion to organic farming.

Increased costs	Reduced costs
Crop production	
• Certified organic seed	• Reduced input of mineral fertilizer
• Dehumidification, cleaning, storage	• Reduced or zero input of pesticides
• Handling and packing for marketing	
• Labour costs for manual weed control, flame weeding	
• Bio-dynamic preparations	
Livestock production	
• Increased forage use per unit	• Prolonged productive life of dairy cattle
• Feeding own cereals (organic premium)	• Reduced forage production costs
• Costs for slaughtering, cutting and processing	• Reduced use of concentrates

- facilities for storage of manures and slurry;
- equipment for marketing;
- adaptation of animal housing to meet organic production standards;
- tractors and cars.

Not all of the investments can be assigned exclusively to conversion. On all of the farms, the largest investments were made in tractors (Freyer, 1991). These cannot be designated as specific to conversion. So, with the decision to convert to organic farming (and thus to stay in farming) the farmers invested in necessary replacement machinery and equipment. However, if during conversion higher prices for products cannot be realized, and in situations where high investment expenditures are necessary, investments in livestock husbandry might be postponed.

In addition, the requirement for leased land (forage crops) and hired labour often increases the fixed costs. Disinvestment in machinery, for example sprayers, can reduce fixed costs. Destocking and the sale of sugarbeet or milk quotas can be used for funding investments.

As Table 14.8 shows, the main income on the dairy farms during conversion came from milk production (Rantzau *et al.*, 1990). In this study the variable costs for forage production are allocated to crop production. The real difference between crop and livestock production therefore is much smaller than shown in Table 14.8. In contrast, farms with field vegetables or beef and pig finishing enterprises derived their main income from crop production (Table 14.9). On these farms,

Table 14.8. Crop/livestock gross margins as a percentage of total farm gross margin on seven cropping farms during conversion to organic farming, western Germany, 1986–1990.

	1	2	5	8	9	12	14
Initial farm size (ha)	32	22	31	39	32	56	25
Other main enterprises[a]	P, DM	DM		P, DM	FV, DM		FV, DM
Start (1986)[b]	50/50	n/a	84/16	21/79	18/82	98/2	60/40
Year 1 (1987)	43/47	50/50	84/16	69/31	84/16	97/3	70/30
Year 2 (1988)	44/56	70/30	91/9	58/42	95/5	98/2	80/20
Target[c] (1990)	50/50	64/36	100/0	66/34	85/15	98/2	60/40

[a] In addition to cash crops: DM – direct marketing; FV – field vegetables; P – processing.
[b] Farms 1, 2 and 5 sold at conventional prices.
[c] Planned organic endpoint.
Source: Rantzau *et al.* (1989).

Table 14.9. Crop/livestock gross margins as a percentage of total farm gross margin on seven dairy farms during conversion to organic farming, western Germany, 1986–1990.

	3	4	6	7	10	11	13
Initial farm size (ha)	20	45	14	13	32	11	24
Other main enterprises[a]	DM	FV, DM	FV, DM	DM	DM	DM	
Start (1986[b]	15/85	25/75	4/96	32/68	17/83	30/70	29/71
Year 1 (1987)	20/80	33/67	18/82	55/45	33/67	20/80	30/70
Year 2 (1988)	37/63	29/71	23/77	62/38	29/71	12/88	13/87
Target[c] (1990)	34/66	39/61	32/68	65/35	27/73	22/78	23/77

[a] In addition to dairy cows: DM – direct marketing; FV – field vegetables.
[b] Farms 6, 7 and 13 sold at conventional prices. Farms 10 and 11 sold some produce at organic prices.
[c] Planned organic endpoint.
Source: Rantzau *et al.* (1989).

premium prices can be obtained more easily for vegetables and cereals than for meat.

On all case-study farms, farm profits increased during the conversion (Figs 14.2 and 14.3). However, on many farms a generation change took place, which generally worked out positively for the farm development. On the other hand, five cases involved the establishment of new businesses, which took place in the period shortly before

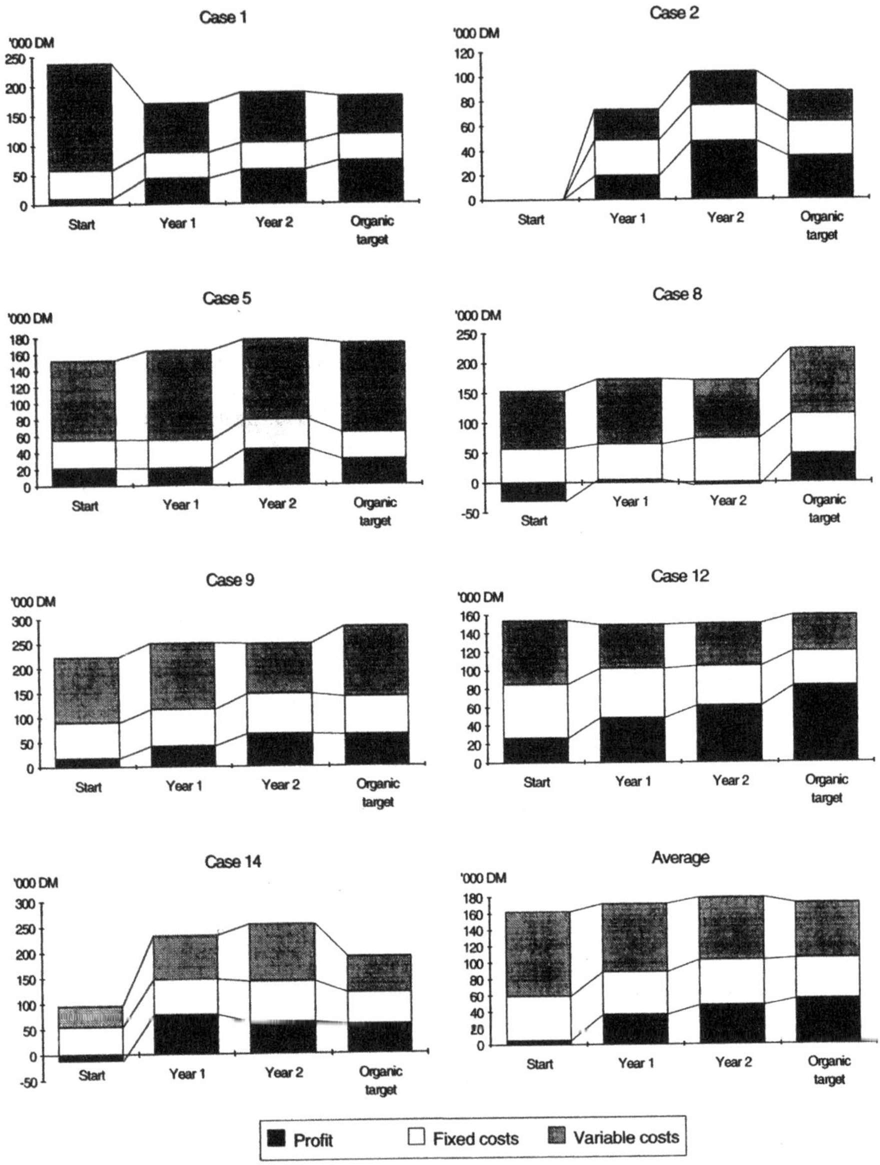

Fig. 14.2. Allocation of total output to variable costs, fixed costs and profit on seven cropping farms during conversion to organic farming, western Germany, 1986–1990. Except where profit is negative, the whole column represents total output (profit plus fixed and variable costs), and the top of the white section represents the gross margin (total output less variable costs). Source: Adapted from Rantzau *et al.* (1989, 1990).

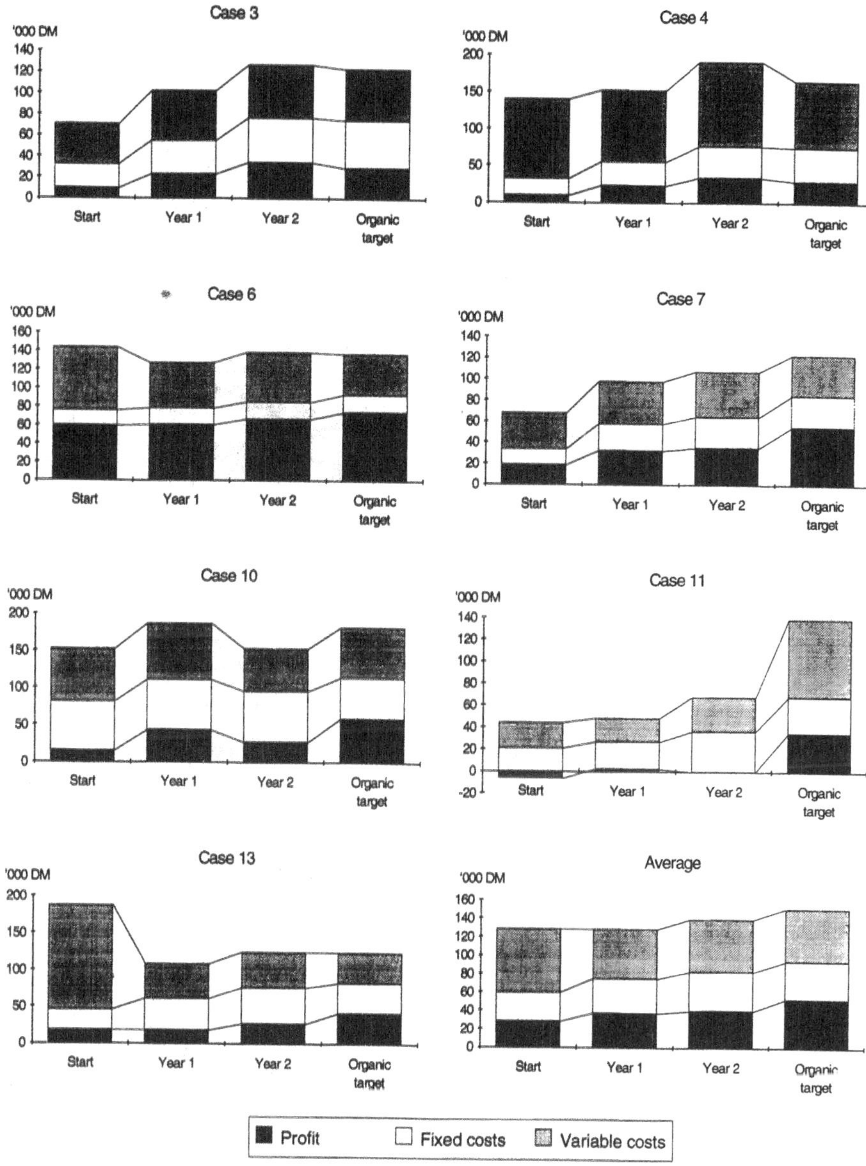

Fig. 14.3. Allocation of total output to variable costs, fixed costs and profit on seven dairy farms during conversion to organic farming, western Germany, 1986–1990. Except where profit is negative, the whole column represents total output (profit plus fixed and variable costs), and the top of the white section represents the gross margin (total output less variable costs). Source: Adapted from Rantzau *et al.* (1989, 1990).

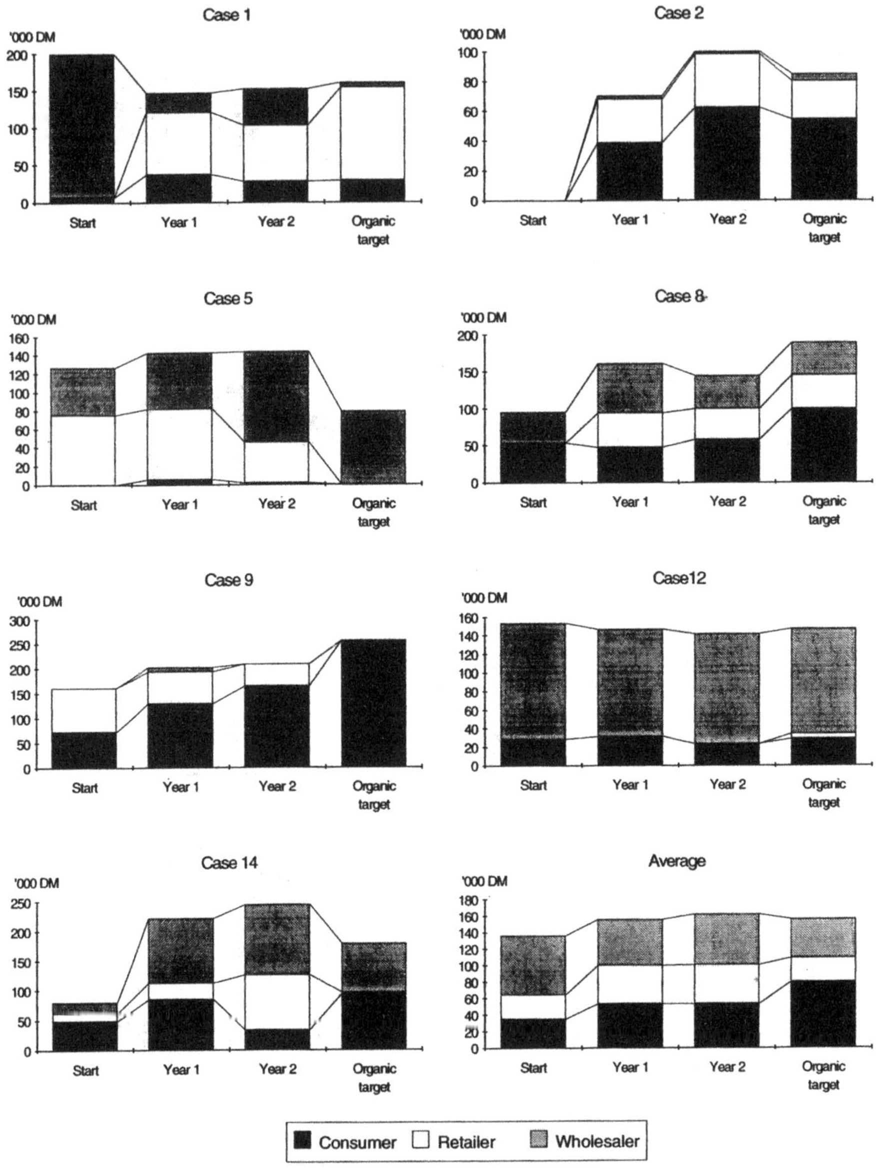

Fig. 14.4. Changes in market outlets used by seven cropping farms during conversion to organic farming, western Germany, 1986–1990. Source: Adapted from Rantzau *et al.* (1990).

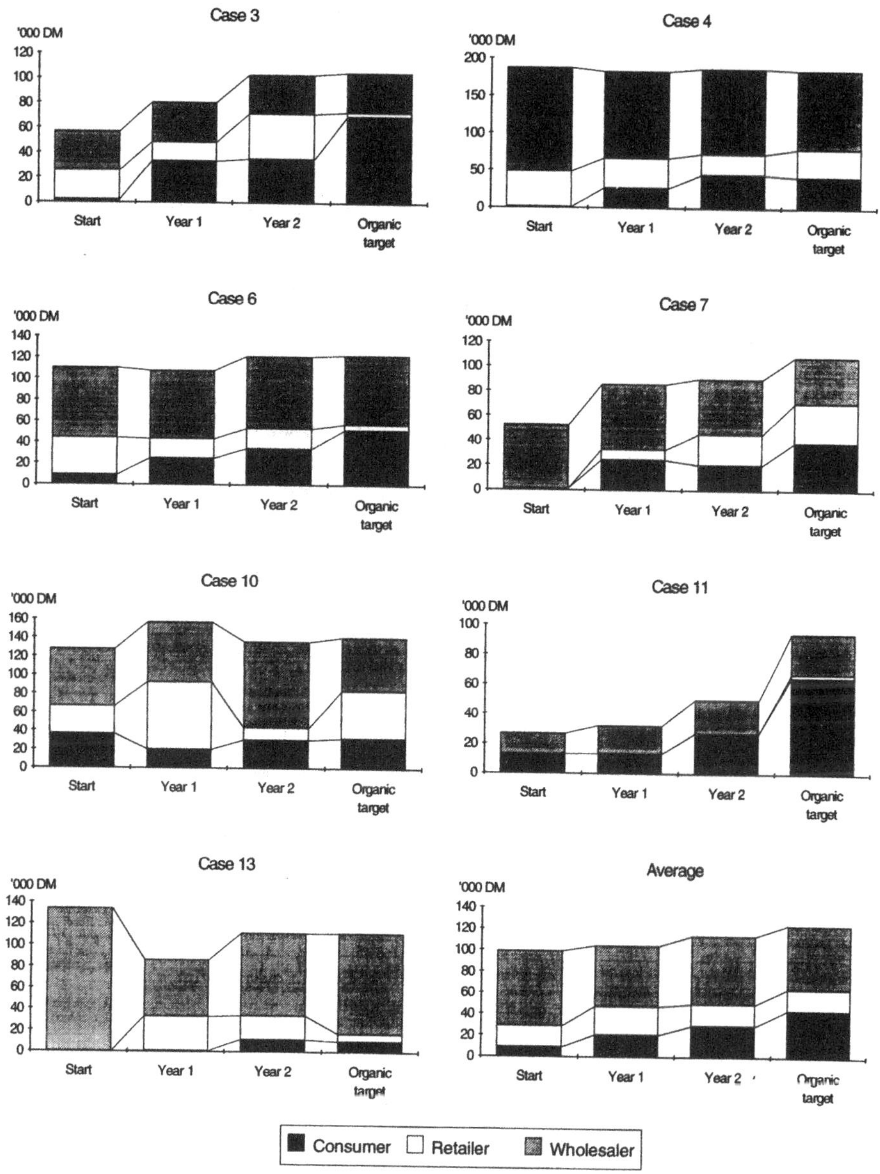

Fig. 14.5. Changes in market outlets used by seven dairy farms during conversion to organic farming, western Germany, 1986–1990. Source: Adapted from Rantzau *et al.* (1990).

conversion. For farmers converting in Germany, these starting points are very typical.

Marketing

On all farms, there was a gradual increase in the quantity and range of products which could be marketed with an organic premium. While the marketing of crop products through alternative channels was possible already in the first year of conversion, only one farm (case 13) succeeded in marketing milk as an organic product, although no premium was obtained. The use of different market outlets, for example consumers, retailers, or wholesalers (Figs 14.4 and 14.5), and the prices realized in each case varied considerably from farm to farm. These are determined primarily not only by locational advantages or disadvantages, but also by the ability of the farm managers and the available labour for the organization of a successful marketing operation.

Conclusions

The financial performance of the case-study farms reported in this chapter is only of limited relevance to other farms now converting to organic farming in Germany. The financial situation during conversion has changed in recent years. Before 1990, income was derived primarily from the marketplace. Now German organic farmers and farmers converting receive a considerable level of subsidy support through the European Union's extensification and agri-environment programmes (see Chapters 25 and 26). Premium prices have fallen in response to the large increase in the number of organic farms. For small farms in western Germany, the new and much larger eastern German farms mean keen competition. Increased imports from France or central European countries such as Hungary and the Czech Republic will aggravate the situation, in particular for the smaller farms during conversion. The larger farms in eastern Germany are, however, in a difficult phase of development because of their high investment costs independent of conversion, although the need to invest in major restructuring anyway may make it easier to take the decision to convert.

Even with the availability of the new subsidy payments, careful planning is essential to avoid financial problems during conversion. This necessitates a careful analysis of the whole farm including the capacities of labour and the need for further training. Successful conversion thus demands the ability to organize and manage a new kind of farm. Planning the conversion process is one of the important measures for personal preparation. It helps to avoid mistakes and unexpectedly high workloads, or financial bottlenecks. Conversion planning also needs to take into account environmental protection and nature conservation factors.

References

Benecke, J., Kiesewetter, B. and Urbauer, H. (1988) *Bauern stellen um.* Alternative Konzepte, 62. Verlag C. F. Müller, Karlsruhe.

Bodenstedt, A. (1988) Ökologischer Landbau als gesellschaftliche Alternative. In: Gödde, H. and Voegelin, D. (eds) *Für eine bäuerliche Landwirtschaft, Materialien zur Tagung in Bielefeld – Bethel, 1988.* Band 14, Schriftenreihe des Fachbereichs Stadtplanung/Landschaftsplanung der Gesamthochschule Kassel, Kassel.

Dabbert, S. (1990) *Zur optimalen Organisation alternativer landwirtschaftlicher Betriebe – Untersucht am Beispiel organisch-biologischer Haupterwerbsbetriebe in Baden-Württemberg.* Verlag Alfred Strothe, Frankfurt am Main.

EC (1991) Council Regulation (EEC) No. 2092/91 of 24 June 1991 on organic production of agricultural products and indications referring thereto on agricultural products and foodstuffs. *Official Journal of the European Communities* 91 (L198), 1–15.

Fischer, R. (1982) *Der andere Landbau.* Verlag Buchhandlung Madlinger-Schwab, Zürich.

Freyer, B. (1991) *Ökologischer Landbau. Planung und Analyse von Betriebsumstellungen.* Verlag Margraf, Weikersheim.

Freyer, B. (1994) Ausgewählte Prozesse in der Phase der Umstellung auf den ökologischen Landbau am Beispiel von sieben Fallstudien. *Berichte über Landwirtschaft* 72, 366–390.

Hotzelt, A. (1985) *Der ökologische Landbau in Mittelfranken – Ablauf, Gunst- und Hemmfaktoren einer Innovation.* Zulassungsarbeit am Institut für Geographie der Universität Nürnberg.

Maire, N., Besson, J.M., Suter, H., Hasinger, G. and Palasthy, A. (1990) *La conversion des domaines agricoles en mode biologique: effect sur l'équilibre physico-chimique et biologique des sols.* Report 43. National Forschungsprogramm 'Boden', Schweizerischer Nationalfond, Liebefeld-Bern.

Rantzau, R., Freyer, B. and Vogtmann, H. (1989) *Betriebliche Erfordernisse und Konsequenzen bei der Durchführung des biologischen Landbaus.* Endbericht im Auftrag des Bundesministers für Ernährung, Landwirtschaft und Forsten. Fachbereich Landwirtschaft, Gesamthochschule Kassel, Witzenhausen.

Rantzau, R., Freyer, B. and Vogtmann, H. (1990) *Umstellung auf ökologischen Landbau. Betriebliche Erfordernisse und Konsequenzen bei der Durchführung des biologischen Landbaus.* Angewandte Wissenschaft, Schriftenreihe des Bundesministers für Ernährung, Landwirtschaft und Forsten, Heft 389, Landwirtschaftsverlag, Münster Hiltrup.

Vogtmann, H., Freyer, B. and Rantzau, R. (1993) Conversion to low external input farming systems: a survey of 63 mixed farms in West Germany. In: Napier, T., Ferro, O. and Paoletti, M.G. (eds) *Socio-economic and Policy Issues for Sustainable Farming Systems.* Coop Amicizia Pubbl. Padua.

15 Conversion to Low-input Farming Systems in Pennsylvania, USA: An Evaluation of the Rodale Farming Systems Trial and Related Economic Studies

STEVEN E. PETERS

Introduction

An alternative farming system may be more profitable than the conventional system it is replacing once it has become established, but the costs incurred during the transition can be considerable. The farmer must adjust to different labour, management, equipment and marketing requirements, and the crop plants must adapt to a changing soil environment. The conditions can result in reductions in crop yields and net profits during the transition period. This could mean increased interest payments on production loans and a longer period before the alternative system becomes economically competitive. Many alternative practices may be cost-effective, but to what extent does the reluctance of lending institutions to invest in non-traditional approaches make them less attractive and riskier to farmers?

Our society's emphasis on short-term profits obscures many of an agro-ecosystem's most valuable assets, including biological resilience, self-regulation, diversity, sustained production, and improvement of the natural resource base. Most economic studies focus on short-term profitability, but to assess the long-term sustainability of a farming system, changes in soil health and quality must be taken into account. Is the soil being regenerated, maintained or degraded? Sustainability is also reflected in the variability of farm profits from one year to the next, which indicates the degree of risk involved. The extent to which farming practices affect the quality of food and feed produced on the farm, and the quality of ground and surface waters within a farm's watershed must also be considered to evaluate the broader costs to society.

The socioeconomic and political environment in which a farm operates has a great influence on the economic viability of a farm. Many of these forces beyond the immediate control of the farmer impose barriers to positive change, such as the current US government commodity programme, which is biased against crop diversity and soil-building crops. To what extent would removing the support (or constraint) of the government programmes change the economics of particular farming systems?

Assessing the real value of a farming operation is therefore a complex task that can be viewed in several ways. This chapter presents a review of the Rodale Farming Systems Trial and a number of economic studies based on the results of the trial. Each of these studies adopts a somewhat different analytical approach. Taken together, they can be used to present a broader and more complete economic picture of alternative and conventional farming practices and the transition process.

Trial Description

Objectives and background

In 1981, a study known as the Conversion Project was initiated on a site adjacent to the Rodale Institute Research Center to examine the process of converting from a conventionally managed to an organically managed farming system (Culik and Liebhardt, 1984). The main objective of the trial was to deal with some practical management questions that farmers changing from one system to another are confronted with, such as identifying the factors that limit crop growth and yield during transition, finding cultural techniques and suitable crops to minimize the transition effect, and determining the net cash returns and other economic constraints (e.g. labour availability and seasonal cash flow during conversion to organic practices). From 1986, the objectives of the study were modified and the study became known as the Farming Systems Trial (FST), since it was assumed that the transition phase (at least biologically) was completed. Emphasis shifted towards evaluating the long-term reliability of organic/low-input techniques and their impact on soil ecology (Peters, 1991; Peters *et al.*, 1992).

Site description

The FST experiment is located in the Great Valley section of the Ridge and Valley province in southeastern Pennsylvania, USA. The gently rolling hills, reasonably fertile soils, and favourable climate in southeastern Pennsylvania have resulted in some of the most productive and profitable agricultural systems in the USA. The area has been farmed continuously since the original hardwood forest was cleared, approximately 250 years ago, to produce pasture, hay, maize and cereals. The dominant farming enterprises involve raising dairy and beef cattle, hogs and poultry. The major crops grown for feed include maize grain and silage, alfalfa and red clover hay, wheat, oats and soyabeans. All of these crops (except maize silage) are also sold off the farm as cash crops.

The 5 ha site is on a gently rolling, southwest facing, 1–5% slope. The silty/clay loam and shaly/silty/clay loam soils are moderately well drained, but in the spring may have a perched water table in low lying areas. Rainfall is about 1045 mm annually, well distributed throughout the year, with about 50% occurring between 1 May and 30 September. Average annual temperature is 12.4°C; summers are hot and humid and winters relatively short and mild. The growing season consists of 180 frost-free days and 3000 growing degree days (based on 10°C).

Management practices

Three farming systems, based on either three- or five-year crop rotations (Fig. 15.1), were established on the site, which had produced maize nearly continuously for at least 20 years with chemical fertilizers and pesticides. Every year three of the rotational entry points from each system were represented in the field for a total of nine distinct treatments. Each of these nine treatments were replicated eight times in a split-plot, randomized block design. The three systems were:

1. A low-input/animal system (LIP-A), which simulated a beef or dairy operation that produced red clover/alfalfa/orchard-grass hay, oats, winter wheat, maize grain and silage, and soyabeans. Nitrogen was provided by cattle manure obtained from the neighbouring Kutztown farm and from third-year hay crops ploughed in prior to the planting of maize.
2. A low-input/cash grain system (LIP-CG), which did not contain an animal component and produced a cash grain every year. This system raised maize (grain only), soyabeans, oats, winter wheat, and spring barley. Ploughed-in legumes were the primary nitrogen source.
3. A conventional/cash grain system (CONV) produced maize and

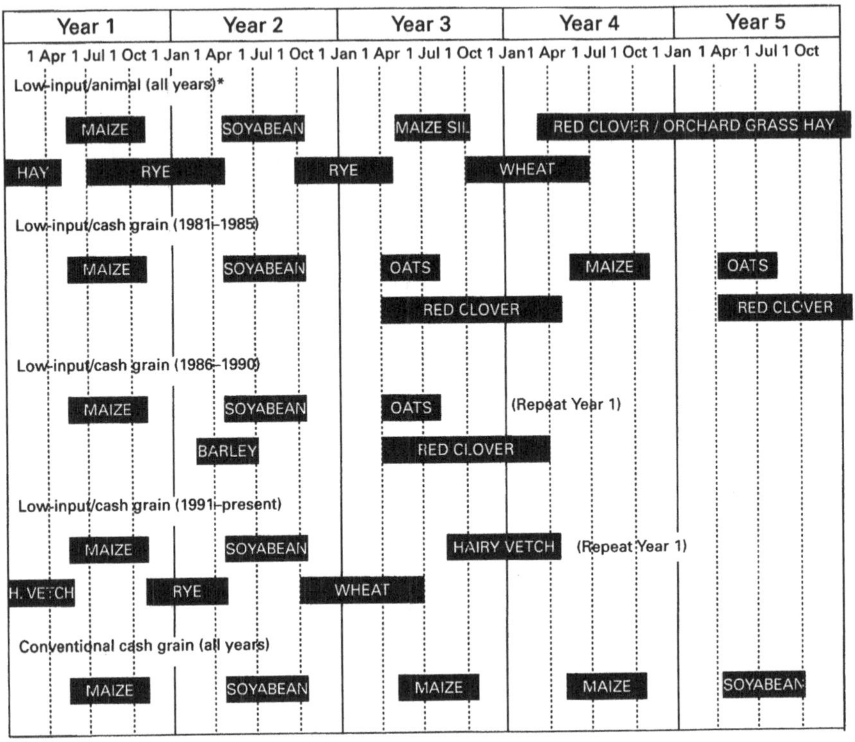

Fig. 15.1. Rotation design, Rodale Farming Systems Trial, Pennsylvania, 1981–1992. *With the exception of a rye cover crop added to the rotation in 1991.

soyabeans using Pennsylvania State University recommendations for commercial fertilizer and pesticide inputs.

Both low-input systems avoided herbicide use, and instead employed rotary hoeing, row cultivation, crop rotation, green manuring, and relay cropping for their weed control. Potassium sulfate and high-calcium limestone were applied to both low-input systems in 1989.

Conventional tillage (mouldboard plough, discs, field cultivator, cultipacker) was used in all three farming systems. The two low-input systems, except for the potassium sulfate application, would be considered 'organic' under most certification programmes. The term 'low-input' is retained to emphasize the reduction of purchased inputs. These low-input systems, however, also rely on more farm-generated resources including a greater range of management skills than does the conventional system.

Agronomic Results

During the first few years of the study, maize grain yields were depressed in the low-input systems relative to the conventional. After four or five years, however, maize grain yields in all systems became more similar (Fig. 15.2 and Table 15.1). From 1981–1984, LIP-A and LIP-CG grain yields averaged 77% and 67% of CONV, respectively. From 1985 to 1992, LIP-A yields were 98% and LIP-CG yields were 95% of CONV. Shifting from chemically-based to more biologically-based crop production practices resulted in a period of biological adjustment when nutrient cycling dynamics and/or other processes were changing and influencing plant productivity.

This 'transition effect' may largely have been caused by a lack of plant-available nitrogen (N), i.e. N-immobilization by microorganisms in the low-input systems during the conversion period and a subsequent release of mineral N after a few years. Maize ear leaf N levels at silking substantiate this conclusion. Average ear leaf N concentration from 1981 to 1984 was 2.73% in the CONV and only 2.43% in both LIP systems. After 1984, ear leaf N levels were nearly the same in all systems: LIP-A 2.76%, LIP-CG 2.74%, and CONV 2.79%.

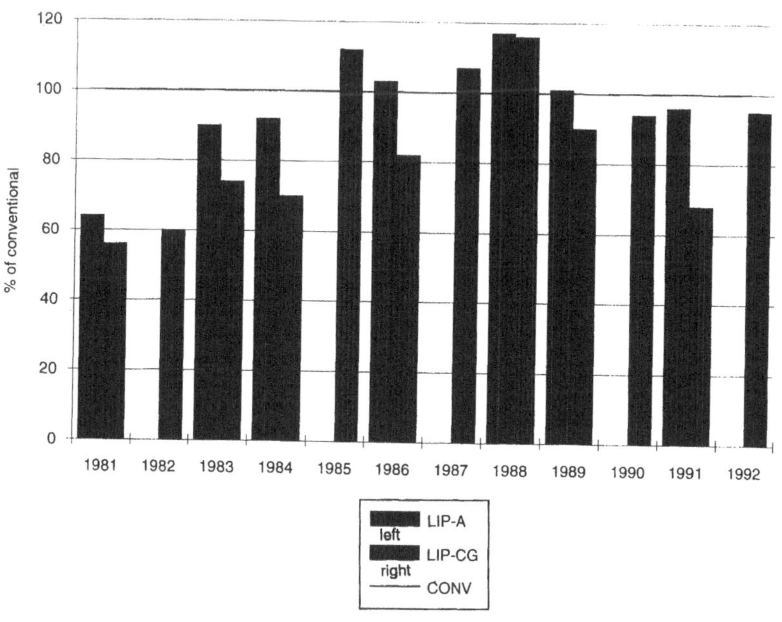

Fig. 15.2. Low-input maize grain yields as a percentage of conventional, Rodale Farming Systems Trial, Pennsylvania, 1981–1992.

Table 15.1. Maize grain yields[a] ($t\ ha^{-1}$), Rodale Farming Systems Trial, Pennsylvania, 1981–1992.

System	1981	1982	1983	1984	1985	1986	1987	1988	1989	1990	1991	1992
LIP-A	1.56y	–	5.52xy	8.70y	–	10.78x	–	6.89x	7.81xy	–	8.21x	–
LIP-CG	1.37y	5.47y	4.60y	6.69z	9.57x	8.62y	9.03x	6.84x	6.95z	9.35x	5.76y	9.66y
CONV[b,d]	2.38x	9.55x	5.92x	9.33x	–	10.51x	8.18x	5.33y	7.25yz	–	8.60x	9.60y
CONV[c,d]	2.54x	8.59x	6.35x	9.69x	8.54x	10.42x	8.71x	6.51x	8.19x	9.90x	8.16x	10.65x

[a] Yields adjusted to 15.5% moisture; values within the same year followed by the same letter are not significantly different from one another at the 5% level.
[b] Maize following maize.
[c] Maize following soyabeans.
[d] Both CONV maize followed a wheat/winter fallow in 1981; and soyabeans in 1986 and 1991.
Source: Peters *et al.* (1992).

Excessive weed growth in the LIP systems during the early years has also been identified as a cause of depressed yields (Liebhardt *et al.*, 1989). However, the data do not always support this conclusion. Annual average weed biomass levels from 1985 to 1992 decreased by 59% in the LIP-CG compared to the 1981–1984 period, but increased in the LIP-A by 54%. The application of relatively fresh beef manure in the LIP-A was probably adding weed seeds and causing a stimulation of weed growth, but crop growth and yield were also increasing. By 1985 the LIP-A soil environment was capable of supporting greater crop and weed biomass. With little plant-available N present during the transition period in the LIP systems, a relatively small amount of weeds was very likely competing significantly with the maize crop, but the importance of this weed competition and the weed–nitrogen interaction is unclear.

Although both low-input systems experienced a transition, they responded differently. By the third year (1983), the animal-based system (LIP-A) already had a maize yield of 90% of conventional (CONV). All subsequent LIP-A maize yields (1984–1992) ranged from 92% to 117% of CONV. In contrast, the transition period for the maize in the legume-based system (LIP-CG) was longer, with yields of 56–74% of CONV from 1981–1984. In 1985, LIP-CG maize followed an excellent cover crop of hairy vetch and the yield exceeded CONV. A fair assessment cannot be made in 1986 because a shorter-season variety of maize with lower yield potential was grown in LIP-CG than in the other systems, resulting in reduced yields. The other anomaly was in 1991 when LIP-CG maize yielded only 68% of CONV. This occurred because the crop rotation sequence was changed that year and a legume cover crop could not be established. The result was N-deficient maize with poor grain fill. In 1992, following an adequate legume cover crop, yields recovered to former levels.

Total above-ground maize dry matter yields for both low-input systems averaged about 75% of CONV from 1981 to 1984, and both became more similar to CONV after 1984 (Table 15.2). The low-input systems differed markedly, however, in that overall (1981–1992) dry matter yields were 7% higher in LIP-CG, while grain yields were 5% higher in LIP-A. Therefore the harvest index (grain to dry matter yield ratio) was higher in LIP-A (0.46) than in LIP-CG (0.41) and similar to CONV (0.45).

Although it was evident that maize growth and yield in the LIP systems was affected by a transition phenomenon, the weather conditions during the early years of the experiment partly caused this result. In 1981 and 1983 there were extended periods of drought during the growing season which severely reduced plant growth. After 1985, the only significant drought year was in 1988. Relative to 1981–1984, maize

Table 15.2. Maize above-ground dry matter yields[a] (t ha^{-1}), Rodale Farming Systems Trial, Pennsylvania, 1981–1992.

System	1981	1982	1983	1984	1985	1986	1987	1988	1989	1990	1991	1992
LIP-A[b]	9.71z	–	7.95z	14.72y	–	18.22x	–	10.40yz	12.94x	–	16.23x	–
LIP-CG	11.51y	9.69y	9.24y	13.72y	17.98x	16.55y	15.90x	12.85x	12.76x	18.18x	13.40x	18.59x
CONV[c]	13.51x	13.95x	13.07x	17.57x	–	18.90x	14.11y	9.39z	12.51x	–	14.67x	17.59x
CONV[d]	14.08x	10.90y	13.65x	17.60x	16.16y	18.74x	17.11x	11.46xy	14.42x	17.80x	15.46x	18.14x

[a] Values within the same year followed by the same letter are not significantly different from one another at the 5% level.
[b] Grain maize only (i.e. maize grown for silage not included).
[c] Maize following maize.
[d] Maize following soyabean.
Source: Peters *et al.* (1992).

Table 15.3. Soyabean seed yields[a] (t ha^{-1}), Rodale Farming Systems Trial, Pennsylvania, 1981–1992.

System	1981	1982	1983	1984	1985	1986	1987	1988	1989	1990	1991	1992
LIP-A	–	3.29x	–	3.56x	3.42xy	–	2.92y	–	2.91x	3.18x	–	3.13x
LIP-CG[b]	1.78x	3.31x	–	–	3.58x	3.20x	2.95y	–	2.22y	1.93z	3.38x	3.18x
CONV	1.78x	2.87y	2.83	3.53x	3.31y	2.79y	3.54x	3.26	3.07x	2.72y	3.37x	3.18x

[a]Yields adjusted to 13% moisture; values within the same year followed by the same letter are not significantly different from one another at the 5% level.

[b]Soyabeans from 1986 to 1990 were relay-cropped with spring barley or winter wheat.

Source: Peters *et al.* (1992).

grain yields from 1985 to 1992 were 40% and 81% higher for LIP-A and LIP-CG, respectively, but yields were also 27% higher in CONV. Therefore, not all of the transition effect can be attributed to management factors such as changes in fertilizer and weed control inputs, and climatic conditions must also be considered.

Soyabean yields were similar in all three systems throughout the study, and thus seemed to be relatively unaffected by conversion to organic practices (Table 15.3). Soyabeans require much less soil and fertilizer-derived nitrogen and withstand drought conditions much better than maize. In addition, soyabean yields are much less affected by reductions in plant population. LIP-CG soyabean yields were reduced in 1987, 1989 and 1990 because these soyabeans were relay-cropped with barley or wheat. Combining the LIP-CG soyabean and cereal yields, however, resulted in higher overall yields than CONV. LIP-A soyabean yields were equal or slightly higher than CONV in all years except in 1987. That year heavy rains in early summer led to excessive weed growth and competition, but soyabean yields were still very respectable.

Hay and cereals were only grown in the low-input systems so a comparison with conventional management could not be made for these crops. Yields of hay and cereals were similar to county average yields during the transition period. Oats and wheat yields, however, increased 40–60% after 1984 indicating that these crops may have also been adversely affected by the transition.

Economic Analysis

Farming systems trial: The transition phase

The first formal economic analysis performed on the Rodale Farming Systems Trial dealt with the early (transition) years of the trial, i.e. 1981–1986 (Duffy *et al.*, 1989). The primary objectives were to estimate the effect of the starting crop on returns over a complete five-year rotation and to compare the returns from the best rotation in each of the three farming systems. The hypothesis being tested was that starting the alternative (low-input or organic) rotations with cereals would provide higher returns than starting with maize or soyabeans. This should be true because cereals require fewer nutrients and the high-density plantings inhibit weed growth without fertilizers, and also because farmers have historically used less chemicals on cereals. Furthermore, the cereals in the study were established with legumes.

A partial budgeting analysis was conducted, and all crops (even forage crops such as silage) were assumed to be commercially marketed

Table 15.4. Returns to land and management by farming system and rotation entry point, Rodale Farming Systems Trial, Pennsylvania, 1981–1986.

		Annual net profit (US\$ ha^{-1})	
Farming system	Rotation entry point	1981–1985	1982–1986[a]
LIP-A	Maize	346	563
LIP-A	Maize silage	474	494
LIP-A	Oats/red clover hay	519	610
LIP-CG	Maize	272	389
LIP-CG	Oats/red clover hay	408	316
LIP-CG	Soyabeans	296	415
CONV	Maize	388	553
CONV	Soyabeans	420	415
CONV	Maize	346	494

[a]1986 data substituted for 1981 to provide complete rotation cycle because extremely dry weather and start-up problems in 1981 resulted in very poor maize yields in all systems.
Source: Duffy *et al.* (1989).

or given an equivalent cash value. Revenues were based on US Department of Agriculture (USDA) values for market prices. Operating expenses were determined using Pennsylvania custom rate surveys, included charges for fixed and variable machinery costs, labour costs and a modest profit margin for the operator. These costs were probably assigned a higher value than on a family farm using family labour. Material expenses were based on the actual prices paid by the Rodale Institute Research Center. Manure was charged on application cost only. A land charge was excluded from the analysis since the type of farming system used would not influence land charges in the short term. The returns to land and management (net profit = revenues minus expenses) are shown in Table 15.4.

Duffy *et al.* (1989) concluded that the starting crop in the conversion from conventional to organic farming had a significant influence on returns to land and management, with the most profitable rotation over the five years (1981–1985) starting with oats/red clover in the case of both low-input systems and soyabeans in the conventional system. Where the rotation has to begin with maize, supplemental fertility and weed management must be provided in the initial years of conversion or loss of potential return will result. Animal manure can effectively replace fertilizer to supply fertility needs during the transition phase and the ability to use forages to feed animals in the LIP-A system can provide returns from crops with limited cash market potential. When the most profitable rotation over the five years 1982 to 1986 from each system was compared, CONV and LIP-A had equal returns and LIP-CG had lower returns.

Whole-farm models: Grain production

A whole-farm analysis of a representative grain farm in the Mid-Atlantic region was conducted using the FST data for the nine-year period from 1981 to 1989 (Hanson *et al.*, 1990). The primary objective of this study was to analyse the profitability, liquidity, solvency and risk of a conventional cash grain and a low-input cash grain farm. The two systems were assessed with and without participation in government commodity programmes.

The cultural practices and crop yields from the FST and input/output prices from central Maryland were used to develop crop budgets for a representative grain farm. The low-input and conventional farms started with the same resources, which included: 300 ha tillable cropland; farmer-owned tillage, planting, cultivating and spray equipment; custom-hired harvesting and fertilizer/lime applications; 250 hours of operator and family labour per month; sufficient building space for machinery and hay storage; no on-farm grain storage; a 22% intermediate and long-term debt-to-asset ratio; and 15 years remaining on the farm mortgage.

Assets were valued at the original cost minus accumulated depreciation. New machinery was purchased at 15% of the current year's opening asset value. Building improvements were made at 6% of the current year's opening asset value. Labour costs were calculated using standard farm management conversion rates that estimated the time required for each field operation. An indirect time requirement of 25% was added to each operation. If the farm required more than 250 hours per month, then help was hired.

The main revenues and expenses are summarized in Table 15.5. Gross income was considerably less in the low-input system because of lower maize and soyabean acreages, and lower yields during the transition. The low-input system net income, however, was only slightly less than the conventional system because cash operating expenses, primarily fertilizers and pesticides, were greatly reduced. As a result, cash expenses were only 72% of total income in the low-input system versus 81% in the conventional.

Hired labour costs were greater in the low-input system because cultivation and haying required more labour, but labour was distributed more evenly throughout the year. Machinery purchases, depreciation and maintenance costs were also higher in the low-input

Table 15.5. Annual income and expenses (US$ per farm) for a representative grain farm under low-input and conventional scenarios,[a] Mid-Atlantic region, 1981–1989.

	Low input	Conventional	Relative (conv=100)
Gross income	140,765	174,433	81
Cash operating expenses	99,715	133,922	74
Depreciation	13,436	10,620	127
Net income	27,614	29,891	92
Fertilizer/pesticides	1,875	37,371	5
Hired labour costs	3,096	2,426	128
Machinery purchases	17,770	11,719	152
Machinery fuel/maintenance	12,862	10,081	128
Interest (excluding land)	10,745	9,625	112
Interest on land	25,200	25,200	100

[a] Excluding government programmes.
Source: Hanson *et al.* (1990).

system because of haying equipment. A farmer who already owned haying equipment could make an easier transition. Interest payments (excluding land mortgage interest) were higher in the low-input system because lower profits during transition meant that more interest was paid on carry-over of current liabilities. Less interest was paid on production loans because less was borrowed, but more interest was paid on machinery.

The income statements and balance sheets were combined into yearly financial summaries to provide profitability, solvency, liquidity and risk analyses (Table 15.6). Profits and losses were carried over from year to year to provide financial trend analyses over the nine-year period. Income taxes and debt refinancing were also considered. Profitability was defined as the profit before-tax return to operator, family labour and equity capital. Solvency was net worth of the farm, and liquidity was reflected in changing levels of current liabilities, assuming zero liabilities at the start of the study. Risk was defined as the year-to-year variation (standard deviation) of profits.

Although results are shown for both systems with and without government subsidy programmes, the low-input government scenario was not possible because of cross-compliance restrictions (i.e. this farmer raised wheat and barley when the farm did not have a base for these crops), but was used to illustrate what could be possible if government policies were changed.

On the basis of these results, Hanson *et al.* concluded that low-input

Table 15.6. Profitability, liquidity, solvency and risk measures (US$ per farm) for a representative grain farm under low-input and conventional scenarios, Mid-Atlantic region, 1981–1989.

	Average annual profit			Current liabilities	Increase in net worth	Risk measures 1981–1989	
	1981–1989	1981–1984	1985–1989			SD[c]	Limit[d]
Low input[a]	27,614	22,027	32,083	144,448	17,124	16,985	16,166
Low input[b]	32,464	17,790	44,202	110,039	50,533	n/a	n/a
Conventional[a]	29,891	39,769	21,990	124,118	15,444	37,811	4,406
Conventional[b]	39,193	36,125	34,738	54,315	83,247	24,416	12,777

[a] Excluding government programmes.
[b] Including government programmes.
[c] Standard deviation.
[d] 75% lower confidence limit.
Source: Hanson *et al.* (1990).

profits trended upwards yet the economic transition was still much longer than biological transition. Equal profits would occur (excluding government programmes) after 11 years if current trends continued. The low-input approach was advantageous for risk-averse farmers, except during transition. Without restructuring debt, all scenarios faced cash flow problems, but all scenarios increased their net worth. Government programmes rewarded the conventional farmer with a large base acreage of maize. Cross-compliance restrictions discouraged rotations.

The analysis did not account directly for the likely changes in soil properties over time. While the increased profits of the low-input system in the later years reflected some of this investment in soil improvement, it is unknown how continued soil improvement (or degradation) would determine profitability in the long term. Focusing on only a few years' data greatly understated the potential profitability of the low-input approach.

Whole-farm models: Dairy farming

Dunbar (1991) used linear programming to evaluate the profitability of a representative dairy farm in south-central Pennsylvania under conventional management and an alternative low-input scenario based on the FST LIP-A rotation. The farm resources assumed to be available included 44 ha land, 65 dairy cows, livestock facilities for housing and milking up to 65 dairy cows plus the necessary number of replacement calves and heifers.

The conventional model included: (i) full utilization of livestock capacity; (ii) a rotation with maize grain, maize silage, alfalfa and a cereal nurse crop; (iii) supplemental feed purchases of alfalfa hay and soyabean meal; (iv) a straw harvest; and (v) no purchased N, P and K fertilizer because manure plus residual N from alfalfa was sufficient. In addition, optimal feed rations were calculated for three lactation periods plus a dry period.

The low-input model included: (i) full utilization of livestock capacity; (ii) a rotation with equal amounts of land (9 ha) for each of the following: wheat/red clover/alfalfa, second year red clover/alfalfa, maize grain, soyabeans, maize silage; (iii) crop sales of wheat and soyabeans; (iv) less purchased soyabean meal and straw, and more purchased alfalfa hay and maize than the conventional system; and (v) more maize grain and less maize silage, soyabean meal, lime, and dicalcium phosphate fed to dairy cows during lactation periods 1 and 3 and the dry period than in the conventional system.

The low-input model (based on the FST LIP-A rotation) had about 13% lower returns over variable costs than the conventional model (an alfalfa hay and maize silage/grain rotation). Low-input maize production was more profitable, but this was offset by more profitable cereal and hay production in the conventional system. Low-input maize production costs were half those of the conventional (US\$ 114 ha^{-1} versus US\$ 222 ha^{-1}) while grain yields were similar (low input 8.5 t ha^{-1} versus conventional 8.2 t ha^{-1}). Low-input maize silage costs were 60% of conventional (US\$ 141 ha^{-1} versus US\$ 232 ha^{-1}) while crop yields were similar (low input 43.5 t ha^{-1} versus conventional 47.1 t ha^{-1}). Low-input cereal and hay yields were lower than conventional. Available labour was better utilized throughout the growing season in the low-input model. In the conventional model, labour was a constraint in September because of silage making, alfalfa hay harvest and cereal establishment.

Linear programming was also used by Dabbert and Madden (1986) to investigate financial changes during the transition to organic farming. In this case, data were obtained from the 130 ha Kutztown family farm adjacent to the Rodale Institute Research Center. The Kutztown farm was predominantly organic, but some restricted inputs were used on parts of the farm. Yields exceeded county averages for most crops. The financial returns from the Kutztown farm were calculated to be 5% higher than a hypothetical conventional farm typical of the region, assuming similar crop production levels and prices, and 10% lower costs overall (Culik *et al.*, 1983). Dabbert and Madden found that income reductions during the transition phase were primarily attributable to assumptions made concerning conversion-specific yield reductions, although the Kutztown farm was already a well-established low-input system and provided no direct evidence for this. The value of the Kutztown farm study was that it provided real-world information for the design of resource-efficient farming systems to be incorporated into a more statistically rigorous, long-term research study (i.e. the FST).

Policy implications

Faeth *et al.* (1991) used data from the Rodale Farming Systems Trial and the University of Nebraska Research Station in a natural resource accounting framework to estimate the depreciation (or appreciation) of the soil under different farming systems. The intention was to illustrate the major influence of government policies and the farm support programme on farm incomes and how these policies mask the negative effects of soil degradation in the hope of influencing future farm bill policy decisions towards promoting sustainable farming practices.

Typical financial analyses only calculate the depreciation of man-made capital assets (i.e. buildings and machinery). In the words of the authors: 'Economic realities are obscured by traditional farm accounting systems, but the facts can be laid bare by environmentally honest accounting practices'. The study included a soil depreciation allowance (SDA) which is an estimate of the present value of future income losses due to the impact of crop production on soil quality. Soil erosion causes loss of plant-available water, soil crusting that restricts seedling emergence and root penetration, and loss of plant nutrients. SDA measures soil productivity and not the productivity of technology. Therefore, the important issue is whether the technology is more productive on a better soil than on a degraded soil to determine whether or not some of the effects of soil degradation have been masked by technological improvements. If this is the case, then there has been a real loss of income.

The estimation of net farm income (NFI) included the value of changes in soil productivity (as in business accounting which incorporates asset formation and depreciation in measures of income). Gross operating margin (crop sales minus variable costs) and government subsidies were added together and then the SDA was subtracted from this. The net economic value (NEV) was calculated by subtracting from the NFI an additional amount for the off-site environmental costs (sedimentation, effects on fisheries and recreational areas, downstream water users). Therefore the NEV takes into account the costs that farmers' activities impose on others, but excludes transfer payments to farmers such as subsidies that are not income generated by agricultural production.

A measure of sustainability of the farming systems in the study was considered at several hierarchical levels. At the field level, a computer simulation model was used to estimate physical changes of soil (erosion) and productivity changes as a result of different agronomic practices. At the farm level, a programming model was developed to assess the impact of commodity programmes on NFI and NEV. Field- and farm-level analyses were then linked to calculate crop sales, production expenses, government deficiency payments, NFI, soil erosion, off-site damages, and the SDA. The farm-level data were then linked to regional estimates of off-site damage per tonne of eroded soil. Finally, at the national level agricultural sector models were used to estimate changes in crop prices under various policies.

The NEV generated by the alternative systems (LIP-CG, LIP-A) was higher than for CONV under all policy options. This was true even during the transitional period (1981–1985) when LIP yields were depressed. This was mainly due to the much greater soil depreciation and off-site costs in the conventional system. The NEV was negative in all CONV scenarios except with multilateral decoupling (i.e. elimination by the USA, EC and Japan of all commodity programmes tied to production). All LIP-A and LIP-CG scenarios had a positive NEV except the LIP-CG with normal crop acreage (i.e. 100% planting flexibility, but no deficiency payments for non-programme crops if harvested).

Despite this result, under the commodity support payment programmes, farmers who would or could not look beyond the five-year transitional period would find the conventional practices more profitable. Farmers who do not account for long-term losses in soil productivity in their financial estimates would find conventional practices more profitable under all policy options except the Sustainable Agriculture Adjustment Act (SAAA) which allows farmers using resource-conserving rotations 100% planting flexibility, maintenance of crop base acreages, a waiver of crop set-aside acreage, and payments on forage crops.

With complete accounting of on- and off-site costs and elimination of baseline agricultural policies, farming systems with diverse rotations that maximize use of biological nutrients are economically competitive compared to CONV systems when environmental costs are low (e.g. in Nebraska) and markedly superior to CONV systems when environmental costs are high (e.g. in Pennsylvania). Shifting towards biologically-based farming systems through appropriate policy changes can raise productivity, reduce the amount of subsidies to maintain farm income, and lower environmental costs.

Summary and Conclusions

An accurate assessment of the economics of a particular farming enterprise requires a clear understanding of the assumptions being made, the parameters measured, and the parameters excluded from the analysis. Each of the studies comparing conventional with alternative management practices had distinct goals, particular biases, and hence certain limitations. Obviously, many of the specific results can be debated endlessly, yet they do provide a framework upon which some general conclusions can be drawn.

The agronomic results of the Rodale Farming Systems Trial show a 'transition' effect in terms of reduced maize yields that probably is related to low availability of N at the beginning of a low-input rotation (Liebhardt *et al.*, 1989). Starting the conversion with a different crop such as soyabeans could help avoid such yield losses. The partial budgeting analysis of the FST results (Duffy *et al.*, 1989) was a straightforward approach that showed the importance of the starting crop during transition to organic farming practices. It also made clear that removing all fertilizer N inputs at once and substituting them with

N derived from legume green manures would mean a significant loss of profits during the early years of transition. In contrast, profits were not reduced during the transition in the animal-based low-input system. This approach does not consider the effects of transition on the whole farm (e.g. labour, management, marketing, buildings, family goals), the risk factors, the environmental impact and the influence of government subsidies.

The whole-farm analysis (Hanson *et al.*, 1990) addressed many of the issues identified as limitations in the partial budgeting study. Although the low-input/organic farm was less profitable (primarily during the transition years), it produced a steadier yearly income (less risk) because of greater crop diversity. In addition, the greater profits received by the conventional farm were largely due to government subsidies rather than being an inherently 'better' system. A key finding was that, although the organic farm established a new biological equilibrium in three to five years (i.e. when maize yields were equivalent to or higher than conventional yields), the economic transition took much longer because of high interest payments on debt incurred in the early transition years. This problem could probably be alleviated with a more gradual removal of synthetic inputs, but research is needed to confirm this.

The linear programming analysis (Dunbar, 1991) integrated the livestock component with the crop component. It was the only study that adequately assessed the low-input/animal (LIP-A) system in the FST. It also illustrated how difficult it can be for a livestock farmer to make the appropriate decisions for maximizing profit. Unfortunately, linear programming techniques are still quite inaccessible to most farmers, who rely primarily on experience and intuition to make the right choices.

The policy study and a recent update (Faeth *et al.*, 1991, 1993) clearly illustrated that the government commodity programmes encouraged dependence on inorganic inputs and discouraged sustainable farming practices. Multilateral decoupling could remove the distorting influence of commodity programmes, encourage agricultural sustainability, and increase net economic value. Perhaps the most significant contribution of this study was its use of natural resource accounting to present a more complete picture of the real value of a particular farming system. Computer simulation models can only provide a rough approximation of changes in soil quality (soil erosion in this case) and subsequent effects on crop productivity. Nevertheless, the results from these models underscore the need for more concerted, long-term efforts to document soil changes resulting from management practices.

References

Culik, M.N., McAllister, J.C., Palada, M.C. and Rieger, S.L. (1983) *The Kutztown Farm Report: A Study of a Low-input Crop/Livestock Farm.* Technical Bulletin. Rodale Research Center, Kutztown, Pennsylvania.

Culik, M.N. and Liebhardt, W.C. (1984) *Conversion Project Progress Report.* Rodale Research Center, Kutztown, Pennsylvania.

Dabbert, S. and Madden, P. (1986) The transition to organic agriculture: a multi-year simulation model of a Pennsylvania farm. *American Journal of Alternative Agriculture* 1, 99–107.

Duffy, M., Ginder, R. and Nicholson, S. (1989) *An Economic Analysis of the Rodale Conversion Project: Overview.* Staff Paper 212. Department of Economics, Iowa State University, Ames, Iowa.

Dunbar, T.V. (1991) An economic evaluation of alternative agriculture production methods in south central Pennsylvania. MSc thesis, Department of Agricultural Economics and Rural Sociology, Pennsylvania State University, University Park, Pennsylvania.

Faeth, P. and Westra, J. (1993) Alternatives to corn and soybean production in two regions of the United States. In: Faeth, P. (ed.) *Agricultural Policy and Sustainability: Case Studies from India, Chile, the Philippines and the United States.* World Resources Institute, Washington, DC.

Faeth, P., Repetto, R., Kroll, K., Dai, Q. and Helmers, G. (1991) *Paying the Farm Bill: US Agricultural Policy and Transition to Sustainable Agriculture.* World Resources Institute, Washington, DC.

Hanson, J.C., Johnson, D.M., Peters, S.E. and Janke, R.R. (1990) The profitability of sustainable agriculture on a representative grain farm in the Mid-Atlantic region, 1981–1989. *Northeastern Journal of Agricultural and Resource Economics* 19(2), 90–98.

Liebhardt, W.C., Andrews, R.W., Culik, M.N., Harwood, R.R., Janke, R.R., Radke, J.K. and Rieger-Schwartz, S.L. (1989) Crop production during conversion from conventional to low-input methods. *Agronomy Journal* 81, 150–159.

Peters, S. (1991) Organic and conventional cropping beyond transition. *Organic Farmer* 2, 13–18.

Peters, S., Janke, R. and Bohlke, M. (1992) *Rodale's Farming Systems Trial 1986–1990.* Research Report. Rodale Institute Research Center, Kutztown, Pennsylvania.

16 Economics of Conversion to Organic Farming: Cross-sectional Analysis of Survey Data in Germany

STEPHAN DABBERT

Introduction

For many farmers the period of conversion to organic farming involves a far-reaching reorganization of many aspects of their farm. This period of reorganization, when many things change simultaneously, is a challenge to farmers – it is also a challenge to researchers who try to describe and explain the process and possibly develop ways to manage the conversion process more efficiently. Cross-sectional studies are but one step in reaching understanding of the conversion process. The reason why these studies are performed can be better understood if some general concepts of the conversion process are developed. Consequently this chapter first discusses some important conceptual issues of the conversion process. This is followed by a discussion of the methods and results of the two existing cross-sectional studies in Germany that, among other areas, also touch upon the conversion process. At the end of the chapter empirical findings and strengths and weaknesses of the cross-sectional approach are summarized.

Conceptual Issues of the Conversion Process

From a legal perspective the process of conversion to organic farming is a definite time span between conventional management and the time when the farm is fully organic. A prominent example of this view is given by the new EC Regulation (2092/91) on organic farming (EC, 1991). The conversion period is a clearly defined period that begins when a state-authorized institution testifies that organic methods of production are used, and it has a length of two years. In principle, only after this period can the products produced be sold as 'organically produced', although continued sale of 'in conversion' products has been allowed on a temporary basis. While this legal perspective severely influences the economics of the conversion process, from an economic viewpoint this definition is not satisfying because the conversion process goes on for a longer period. It may possibly start before the farm uses organic production methods on the total operation and it will in most cases not be finished after two years. From an economic viewpoint the term 'conversion' marks a period that is characterized by severe changes in the farm organization and other parameters that influence economic success if these changes are caused by the decision to convert to organic farming. Of course, what is meant by 'severe' in this definition is open to interpretation.

A farmer who wishes to consider whether a conversion to organic farming methods will be profitable or not has to compare (i) the development path his farm will take if he stays with conventional methods; and (ii) the development path his farm will take if he converts to organic methods. At the moment of decision neither option can be specified with certainty. Both development paths will be influenced by a wide range of factors, from changing technologies to the impact of GATT negotiations. Some hypotheses relating to typical developments of farm parameters over time connected with the transition process are shown in Fig. 16.1. While the general tendencies described in Fig. 16.1 are typical in many cases in Germany there also are many open questions, especially with respect to quantitative assessment of the development of the parameters.

Product prices increase because in the past it was, and for the time being still is, possible to market part of the products under trade marks either 'from a farm in conversion' or 'from a fully organic farm', with the premium received for organic production being higher in the second case. The yield developments (which will be elaborated on later in the next section) have been explained partly by a 'biological transition effect' (Dabbert and Madden, 1986), that is, the adaptation of the agro-ecosystem and especially soil conditions to the new methods of production takes time during which increased weed and pest pressure and other factors negatively influence yields. Variable costs per hectare are lower and labour input is higher on organic farms (Dabbert, 1990). The studies reported elsewhere in Part 3 and in Chapter 25 of this volume have shown that additional investment in machinery and in some cases in buildings takes place. Taking all these factors into

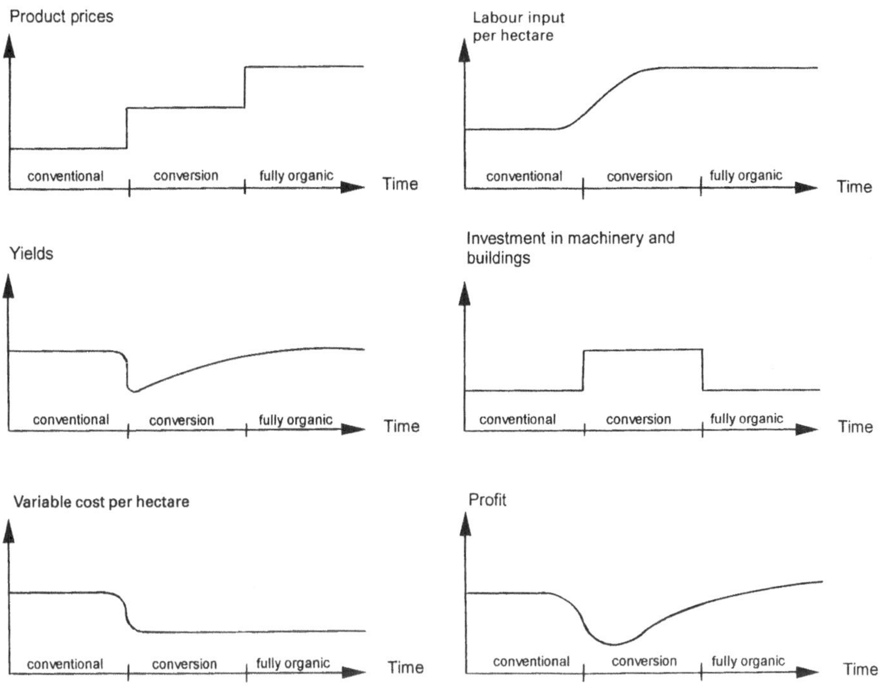

Fig. 16.1. Schematic analysis of the development of important parameters over the period of conversion to organic farming.

account, it is possible to estimate profit development over the conversion period. However, despite the plausibility of the profit curve depicted in Fig. 16.1 (see, for example, Dabbert and Madden, 1986) representative empirical results on these trends are still missing.

In the past there have been two lines of research on the economic aspects of the transition to organic farming. The first intended to provide information on what actually happens during this time. How are farmers making the conversion? What are the results with respect to parameters relevant to economic success? These were the questions asked. The second line of research aims at the development of tools to provide better and more systematic ways to judge the consequences of the transition, namely specific farm planning software.

In this chapter I concentrate on one aspect of the first line, specifically, on the use of cross-sectional studies to obtain information on the development of yields during the conversion process.

Cross-sectional Studies in Germany and their Results

The cross-sectional studies in Germany which covered a larger number of farms and for this reason are relevant to the topic discussed here were not primarily concerned with developments during the conversion. However, in two of these studies (Böckenhoff *et al.*, 1986; Dabbert, 1990) the development of yields as a function of the number of years since the beginning of the conversion process turned out to be an important sideline.

Böckenhoff *et al.* surveyed a little under 200 organic farms in Baden-Württemberg. In winter 1983/84 when the survey was done these operations farmed 61% of total organic acreage in the southern German state of Baden-Württemberg. While the central aim of their survey was to obtain reliable information on the general picture on organic farms, they also tried to explain yield differences between organic farms by the length of time of organic management.

Using cross-tabulation Böckenhoff *et al.* (1986) found winter wheat, spelt and winter rye to be dependent on the length of time under organic management, with the strongest increase in yields within the first five years after the conversion (Table 16.1). For spring cereals they were not able to establish such a relationship, which they attribute to regional differences in weather conditions in the summer of 1983. This argument, they also suggest, explains the reason why they could not find increasing yields over time for potatoes.

For carrots, which are an important vegetable crop on many organic farms in Germany, Böckenhoff *et al.* (1986) established a very

Table 16.1. Average organic and conventional (conv) yields (t ha^{-1}) and organic yields by the number of years of organic management, Baden-Württemberg, 1983.

			Organic yields by number of years under organic management			
	Organic	Conv	<2	3–5	5–10	>10
Winter wheat	3.34	4.75	2.74	2.85	3.48	3.53
Spring wheat	2.81	3.87	2.76	2.29	3.11	2.89
Spelt	2.63	–[a]	1.40	2.50	2.75	2.73
Winter rye	3.17	3.80	2.07	2.88	3.27	3.37
Potatoes	16.19	22.20	20.98	11.39	17.28	15.56
Carrots	40.53	42.25	13.8	23.0	40.9	43.1

[a] Included in winter wheat.
Source: Böckenhoff *et al.* (1986).

strong increase with the time since conversion. They hypothesize that this observation is heavily influenced by the fact that carrots are not a common crop on conventional farms, so the results show (at least partly) that the learning process of how to grow carrots needs time.

Böckenhoff *et al.* (1986) stress the importance of soil and climatic conditions for yield level. However, their data did not include indicators of these factors, so they could not test for this influence. Also the method of cross-tabulation is not well suited to take account of multi-factor influences on yield.

For these reasons a study by Dabbert (1990) uses a different approach in order to investigate the influence of the length of time under organic management on yield level. In an attempt to cover every farm of the Bioland organization in Baden-Württemburg that farmed full-time and was not mainly engaged in fruit or vine production, 44 organic farms were surveyed in this state in 1987/88. Because the goal of this study was to gain information on organic farms after the main phase of reorganization had been overcome, all farms which had been under organic management for less than three years were excluded. A simple theoretical yield-explaining model was developed with the following explanatory exogenous variables:

- soil quality index;
- cereal as a proportion of arable land;
- number of animals per hectare;
- farm size;
- proportion of permanent grassland;
- the number of years under organic management.

In simple regression runs, the influence of these variables was individually tested. For most cereals and potatoes, soil quality index and, for cereals only, the number of years under organic management and (to a much lesser degree) the area of cereals as a percentage of total arable land, were found to have some explanatory value. The other variables tested did not have any significant impact on yields. These three variables were then included in multiple regressions and the functions reported in Table 16.2 were found to have the best fit.

The results shown in Table 16.1 and Fig. 16.2 indicate that:

- the yields of winter cereals increase with time since conversion; for spring cereals such an influence was found only for oats;
- except for spelt wheat and oats, there is a strong influence of the soil quality index on yield level;

Table 16.2. Multiple regression analysis of organic cereal yields, Baden-Württemberg.

Endogenous variable, yield (t ha^{-1}) of	Constant	Exogenous variables: (Soil quality index)$^{-1}$ [a]	Exogenous variables: ln (years under organic management) [b]	Exogenous variables: Cereals as % of arable land [c]	R^2	n
Winter wheat	3.986**	−51.122*	0.53710**	–	0.36	32
Winter rye	1.085**	−28.643**	0.51389**	–	0.32	30
Spelt wheat	1.434	–	0.96632**	–	0.16	25
Spring wheat	8.547**	−135.386**	–	−0.000525**	0.71	10
Oats	2.584**	–	0.54490*		0.30	10
Mixed cereals (mainly oats)	5.342**	−39.329**	–	−0.000228*	0.26	20

*Significant at the 90% level.
**Significant at the 95% level.
[a] Range 20–67 points on 0–100 scale.
[b] Range 3–17 years.
[c] Range 0–75% (range = range of values over which functions were estimated for each variable).
Source: Dabbert (1990).

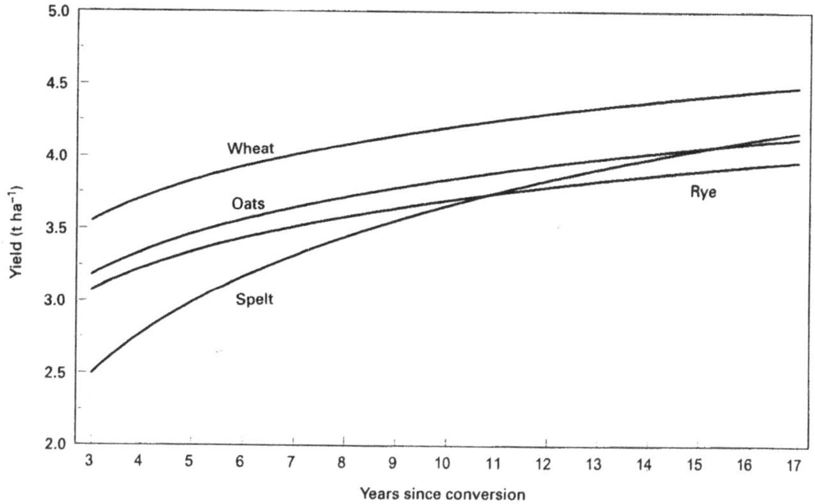

Fig. 16.2. Crop yields as a function of the number of years since conversion to organic farming, Baden-Württemberg. A soil quality index of 50 and a cereal percentage of 50 were assumed. Source: Dabbert (1990).

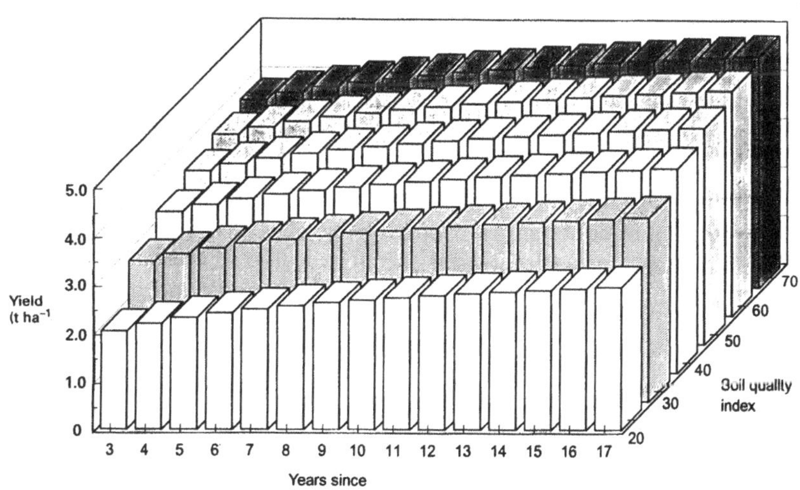

Fig. 16.3. Wheat yields as a function of soil quality index and the number of years since conversion to organic farming, Baden-Württemberg. Source: Dabbert (1990).

- the cereal percentage of total arable land was only found to be significant in two cases.

The simultaneous influence of soil quality index and the number of years since the conversion on wheat yields is illustrated in Fig. 16.3.

The results of the study by Dabbert (1990) have to be interpreted with a certain amount of caution because of the small sample used. However, as these results, which were gained with a different methodology and from a different sample than those of Böckenhoff *et al.* (1986), point in many respects in the same direction, both support each other.

Conclusions

The main results of both studies are as follows:

- For some crops increasing yields with time under organic management can be found.
- In these cases yields increase at a decreasing rate.
- Time under organic management is only one factor that explains yield differences; at least of similar importance are soil and climatic differences, and a large part of yield variation cannot be explained.
- For winter cereals, the dependence of yield level on the time under organic management can be firmly established; for spring cereals, the results remain ambiguous.
- As crops that are newly introduced into the organic farms with the conversion (spelt wheat, carrots) exhibit a relatively strong increase in yields over time, a process of learning how to handle these new crops seems to be a part explanation for the results.
- Two other major factors that play a role in explaining yield developments are: a biological transition effect (Dabbert and Madden, 1986); and the process of learning organic management in general.

On methodological grounds the advantages of cross-sectional studies are the following:

- Results can be gained rapidly and with limited cost.
- Changes in technology and prices that might interfere with the results of long-term time-series observations are absent.

The disadvantages include the following:

- The fact that some of the important parameters (e.g. profit) cannot usually be observed with one-off surveys on many farms with a sufficient degree of certainty limits the scope of cross-sectional studies mostly to developments in yields (and possibly product prices); this, of course, severely limits the inferences that can be drawn from these studies.

- The conclusions are based an statistical inferences. It might be argued that yield differences between farms are not due to the factor 'time under organic management' but to some other unobserved factor. That means cross-sectional studies infer the existence of a conversion process, they do not actually observe it.

For these reasons cross-sectional studies can only form part of our picture of the conversion process. They have to be supplemented by case studies of the conversion over time (see Chapters 13 and 14) and by model calculations. However, the usefulness of cross-sectional studies of the conversion process should not be underestimated. These studies were the first firmly to support the hypothesis that profits decrease after the transition and later increase, a hypothesis that has had important consequences on the type of political support for organic farming (see Chapter 25).

References

Böckenhoff, E., Hamm, U. and Umhau, M. (1986) Analyse der Betriebs- und Produktionsstrukturen sowie der Naturalerträge im alternativen Landbau. *Berichte über Landwirtschaft* 64(1), 1–39.

Dabbert, S. (1990) *Zur optimalen Organisation alternativer landwirtschaftlicher Betriebe. Untersucht am Beispiel organisch-biologischer Haupterwerbsbetriebe in Baden-Württemberg.* Agrarwirtschaft Sonderheft 124. Verlag Alfred Strothe, Frankfurt am Main.

Dabbert, S. and Madden, P. (1986) The transition to organic agriculture: a multi-year simulation model of a Pennsylvania farm. *American Journal of Alternative Agriculture* 1(3), 99–107.

EC (1991) Council Regulation (EEC) No. 2092/91 of 24 June 1991 on organic production of agricultural products and indications referring thereto on agricultural products and foodstuffs. *Official Journal of the European Communities* 91(L198), 1–15.

17 Conversion to Organic Farming: An Overview

Susanne Padel and Nicolas H. Lampkin

Introduction

A period of several years, known as the conversion or transition period, is needed to change a farm from conventional to organic management, during which the farmer should aim to:

- improve soil fertility by establishing a rotation with legumes so that crops can be produced without synthetic nitrogen fertilizer or large amounts of bought-in manures;
- adjust the stocking rate to the natural carrying capacity of the farm, so that livestock can be produced without large amounts of purchased concentrates and/or forage;
- change the management to maintain animal and plant health with the limited inputs available according to organic production standards.

In several aspects this period differs from the established organic farming system. The time required and the difficulties associated with the necessary changes will depend on the intensity of conventional management and the condition of the farm before conversion. Usually some investment in machinery and/or buildings will be required in order to meet organic standards. After the conversion period, the farmer can apply for full organic certification and will usually be allowed to use a symbol and gain access to premium prices where available.

The certification aspects of conversion are regulated by production standards and legislation (Chapter 2). From a socioeconomic perspective other aspects are of interest. For farm planning or policy-making purposes an understanding of the process of conversion and its costs to the farmer, in terms of reduced yields and incomes as well as changes in inputs and fixed costs, is required. For policy-making purposes an understanding of how and why farmers convert to organic farming is also necessary. This chapter attempts to answer some of these questions by drawing together some of the conclusions from previous chapters and discussing them in the context of other related studies.

Adoption of Organic Farming: Characteristics, Motives and Barriers

Farm size

In a number of countries, for example Britain (Chapter 5), Denmark (Chapter 7), Canada (Chapter 9) and the United States (Chapter 10), organic farms are typically smaller than conventional farms, due to a high number of specialist horticultural and 'lifestyle' holdings. However, in other countries such as Germany (Chapters 6 and 25) and Switzerland (Chapters 8 and 24) organic farm sizes are similar to or above the conventional average. These countries typically have an older organic farming tradition with stronger roots in the farming community (Chapter 2). There is some evidence that in countries where the organic farming sector has been dominated by very small holdings, average size is increasing as more established farmers convert (Chapter 5).

Education, individual background and social relationships

Organic farmers are typically well educated or better educated than conventional farmers. Many have an urban background, general academic education and little or no previous farming experience (Chapters 5, 9, 10 and 14). Organic farmers are also typically younger than average (Chapters 9, 10 and 25). Early studies reported a lack of social acceptance as a consequence of the conversion (Wernick and Lockeretz, 1977; Fischer, 1982). Later studies regarded this issue as less important (Lockeretz and Madden, 1987; Vogtmann *et al.*, 1993). Richter (1990) suggested that urban people who start farming organically are less dependent on acceptance in the rural community.

Motivation

The question why farmers go organic has been the subject of several studies since the mid-1970s (Wernick and Lockeretz, 1977; Vine and

Bateman, 1981; Fischer, 1982; Dettmer, 1986; Lockeretz and Madden, 1987; Rantzau *et al.*, 1990; Wynen, 1990; Svensson, 1991). In general, farmers choose to adopt organic farming for a number of reasons in addition to financial ones, ranging from husbandry concerns, in particular soil degradation and livestock health, to personal concerns including family health links to pesticide use, and general political, ideological, philosophical and religious concerns such as the environment, developing countries and food quality.

Financial motives include attempts to solve existing financial problems and to secure the future existence of the farm by exploiting opportunities for cost savings and premium price marketing (Vine and Bateman, 1981; Lockeretz and Madden, 1987; Fisher, 1989; MacRae *et al.*, 1990a; Wynen, 1990; Svensson, 1991). Financial motives are more frequently mentioned in recent studies, but usually in combination with other motives (see Chapters 14 and 25). MacRae *et al.* (1990a) relate this to the general economic situation of farming in the developed world, but it can also be argued that it is linked to the increasing financial attractiveness of organic farming in some countries.

Although financial factors play a role, there are clearly a large number of other objectives or motivations in a farmer's decision to convert to organic farming. Non-financial aspects and farmers' objectives should therefore be included in any analysis of the implications of conversion.

Barriers to conversion

As important as why farmers convert is why farmers do not convert to organic farming, particularly given the stark contrast between the numbers of farmers who have actually adopted organic farming and the numbers who have expressed interest in adopting organic methods (e.g. 38% in Britain (NatWest, 1992); 27% in Victoria, Australia (Clarke, 1991)).

Conventional farmers' *perceptions and access to technical and financial information* have been identified as significant barriers to conversion (Blobaum, 1983; Fisher, 1989; Henning *et al.*, 1990; Wynen, 1990; Chadwick and McGregor, 1991; Clarke, 1991; Olbrich, 1992; Vogtmann *et al.*, 1993). These include:

- lack of information and unhelpful or disparaging extension agencies, leading to misconceptions concerning practices, yield expectations, financial performance and risk;
- concerns about potential problems with weeds, pest and disease control, the risk of yield losses or crop failure, feed shortages and a large increase in manual labour;

- difficulties gaining access to information which does exist but is only available through non-traditional sources, such as books, magazines, neighbours, family and friends and particularly other organic farmers;
- negative images of organic farmers as hippies or hobby farmers (this view is not as widespread as it once was, with the increasing recognition of organic farming as a viable option, and enhanced credibility through the involvement of established agricultural institutions, in particular policy-makers);
- concern that the market for organic produce is limited and that large-scale conversion would lead to massive oversupply, a collapse in prices and reduced incomes, with the lack of technical, marketing and financial data contributing to an enhanced perception of limited feasibility and high risk which risk-averse farmers under financial pressure or high levels of indebtedness are not prepared to take on board.

Fisher (1989) comments that the major difference between adopters and non-adopters was one of attitude. Once the decision to adopt organic farming had been taken, the farmers became fully committed, seeing further problems as challenges and no longer as barriers. This may be helped by improvements in the availability of information, since the problem of lack of information and advice does not feature so highly in recent studies from countries such as Sweden where advisory services for organic producers are better developed (Svensson, 1991).

Institutional barriers to conversion have been identified in a number of studies, particularly from Canada and the United States (Blobaum, 1983; Fisher, 1989; MacRae *et al.*, 1989, 1990b; National Research Council (NRC), 1989; Henning *et al.*, 1990). These include: landlord objections, refusal of loans or insurance for organic production; problems with grant applications; and legislative and certification constraints.

Social barriers, such as the fear of becoming an 'outsider' or intergenerational conflicts, are also cited frequently (Wernick and Lockeretz, 1977; Fischer, 1982; Blobaum, 1983; Dettmer, 1986; Lockeretz and Madden, 1987; Chapter 14, this volume), but these can also be seen as examples of a general phenomenon associated with the social-change process, particularly in tightly-knit rural communities.

Conversion to organic farming and the adoption/diffusion model

Differences in the social status of early adopters of a specific innovation are an important part of the adoption/diffusion model described by Rogers (1983). In the early stages of diffusion, so-called *innovators* are the first to adopt the innovation and adoption rates are low. Innovators are venturesome, have cosmopolitan relationships and communicate with a clique of other innovators, often in spite of considerable distance. They must be able to cope with a high level of uncertainty. Very often they are not respected by other members of the social system. The innovation (and innovators) might even be viewed as a threat to the established system and therefore be rejected initially by the majority. However, innovators have an important gate-keeping role in the diffusion process. *Early adopters* are a more integrated part of the social system than innovators. They usually have a degree of opinion leadership and have intensive contact with information sources. Their role in the diffusion process is to make the innovation acceptable in the community. The next group, the *early majority*, adopts new ideas just before everyone else does, whereas the *late majority* remains rather sceptical. A relationship between innovativeness and socioeconomic status, including education, income level, farm size and commercial orientation, can be generalized as a result of many adoption studies (Rogers, 1983).

Most of the studies of the socioeconomic situation of organic farmers were undertaken when organic farming was at a very early stage of diffusion (less than 0.5% of all farmers). The organic farmers studied can therefore be placed in the category of innovators. The conversion studies match the adoption/diffusion model with regard to education and age, whereas relative farm size varies between countries and over time. Social acceptance problems experienced by many early organic farmers around the world are typical of innovators' experiences with other innovations. Farmers' reasons for adopting organic farming are in line with the adoption/diffusion model if it is modified as suggested by Taylor and Miller (1978) to suit environmental and non-profit-oriented innovations. The relatively slow, but increasing, rate of adoption of organic farming (see Fig. 1.1, p. 4) is also consistent with the model, which predicts that the more complex an innovation is, the slower the adoption rate will be. The adoption/diffusion model therefore provides a useful framework for understanding the social aspects of conversion to organic farming, and for developing strategies to encourage more widespread adoption of organic farming (Padel, 1994).

The better social acceptance of more recent adopters in Germany, increasing average farm size in Britain, and the increased importance of financial motives suggests that, in some countries at least, the early adoption stage has been reached. Fisher (1989) tried to identify different characteristics for later and earlier adopters of organic farming in New Zealand, but did not succeed. If the small number of organic farms in New Zealand (200 certified farms in 1992) is taken into account, the lack of success is not surprising. Different results might be found if a similar study were to be conducted in one of the northern European countries, where more farms have converted, and more financial and other support has been given to organic farming.

However, the most important conclusion from the adoption/diffusion model is that although the first organic farmers are somewhat different from the farming community as a whole, this does not mean that organic farming is not suitable for the majority of farmers. In every diffusion process, the innovators and early adopters differ from the majority adopting at a later stage. Organic farming is, in most countries, at a very early stage on the adoption curve with all the associated social problems for the innovators and early adopters.

The Process of Conversion

The process of conversion to organic farming starts with a change in attitudes. Before the management system of the farm changes, the farmer has to be convinced that the new system is better than the old one. The process outlined by Padel (1988), Andrews *et al.* (1990) and Rantzau *et al.* (1990) is similar to the individual adoption process in the adoption/diffusion model (Rogers, 1983). There are four stages:

1. initial knowledge,
2. acceptance as a good idea,
3. acceptance on a trial basis,
4. adoption.

There may, however, be greater emphasis on individual goals, preferably set with the involvement of all members of the farming family, and the adaptation of the organic approach to the specific situation of the farm in order successfully to manage the conversion.

Stages 1 and 2 will usually involve visits to and discussions with other organic farmers, attendance at seminars and conferences, discussions with extension agents and reading available literature (Lampkin, 1990). Some difficulties are associated with stage 3, the 'trial' phase. Trying organic management on one field allows direct experience to be gained, but the use of just one field does not allow for establishment of the whole farming system, including the development of a suitable rotation with fertility-building crops, the integration of livestock, and the management of the farm ecosystem for crop protection. The conversion strategy of gradual deintensification (see below) can also be interpreted as a way of experimenting with the withdrawal of inputs, but again does not provide experiences of organic farming as a complete system. The 'trial' phase may be replaced in some cases by conversion planning, aiming to assess the feasibility of a conversion for the individual farm. Conversion planning has also has been mentioned by several authors as an important strategy to reduce the risk of the conversion (Padel, 1988; MacRae *et al.*, 1990a; Rantzau *et al.*, 1990; Lampkin, 1993) and is discussed in more detail below.

The final adoption phase (stage 4) differs from farm to farm, but three main approaches can be identified. *Staged* conversion involves the conversion of parts of the farm, typically 10–20%, in successive years, using a fertility-building legume crop as an entry into organic management. The learning costs, capital investments and risks can be spread over a longer period of time, sometimes up to ten years or the full length of a rotational cycle, and are more easily carried by the remainder of the farm business (Lampkin, 1993). However, staged conversions are not accepted by some certification bodies and were not allowed under the German extensification scheme (Chapter 25) on the grounds that the subsidy payment should compensate for the increased costs and risks in the early stages of conversion.

Single-step conversion involves converting all the land on the farm at one time, which enables the farm to gain access to premium prices sooner, but means that all the risks, learning costs and financial impacts of conversion are concentrated into a short period of time, while rotational disadvantages can arise because not all of the farm can be put down to fertility-building crops at the same time.

Both the staged and the single-step conversion approaches involve compliance with organic production standards on the land which is being converted from the start of the conversion period, although access to premium markets may not yet be possible. *Gradual de-intensification* is also an option, but is not usually allowed under organic production standards. However, this approach may be appropriate in situations where certification is not important, and is compatible with an alternative structure for the conversion process suggested by MacRae *et al.* (1990a), involving: (i) increasing *efficiency* of the conventional system, (ii) *substitution* of certain inputs by biological or more environmentally friendly ones, and (iii) *redesign* of the system. This alternative model emphasizes changes in *attitudes and practices* before the different character of the organic *system* is recognized and achieved in the redesign stage.

Technical problems and yield changes

Farmers converting report conversion-specific problems with weed control and nutrient supply that often result in higher labour requirements (Wernick and Lockeretz, 1977; Blobaum, 1983; Lockeretz and

Madden, 1987; Rantzau *et al.*, 1990). It has been suggested that some of these problems may be due to mistakes arising from lack of knowledge, experience or forward planning (see Chapters 13 and 14).

The existence of a conversion-specific crop yield decline (additional to that which might be expected under an established organic system) has been discussed in a number of studies (e.g. Vine and Bateman, 1981; NRC, 1989), reports of farmers experiences (e.g. Benecke *et al.*, 1988) as well as by several authors in this part of this volume. Dabbert (Chapter 16) concluded that time under organic management is one of several factors explaining yield differences between organic farms. Freyer *et al.* (Chapter 14) discussed yield decline in relation to the absence of sufficient legumes as preceding crops in the early years of conversion. Peters (Chapter 15) showed that the choice of the first crop during conversion had a significant influence on the yield and financial return of the whole management system.

Lampkin (Chapter 13) concluded that external factors were more important influences on levels of output than any conversion-specific yield decline. Two documented conversions of experimental farms also do not support the hypothesis of a conversion-specific yield decline – the Boschheide Hof in Germany (Vollmer and Buchner, 1985; Reinken and Buchner, 1987) and the Swiss agricultural college of Ebenrain (Lienhard, 1988) – although there is some indication that forage yields declined during the conversion period in the latter case. Stanhill (1990) found no evidence in his review of a large number of comparative studies to support the hypothesis that yield levels are related to the length of time under organic management.

It can be concluded that conversion-specific crop yield reductions are not found universally and can often be avoided: (i) where the farmer has access to information and the opportunity to learn about organic management before starting conversion; (ii) where the conversion is undertaken in stages and starts on each field with a legume crop; and (iii) where the change to an organic rotation is carefully planned to avoid yield decline as a result of rotational mistakes.

Forage crops are the one area where conversion-specific yield decline may be inevitable, with production lost as a result of reseeding grassland with new mixtures containing legumes, or waiting for clover to establish naturally following the withdrawal of nitrogen fertilizer. The loss of output can place significant pressure on stocking rates for livestock on predominantly grassland farms, particularly where permanent grassland is involved (Chapters 13 and 14). This issue has not been investigated in detail in other studies and further research is needed to quantify the effects.

Innovations

The conversion process requires a large number of innovations on the part of the farmer (Fisher, 1989; MacRae *et al.*, 1990a; Rantzau *et al.*, 1990; Woodward in Lampkin, 1990), including the introduction of:

- new marketing approaches, such as selling products direct to the consumer, local shops or specialist organic retailers, preparation before selling (e.g. cleaning cereals) and processing (e.g. milk into yoghurt or cheese) in order to obtain a premium;
- new crop production practices and inputs, such as mechanical weed control, rock dusts, biological controls, clover/grass mixtures, undersowing, intercropping, and new tillage practices;
- new livestock management practices, such as more natural rearing systems, family or other appropriate small groups, appropriate housing, complementary animal therapies such as homoeopathy, and integration of livestock into the whole farming system;
- new manure management practices, such as composting and slurry aeration;
- new crop and livestock enterprises, such as potatoes, grain legumes, field-scale vegetables, integration of sheep and cattle or sheep and poultry;
- new labour sources, such as student apprentices or casual labour.

The range of innovations associated with conversion to organic farming makes the adoption complex and demands an increased level of observation and flexibility as well as management expertise and confidence on the side of the farm personnel. The need for training and advice in preparation for and as part of the conversion process becomes obvious.

Labour requirements during conversion

Both Lampkin (Chapter 13) and Freyer *et al.* (Chapter 14) reported changes in labour requirements on their case-study farms ranging from an increase of 100% to a decrease of 50%. The increases in labour use were typically in the range 10–25%, and were associated with the introduction of more labour-intensive (but high-value) crops and/or production techniques, on-farm cleaning, grading, processing and marketing of produce, small-scale experimentation with new crops as well as increases in farm size. There may be conversion-specific increases in labour requirements associated with market development, innovations, and delays in labour-saving investments such as mechanization due to liquidity constraints, but little evidence is available on this issue (Chapter 14). There is no evidence that labour requirements for *existing* enterprises increased during conversion. In some cases, the abandonment of labour-intensive enterprises such as pig finishing helped to reduce labour requirements significantly (Chapter 14).

Changes in farm income and costs of conversion

Farm income decreases during conversion are reported by Peters (Chapter 15) and Dabbert (Chapter 16) as a consequence of reduced output, while Lampkin (Chapter 13) attributed the farm income decreases primarily to external factors such as milk quotas and lack of access to premium prices. Freyer *et al.* (Chapter 14) reported increased incomes due in part to the availability in Germany of premium prices during conversion. Increased incomes during conversion are also reported by Schulze Pals *et al.* (Chapter 25) due to the availability of subsidies for conversion as well as premium prices. Despite these differing conclusions, several common factors can be identified that influence farm income levels during conversion.

- *Output reductions* due to changes in husbandry practices, enterprise mix (including a possible conversion-related emphasis on fertility-building legumes at the expense of cash cropping (Chapter 14)), or as a result of mistakes or inappropriate actions which could be avoided through improved information and planning.
- *New investments* in machinery, livestock, buildings and land, which may be financed by capital released through reductions in livestock numbers and sale of quotas (Chapters 13 and 14).
- *Information and experience gathering* including direct costs, e.g. for literature, training courses, advisory services, study tours, seminars and conferences, and indirect costs, such as replacement labour to cover for absences on training courses and crop failures/poor performance where new enterprises or techniques are being trialled.
- *Variable cost reductions* as prohibited inputs are withdrawn and other inputs such as purchased livestock feeds are restricted, although some additional costs may be associated with reseeding grassland, establishing green manures and other fertility-building measures (Chapters 13 and 14).
- *Fixed cost increases* normally restricted to labour use, depreciation of conversion-related investments, and certification charges (Chapters 13 and 14). Higher depreciation costs may also be associated with the writing-off of investments in discontinued enterprises.
- *Lack of access to premium prices,* potentially the largest financial penalty during conversion in the northern European context (Chapters 13, 14 and 25). Where premium prices are available, significant

market development costs may be associated with obtaining them.
- *Eligibility for subsidies,* including loss of eligibility or cross-compliance restrictions on subsidies available to conventional producers (Chapters 15 and 26), and support for conversion and environmental improvements, where available (Chapters 23–26).

Most studies have focused on changes in income levels over the conversion period, although in many situations changes in net worth (the value of the farm family's own stake in the business) might be a more appropriate measure of the impact of transition as it reflects changes in capital stocks as well as income flows (Hanson *et al.*, 1990; Ramsay and O'Mara, 1991). Net worth is also influenced by the level of private drawings from the business, which can vary not only in response to income pressures resulting from the conversion, but also in response to external factors, such as school or university fees. Whether income or net worth is used, careful examination of the reasons for any changes is needed to ensure that they can really be attributed to the conversion.

Conversion planning

Conversion planning with the assistance of specialist advisers can help the individual farmer to identify potential problems relating to labour requirements, other resource constraints, financial returns and cash flow. It involves an assessment of the current situation of the farm, proposals for a 'target' organic endpoint, and a strategy for getting from the current situation to the target (Schmid, 1987; Padel, 1988; Lampkin, 1990; Measures, 1990). A full conversion plan therefore involves a multi-period analysis of financial and physical aspects of the whole farm. Techniques for conversion planning were first developed in Germany and Switzerland as student projects during the early 1980s (reviewed in Lampkin, 1993), and were subsequently adopted by organic farming advisers. Conversion planning also formed part of the case studies reported in Chapters 13 and 14.

The assessment of the current farming system and management is an important element of the conversion planning process, providing the opportunity for identifying the personal and resource limitations that might be faced during conversion (Schmid, 1987). The planning of the organic target or endpoint requires the design of an appropriate rotation and stocking plan, in line with the farmer's objectives and marketing options, and the testing of the technical and financial feasibility of the proposed production activities. The planning of the transition from the current to the target system has to consider the conversion strategy, i.e. staged or single-step conversion, as well as the timing of changes in the management and extent of crop and livestock enterprises, major soil improvements, changes in land and labour resources and capital investments. The plan should include steps to enhance the animal welfare, environmental and nature conservation characteristics of the farm. Financial aspects, in particular cash flow during the transition phase, also need to be considered. Effective planning of the transition phase requires linkages between years and, ideally, replanning during the conversion period to accommodate deviations from the original plan.

Conversion planning is likely to become increasingly important as more farmers convert to organic production, making the development of appropriate techniques and computing software a key area for research.

Methodological Issues

The analyses of conversion to organic farming reported in this part of the book involved a range of methodological approaches, including case studies, comparison trials and cross-sectional analyses of survey data. Conversion planning techniques and analyses of the impact of widespread adoption of organic farming presented in later chapters rely heavily on modelling techniques. The discussion in Chapter 3 is also relevant in the conversion context, but the aim of this section is to consider the methodological issues specifically in the context of the conversion process.

Empirical approaches

The *case-study* approach used in Chapters 13 and 14 allows the process of conversion to be observed on a limited number of (usually) commercial farms over a period of time. If the study period is long enough (up to ten years for the staged conversions reported in Chapter 13), the full process of conversion and its physical and financial implications can be observed. Often, the farms can be investigated in considerable detail, but some aspects such as forage crop yields and allocation of available labour remain difficult to record and assess. The main disadvantage of case studies is that the sample is usually very small and non-representative. Case-study research is interactive; frequent visits by the researchers to the farms are likely to influence the process of conversion. Qualitatively, case studies can help to identify and understand different strategies for conversion and their impact on the farm and the farm family, but they do not provide information on the extent to which particular strategies are applied. Ideally, case studies should be multi-disciplinary, and not restricted solely to financial factors. For example, Patriquin *et al.* (1986) studied a farm in Nova Scotia, with the emphasis on the effects of conversion on soils, nutrient availability, nitrogen fixation, yields, weed–crop interactions and costs.

The *cross-sectional analysis of survey data* approach presented in Chapter 16 has advantages in terms of low costs and the absence of disturbances through changes in technology and prices, particularly when compared with long-term observations in case studies. Random selection of the sample provides additional advantages in terms of representativeness and generalizability. However, difficulties in obtaining data on important parameters, such as farm profit, in large-sample surveys, and the possible interference of unobserved factors with the 'time under organic management' are disadvantages of the method (Chapter 16).

The detailed observation of the conversion process on an *experimental unit* was described by Peters (Chapter 15). The conversion process has also been monitored in detail on experimental farms in Germany (Vollmer and Buchner, 1985; Reinken and Buchner, 1987) and Switzerland (Lienhard, 1988). Other comparative farming systems trials in Europe (e.g. Besson *et al.*, 1978; Zadoks, 1989) did not include a specific analysis of the conversion process. As with case studies, experimental trials have to be carried out over a long enough period to capture the full impacts of transition. The main advantage of the experimental unit approach is the availability of accurate data on physical aspects of the conversion. However, experimental units may differ significantly from commercial holdings in terms of management objectives and structure, livestock integration, labour allocation, capital investments and social circumstances, complicating the assessment of physical and financial performance. Any results obtained are very site-specific and therefore unrepresentative, and may well have been predetermined by decisions made by researchers at the start of the trial.

Modelling approaches

Models of the physical and financial impacts of conversion on a farm scale have been developed in the context of research as well as for the purpose of conversion planning using three different modelling techniques. There is clearly some ground for debate as to which approach is preferable for which purpose. A research project will have different requirements than the development of a model for the purpose of conversion planning. For the latter, a model needs to be able to capture the specific characteristics of organic farming, to be transparent to the farmer and must suit the working situation of advisers (Dabbert *et al.*, 1991).

Budgeting techniques are widely applied to farm planning problems and form the basis for several models of organic farming and the conversion process (e.g. Lampkin *et al.*, 1987; Hermanowski, 1989; Hanson *et al.*, 1990; Freyer, 1991). In several cases the budgeting approach has been expanded to include other factors such as crop nutrients and livestock feeds. The budgeting approach forms the basis of spreadsheet-based conversion planning software, such as ECO-PLAN (Dabbert *et al.*, 1993), KOPLAN (Freyer *et al.*, 1993) and BIOPLAN (Zehr and Zerger, 1993). Budgeting approaches have the advantage that they are readily comprehensible and transparent to farmers, and they can be relatively easily applied by advisers. The major disadvantage of the budgeting approach is that it is non-optimizing, i.e. does not give solutions which yield the highest income levels or net worth growth.

Optimization techniques such as programme planning and linear programming have been used to investigate the implications of conversion (Dabbert, 1990; Dunbar, 1991). Linear programming can incorporate the same variables as budgeting, and in a single matrix rather than separate budgets, if suitable constraints are included. For example, a nitrogen or soil organic matter constraint with supply coefficients representing the required or acceptable level of surplus or deficit can replace separate nitrogen and organic matter budgets (Illi, 1990; Dabbert *et al.*, 1993). The main disadvantage of linear programming is its lack of acceptance by advisers and farmers, due to the perceived complexities of matrix construction and the lack of transparency of the results. Other difficulties include the accommodation of non-financial objectives and multi-period analyses, particularly during the transition phase. One linear programming-based conversion planning package, GENLP (Dabbert *et al.*, 1993), is currently available, and includes facilities to assist with matrix construction, easy manipulation of assumptions, and data capture from standard data banks.

Braun (1990) contrasts the application of budgeting and optimization techniques with the planning of organic farms and concluded that, with an experienced user, the outcome of both approaches might not be very different. Some combination of approaches may be sensible, but would require integration of both into one software package (Dabbert *et al.*, 1991). Such a combination might include the use of multi-period and multiple-objective optimization techniques to achieve the best balance between maximum income, maximum rate of conversion, minimum risk and minimum cost. There is also a need to combine modelling software with software for the analysis of organic farms, such as AUSWERT (Zerger and Padel, 1990), to improve the available data sets. The acceptance of linear programming, and its integration with budgeting approaches, might increase in future now that some common spreadsheet packages include optimization tools.

Two German models use a *dynamic simulation* approach: FELDSIM (Bossel *et al.*, 1988) and ECOSIM (Möller, 1993). Since several aspects of the conversion of livestock farms appear to be under-researched, complete simulation might be difficult to implement. The time required for model development and a lack of transparency are major disadvantages of the simulation approach, but without published results the feasibility of simulation techniques is difficult to judge.

All modelling techniques can be criticized on the grounds that the assumptions may be unrealistic and that insufficient reliable data are available. Case studies, experimental units and survey data can be sources of such data, although the use of the first two is restricted by representativeness and generalizability considerations. Data availability for model building is improving with the publication of *standard data collections* specifically for organic farming (Redelberger *et al.*, 1990; KTBL, 1991; Lampkin and Measures, 1994). These data collections are usually derived from established organic systems, but can be used to define the organic 'target' and some aspects of the transition phase in conversion planning.

Summary and Conclusions

The literature relating specifically to physical and financial aspects of conversion to organic farming is more limited than that relating to established organic systems. However, on the basis of this literature, it is possible to draw some tentative conclusions:

- Farmers convert to organic farming for a variety of reasons; although financial motives have become more important in recent years, they are still not dominant.
- Farmers who have adopted organic farming show similar characteristics to innovators and early adopters of other innovations in agriculture as described by the adoption/diffusion model.
- The process of conversion to organic farming involves changes in attitudes as well as changes and innovations in the enterprise mix, marketing strategies and production practices.
- Testing organic farming on a small scale is difficult because of its system character: alternatives include staged conversions, which allow learning costs and risk to be spread over a period of years, and conversion planning, ideally using a multi-period combination of optimization and budgeting techniques, making the development and assessment of appropriate techniques and software a key area for research.

- The evidence for a conversion-specific decline in crop yields below the yield of established organic systems is weak, except in the case of forage output from grassland and associated stocking rates; careful preparation and planning can help reduce yield losses due to husbandry errors.
- Additional labour requirements during conversion are associated with new enterprises and changes in marketing rather than increased labour demand for existing enterprises.
- The lack of premium prices, conversion-related investments and information-gathering costs have been identified as the most important financial constraints during conversion.

Acknowledgement

The work on this chapter has been partly funded by the EC (AIR CT 92 0776).

References

Andrews, R.W., Peters, S.E., Janke, R.R. and Sahs, W.W. (1990) Converting to sustainable farming systems. In: Francis, C.A., Flora, C.B. and King, L.D. (eds) *Sustainable Agriculture in Temperate Zones.* John Wiley, New York. pp. 281–313.

Benecke, J., Kiesewetter, B. and Urbauer, H. (1988) *Bauern stellen um.* Alternative Konzepte, 62. Verlag C.F. Müller, Karlsruhe.

Besson, J.-M., Vogtmann, H., Lehmann, V. and Augstburger, F. (1978) DOK: Versuchsplan und erste Ergebnisse eines Projektes zum Verlgleich von drei verschiedenen Anbaumethoden. *Schweizerische landwirtschaftliche Forschung* 17(3/4), 191–209.

Blobaum, R. (1983) Barriers to conversion to organic farming practices in the mid-western United States. In: Lockeretz, W. (ed.) *Environmentally Sound Agriculture.* Praeger, New York.

Bossel, H., Kramer, P., Krieger, H., Schuller, U., Trost, N. and Zerger, U. (1988) *Dokumentation zum Fruchtfolgeberatungssystem FELDSIM.* AG Resourcen- und Systemforschung RSF 88-1, Gesamthochschule Kassel.

Braun, J. (1990) Betriebsvoranschlag, Programmplanung und lineare Programmierung im alternativen Landbau. Diplomarbeit, Institut für landwirtschaftliche Betriebslehre, Universität Hohenheim, Stuttgart.

Chadwick, L. and McGregor, M. (1991) Non-organic farmers' perceptions of and attitudes towards organic farming. In: *Organic Farming Centre Annual Report 1990.* School of Agriculture, Edinburgh.

Clarke, R. (1991) *Farmers' Attitudes to Organic Farming Practices in Victoria.* Department of Agriculture, Victoria.

Dabbert, S. (1990) *Zur optimalen Organisation alternativer landwirtschaftlicher Betriebe. Untersucht am Beispiel organisch-biologischer Haupterwerbesbetriebe in Baden-Württemberg.* Agrarwirtschaft Sonderheft 124. Verlag Alfred Strothe, Frankfurt am Main.

Dabbert, S., Freyer, B. and Braun, J. (1991) Betriebsvoranschlag und lineare Programmierung als Hilfsmittel für die Beratung alternativer landwirtschaftlicher Betriebe. In: Budde, H.-J., Geidel, H. and Schiefer, G. (eds) *Referate der 12. Jahrestagung der Gesellschaft für Informatik in der Land-, Forst- und Ernährungswirtschaft, Göttingen, Oktober 1991.* Agrarinformatik, 21. Ulmer, Stuttgart, pp. 1–7.

Dabbert, S., Braun, J. and Müller, U. (1993) *Betriebsvoranschlag und lineare Programmierung als Hilfsmittel für die Betriebsplanung alternativer landwirtschaftlicher Betriebe.* Arbeitsbericht. Institut für landwirtschaftliche Betriebslehre, Universität Hohenheim, Stuttgart.

Dettmer, J. (1986) *Motive und sozial-ökonomische Probleme der Umstellung auf ökologischen Landbau aus der Sicht von ökologisch wirtschaftenden Bauern.* Studien zur ländlichen Humanökologie, Institut für Rurale Entwicklung, Georg-August-Universität, Göttingen.

Dunbar, T.V. (1991) An economic evaluation of alternative agriculture production methods in south central Pennsylvania. MSc thesis, Department of Agricultural Economics and Rural Sociology, Pennsylvania State University, University Park, Pennsylvania.

Fischer, R. (1982) *Der andere Landbau: Hundert Bio-Bauern und Gärtner berichten über ihre Beweggründe, Arbeitsweisen und Erfahrungen.* Verlag Buchhandlung Madliger-Schwab, Zürich.

Fisher, P. (1989) Barriers to the adoption of organic farming in Canterbury. MSc thesis Centre for Resource Management, Lincoln University, Lincoln, New Zealand.

Freyer, B. (1991) *Ökologischer Landbau. Planung und Analyse von Betriebsumstellungen.* Verlag Margraf, Weikersheim.

Freyer, B., Hartnagel, S., Pericin, C. and Leyerer, A. (1993) *KOPLAN, EDV gestützes Kalkulationsprogramm für die Betriebskontrolle und Betriebsplanung.* Forschungsinstitut für biologischen Landbau, Oberwil/BL, Switzerland.

Hanson, J.C., Johnson, D.M., Peters, S.E. and Janke, R.R. (1990) The profitability of sustainable agriculture on a representative grain farm in the mid-Atlantic region, 1981–1989. *Northeastern Journal of Agricultural and Resource Economics* 19(2), 90–98.

Henning, J., Thomassin, P. and Baker, L. (1990) *Organic Farmers in Québec: Results of a Survey.* Department of Agricultural Economics, McGill University, Macdonald Campus, Quebec.

Hermanowski, R. (1989) *Vergleich alternativer und konventioneller landwirtschaftlicher Betriebe in Hessen.* Regionalpolitik und Umweltschutz im ländlichen Raum. Wissenschaftlicher Fachverlag Dr Fleck, Niederkleen, Germany.

Illi, S. (1990) Möglichkeit und Bedeutung der Humusbilanzierung im ökologischen Landbau. Diplomarbeit, Fachbereich Landwirtschaft, Gesamthochschule Kassel, Witzenhausen.

KTBL (1991) *Datensammlung alternative Landwirtschaft.* Kuratorium für Technik und Bauwesen in der Landwirtschaft, Darmstadt.

Lampkin, N.H. (1990) *Organic Farming.* Farming Press, Ipswich.

Lampkin, N.H. (1993) The economic implications of conversion from conventional to organic farming systems. PhD thesis, Department of Agricultural Sciences, University of Wales, Aberystwyth.

Lampkin, N.H. and Measures, M. (eds) (1994) *1994 Organic Farm Management Handbook.* Department of Agricultural Sciences, University of Wales, Aberystwyth and Elm Farm Research Centre, Newbury.

Lampkin, N., Jones, W., Mann, M. and Midmore, P. (1987) *Proposal for an Integrated Rural Development Programme Focusing on Organic Agricultural Production in the Teifi Valley, Dyfed.* I. Department of Economic and Agricultural Economics, University College of Wales, Aberystwyth.

Lienhard, G. (1988) *Wahrheitssuche im biologischen Landbau.* Verlag des Kantons Basel-Land, Liestal.

Lockeretz, W. and Madden, P. (1987) Midwestern organic farming: a ten-year follow-up. *American Journal of Alternative Agriculture* 2, 57–63.

MacRae, R.J., Hill, S.B., Henning, J. and Mehuys, G.R. (1989) Agricultural science and sustainable agriculture: a review of the existing scientific barriers to sustainable food production and potential solutions. *Biological Agriculture and Horticulture* 6, 173–219.

MacRae, R.J., Hill, S.B., Mehuys, G.R. and Henning, J. (1990a) Farm-scale agronomic and economic conversion from conventional to sustainable agriculture. *Advances in Agronomy* 43, 155–198.

MacRae, R.J., Hill, S.B., Henning, J. and Bentley, A.J. (1990b) Policies, programs and regulations to support the transition to sustainable agriculture in Canada. *American Journal of Alternative Agriculture* 5, 76–92.

Measures, M. (1990) Converting to organic farming. In: Clunies-Ross, T. and Weisselberg, T. (eds) *Organic Farming: An Option for the Nineties.* British Organic Farmers/Organic Growers Association, Bristol.

Möller, D (1993) ECOSIM – Ein dynamisches Systemsimulationmodell zur mittelfristigen Betriebsplanung im ökologischen Landbau. In: Zerger, U. (ed.) *Forschung im ökologischen Landbau.* Sonderausgabe 42. Stiftung Ökologie und Landbau, Bad Dürkheim, pp. 353–359.

NatWest (1992) *National Farm Survey.* National Westminster Bank, Coventry.

NRC (1989) *Alternative Agriculture.* National Academy Press, Washington, DC.

Olbrich, M. (1992) Akzeptanz von Umstellungsbeihilfen: Beurteilung des EG-Extensivierungsprogramm durch saarländische Landwirte. *Berichte über Landwirtschaft* 70, 181–212.

Padel, S. (1988) Umstellungsplanung: Risikoabschätzung für zukünftige Bio-Bauern. *Bioland* 88/3, 16–19.

Padel, S. (1994) Adoption of organic farming as an example of the diffusion of an innovation: a literature review. Discussion Paper 94/1. Centre for Organic Husbandry and Agroecology, University of Wales, Aberystwyth.

Patriquin, D.G., Hill, N.M., Baines, D., Bishop, M. and Allen, G. (1986) Observations on a mixed farm during the transition to biological farming. *Biological Agriculture and Horticulture* 4, 69–154.

Ramsay, D.A. and O'Mara, M.M. (1991) *Economics.* Final Report, III. Organic Farming Centre, University of Edinburgh, Edinburgh.

Rantzau, R., Freyer, B. and Vogtmann, H. (1990) *Umstellung auf ökologischen Landbau. Betriebliche Erfordernisse und Konsequenzen bei der Durchführung*

des biologischen Landbaus. Angewondte Wissenschaft, Schriftenreihe des Bundesministers für Ernährung, Landwirtschaft und Forsten, Heft 389, Landwirtschaftsverlag, Münster Hiltrup.
Redelberger, H., Scharf, A. and Voelkel, G. (1990) *Datenmaterial für die Betriebsplanung – ökologischer Landbau – Pflanzenproduktion.* Information für Beratung und Verwaltung, 115/90. Hessisches Landesamt für Ernährung, Landwirtschaft und Landentwicklung, Kassel.
Reinken, G. and W. Buchner (eds) (1987) *Biologisch-dynamischer Land- und Gartenbau. Vortragstagung zu den Forschungsvorhaben 'Alternative Erzeugung von Äpfeln und Gemüse', 'Alternativer Landbau Boschheide Hof'.* Forschung und Beratung, C, 44. Landwirtschaftsverlag, Münster-Hiltrup.
Richter, F. (1990) Du merkst aufeinmal, dass es nicht nur ein Traum ist. Diplomarbeit. Fachbereich Landwirtschaft, Gesamthochschule Kassel, Witzenhausen.
Rogers, E.M. (1983) *Diffusion of Innovations.* The Free Press, New York.
Schmid, O. (1987) *Besonderheiten bei der Umstellung auf Biolandbau: Konsequenzen für die Planung.* Kursunterlagen zum Bioberaterseminar 15.9.87. Forschungsinstitut für biologischen Landbau, Oberwil/BL, Switzerland.
Stanhill, G. (1990) The comparative productivity of organic agriculture. *Agricultural Ecosystems and Environment,* 30, 1–26.
Svensson, I. (1991) *Statligt stöd till alternativ odling 1989.* Alternativ odling, 7. Research Committee for Alternative Agriculture, Uppsala.
Taylor, D.L. and Miller, W.L. (1978) The adoption process of environmental innovations: a case study of a government project. *Rural Sociology* 43, 634–648.
Vine, A. and Bateman, D.I. (1981) *Organic Farming Systems in England and Wales: Practice, Performance and Implications.* Department of Agricultural Economics, University College of Wales, Aberystwyth.
Vogtmann, H., Freyer, B., and Rantzau, R. (1993) Conversion to low external input farming systems: A survey of 63 mixed farms in West Germany. Paper presented at the 'Agroecology and Conservation Issues in Temperate and Tropical Regions' conference, Padua.
Vollmer, F.-J. and Buchner, W. (eds) (1985) *Zur Umstellung eines landwirtschaftlichen Betriebes der Rheinniederung auf biologisch-dynamische Wirtschaftsweise: Forschungsvorhaben 'Alternativer Landbau Boschheide Hof' (1979-1983).* Forschung und Beratung, C, 41. Landwirtschaftsverlag, Münster-Hiltrup.
Wernick, S. and Lockeretz, W. (1977) Motivations and practices of organic farmers. *Compost Science* 77, 20–24.
Wynen, E. (1990) *Sustainable and Conventional Agriculture in South-eastern Australia: A Comparison.* Economics Research Report No. 90.1. School of Economics and Commerce, La Trobe University, Bundoora, Australia.
Zadoks, J.C. (ed.) (1989) *Development of Farming Systems – Evaluation of the Five-year Period 1980–1984.* Pudoc, Wageningen.
Zehr, M. and Zerger, U. (1993) *Handbuch für das Betriebsplanungsprogramm BIOPLAN.* Sonderausgabe 47. Stiftung Ökologie und Landbau, Bad Dürckheim.
Zerger, U. and Padel, S. (1990) *AUSWERT – ein EDV-Programm zur Betriebsauswertung ökologisch wirtschaftender Betriebe.* Fachbereich Landwirtschaft, Gesamthochschule Kassel, Witzenhausen.

Part 4

Widespread Conversion to Organic Farming: Modelling the Regional Impacts

18 Comparative Analysis of Future Development Paths for Agricultural Production Systems in Germany

ULI ZERGER AND HARTMUT BOSSEL

Introduction

This chapter presents some results from a study commissioned by the (West) German federal parliament (*Bundestag*) and completed in 1986 (Bossel *et al.*, 1987). The objective of the study was the development of a systems analysis instrument suitable for agricultural policy development. The instrument was designed to assess the range of the possible development paths for agricultural production in West Germany and their agricultural, ecological, economic and social consequences under the influence of new technical and agricultural policy developments over a period of several decades.

The most relevant development perspectives were defined in three scenarios: intensification (A); business as usual (B); and ecologization (C). Organic farming, as defined by IFOAM (1992) or the EC (1991), did not form an exclusive scenario, but it was seen as one of the most important means to realize scenario C. The main results for this scenario can therefore be taken as an indication of the effects that widespread adoption of organic farming in Germany would have.

Essential Parts of a Technology Assessment Study

A technology assessment study for a complex system requires four distinct components:

1. A systems description appropriate to the given problem, which must contain all significant systems elements and their causal linkages. Normally, the system structure will be translated into a computer simulation model.
2. A complete description of the initial system state, i.e. the initial conditions necessary for starting the calculations.
3. Scenarios for the temporal development of external influences over the time period considered. If necessary, the scenario data will consist of (mutually consistent) assumptions concerning plausible and likely development conditions. Several scenarios characterized by different guiding aspects should cover the total spectrum of possible exogenous influences in order to assess the resulting consequences and developments.
4. A criteria set based on the viability requirements of the system considered which will be used for comparative evaluation of the different development paths.

Components of the Method and their Linkages

The basis for the assessment approach is the description of the prototype farm by the sum of its individual processes of plant and animal production with their respective physical and monetary inputs and outputs. Prototype farms are defined for each production region; their development over the time period under consideration is controlled by region-specific scenarios. By aggregating the farm data to the region and finally to the nation, the corresponding regional and national indicators of agricultural production and their consequences (e.g. for food production and environment) can be obtained. The production results are compared with the food demand resulting from the population development and nutrition scenarios. The final results are evaluated using a comprehensive set of evaluation criteria, which are chosen to reflect national supply security and viability objectives.

Production activities

Twenty-three different crop production and five different animal production activities are presently implemented. These include identifiable future developments (e.g. with regard to renewable resources). Each of these activities is characterized by some 30 parameters describing the specific material and monetary inputs and outputs. Since each of these activities can be operated at different levels of intensity, five different intensity levels are defined and applied to most activities. For each of these intensity levels the corresponding input and output parameters are defined for each production activity on the basis of empirical data. It is assumed that these technical and monetary data change over time, and corresponding temporal development scenarios are incorporated in the model.

Description of farm prototypes

The regional prototype farm is described by a region-specific mix of production activities and intensities of production. The data for the prototype farms are defined to be as close as possible to the data of the official agricultural statistics (*Agrarbericht* and *Statistisches Jahrbuch*). These initial conditions are assumed to change over time, where the changes in the type and the amount of production of each of the prototype farms are laid down in the scenarios. At the farm level the material, monetary, ecological and other consequences of the corresponding agricultural production are computed via some 110 intermediate variables, among them a set of indicator variables which are used to evaluate the current system state of the regional prototype. The most important results are recorded by some 75 state indicators for each scenario point in time (every five years).

Aggregation

The production results for each agricultural region follow by aggregation of the data of the regional prototype farm via the regional number of farms and the regional agricultural area. The results are collected in some 35 state indicators. The production data are obtained by summation of the regional data over all of the 15 agricultural regions.

National food demand

The demand for food products is computed on the basis of population development and considering age-specific food consumption data. The results are very much dependent on the scenario assumptions for the average national diet. Three scenarios are assumed: maintenance of 1983 per capita consumption levels (0); increasing per capita meat consumption (1); and reduction in meat consumption to 1960 levels (2). Comparing these demand data with the production data leads to specific indications concerning production deficits or surpluses to be expected. It is further assumed that demand scenarios 0, 1 and 2 are linked to supply scenarios B (business as usual), A (intensification) and C (ecologization), respectively, given the likelihood that production and consumption trends will be influenced in similar directions by policy changes and environmental or other pressures.

Evaluation method

The results of the computations of the state indicators at the regional and national level are compared to a set of evaluation criteria which are selected to allow comprehensive assessment of the development path with regard to food supply, and the viability of farms and of society.

Application of the Method

The application of the method requires several preparatory steps.

Regional disaggregation

In order to be able to consider the regional differences based on natural and historical development in the Federal Republic of Germany, the total agricultural area is disaggregated into 15 agricultural regions. These regions were obtained by aggregation of some of the 42 agricultural areas of the official agricultural statistics. For each agricultural region the initial data of a prototype farm were defined on the basis of data from official agricultural statistics and other sources.

General development scenarios

In order to describe the systemic alternatives and their consequences, the spectrum of the possible external developments (technical progress, agroeconomic conditions and agropolitical policies) must be collected in corresponding, plausible and consistent scenarios. In this study three scenarios are used:

- *Intensification* (A): The essential element is the reduction of trade restrictions and market distortions which would lead to stronger intensification of agricultural production in particular regions for economic reasons. Regions which are no longer competitive under these conditions are taken out of agricultural production.
- *Business as usual* (B): The major theme is the assurance of continued operation of the most productive farms in their current form. This means that markets cannot be liberalized in the way planned in scenario A. For this reason, a similar intensification and concentration of production cannot take place.
- *Ecologization* (C): The major impulse is a more conscious orientation of agricultural production towards decreasing ecological impacts and supporting natural nutrient cycles, coupled with an attempt to

Table 18.1. Qualitative description of production trends in three scenarios for the development of western German agriculture, 1985–2025.

Product	Scenario A (intensification)	Scenario B (business as usual)	Scenario C (ecologization)
Cereals	– –	–	–
Oilseeds	–	+	+
Legumes	–	+	+
Row crops	–	0	–
Renewable resources	++	+	0
Fodder crops	–	0	+
Grassland	–	0	+
Fallow area	++	+	0
Cattle	– –	0	–
Pigs	+	0	–

++, Large increase; +, small increase; 0, no change; –, small decrease; – –, large decrease.
Source: Adapted from Bossel *et al.* (1987).

support the traditional small family farm structure. Organic farming is one of the techniques that can be used to realize this scenario.

The production trends for all three scenarios are shown in Table 18.1. The qualitative scenario descriptions have to be translated into corresponding quantitative parameters or time functions to allow computation. For the technical and economic parameters, identical conditions were assumed for all of the regions for a given scenario.

To assess regional farm development, certain parameters must be quantified and specified for each region and for each scenario (farm production area, extent of each production activity, introduction of new activities, etc.). Following the prescription and quantification of the scenario assumptions, the temporal developments of production data and production consequences are computed for each regional prototype, aggregated to the regional and national level and compared to the demand data and finally evaluated using the comprehensive set of criteria.

General Results of the Development Path Study

Despite the preliminary nature of the study, some general results can be obtained.

Table 18.2. Relative changes in human nutritional consumption patterns, western Germany, 1985 and projected to 2025 (1950 = 100).

Nutrient	1985	Scenario 0[a]	Scenario 1[b]	Scenario 2[c]
Animal proteins	157	147	159	137
Total proteins	113	105	103	108
Fats	140	131	136	127
Carbohydrates	77	72	68	77
Energy	105	98	98	98

[a] Static meat consumption – corresponds to production scenario B.
[b] Increased meat consumption – corresponds to production scenario A.
[c] Reduced meat consumption – corresponds to production scenario C.
Source: Adapted from Bossel *et al.* (1987).

- Food consumption in the Federal Republic of Germany will fall by some 20% in the next few decades, mainly as a result of demographic change (Table 18.2).
- Reduction in demand for animal products has a proportional effect on agriculture, since this affects grain production (for livestock feed) significantly.
- It is not possible to secure the operation of the current numbers of farms under the conditions of a free agricultural market, with increasing yields and decreasing demand.
- There are only a few ways of alleviating this dilemma: to reduce the acreage for plant production and the number of farms significantly; to extensify production, or to subsidize production and its subsequent destruction; to subsidize food exports; or to decouple the internal market prices for agricultural products from the world market, bringing them to a level high enough to guarantee the continual existence of a greater number of extensively producing farms.
- Scenario A, which assures survival only for the most productive farms using intensive production while large unproductive areas would be taken out of production, certainly has the most severe social and ecological impacts. On the other hand, this is a scenario which under the current conditions seems to have the best economic and political prospects. Here, a problem area with high decision pressure arises if the development of socially unacceptable conditions is to be avoided.
- Scenario B, which assumes only incremental policy changes to avoid the most obvious negative impacts, must also be assessed as having

overwhelmingly negative consequences. In this case, subsidized production of agricultural products continues to be a necessity.

- Scenario C, in which an attempt is made to preserve the current small-farm production patterns as much as possible by extensification, does show the most favourable ecological and social impacts, but it also has to deal with difficult problems of economic and social acceptance. The technology assessment for this scenario again shows the necessity for timely decisions and action in order to favour desirable developments by switching policies, and by blocking undesirable developments in time.

In a subsequent study, Bossel *et al.* (1989) were able to revise the original model. The main question in this later study was whether the scenario assumptions were conclusive and consistent, and which consequences resulted from each of the development paths. In addition to the three original paths, a fourth path, *pluralization*, was developed. This was based on the concept that agriculture will not follow a single, uniform line of development, but instead, due to varying regional circumstances, clearly differentiated farm developments may be expected. A trend away from mass production to the production of specialist regional products might be expected. A summary of the two studies, as well as an economic evaluation, can be found in Enquete-Kommission (1990).

Assessment of the Impacts of Widespread Organic Farming on the Basis of this Study

As already mentioned, organic farming was not defined as one of the scenarios, but it is one of the most important approaches to realizing scenario C. In the following section an attempt has been made to assess the effects of widespread adoption of organic farming on the basis of the results for this development path by contrasting the situation in 1985 with that in 2025 (Tables 18.3–18.6):

- Grain production will be reduced by about 40% due to reduced area and a lower level of intensity; the share of wheat will decrease while the share of oats and rye will increase.
- Oilseed production will increase slightly, but most important will be the increase of grain legumes because of crop rotation demands.
- The production of sugarbeet will clearly be reduced while more potatoes will be harvested.
- Renewable resources (e.g. bio-fuels) will not be grown, because the

arable land will be needed for food production purposes.

- Cattle numbers will be reduced by 20% and will only be half as much compared to 1985; milk production will also be reduced by about 40% because of the reduction in use of imported feed grains.
- Erosion levels will be reduced by about 50%, which will be one of the major environmental benefits.
- This scenario will result in a 40% decline in the number of farms, much less than the other scenarios, with a similar impact on the number of persons employed in agriculture.
- Farm incomes will be slightly lower than in 1985 as long as farmers are not able to achieve premium prices for their products.

Bechmann (1987) presented a study which examined whether a 100% conversion of German farmland is possible, and what consequences might result from this scenario. On the basis of average yields as well as specific rotations, he calculated the total output of crops and livestock which might be expected. He estimated that total crop output would fall (cereals by 45%, wheat by 38%, oilseed rape by 79% and sugarbeet by 29%), although there would be an increase in some individual crops (71% more potatoes, 700% more grain legumes). This would still be sufficient for human requirements, although the availability of feeds for livestock would decline substantially, also as a result of the restrictions on imported feedstuffs. While the number of ruminant livestock would fall by only 11%, the numbers of pigs would fall by 63% and poultry by 39% as a result of the feed restrictions, and assuming no improvement in the productivity of organic production methods. Bechmann assumes that a reduction in meat consumption is in any case nutritionally desirable, and therefore concludes that no one

Table 18.3. Relative changes in land use, western Germany, projected to 2025 (1985 = 100).

Land use	Scenario A (intensification)	Scenario B (business as usual)	Scenario C (ecologization)
Agricultural land	89	89	89
Grassland	59	84	91
Arable	68	86	86
Cereals	39	69	68
Row crops	29	71	86
Fodder crops	55	65	100

Source: Adapted from Bossel *et al.* (1987).

Table 18.4. Land use by type as a percentage of total agricultural and arable areas, western Germany, 1985 and projected to 2025.

Land use	1985	Scenario A (intensification)	Scenario B (business as usual)	Scenario C (ecologization)
Agricultural land	100	100	100	100
Grassland	37	25	35	38
Permanent crops, woodland, natural	2	28	4	2
Arable	62	47	60	60
Arable land	100	100	100	100
Cereals	69	40	54	53
Oilseeds	3	7	7	6
Grain legumes	–	–	4	11
Row crops	11	5	9	11
Bio-fuels	–	33	11	–
Fodder crops	16	13	12	19

Source: Adapted from Bossel *et al.* (1987).

Table 18.5. Relative changes in livestock output, western Germany, projected to 2025 (1985 = 100).

Product	Scenario A (intensification)	Scenario B (business as usual)	Scenario C (ecologization)
Dairy cows	33	65	54
Finishing cattle	58	63	86
Sows	40	80	60
Finishing pigs	52	75	51
Milk	53	82	63
Cattle meat	52	74	81
Pig meat	70	93	57

Source: Adapted from Bossel *et al.* (1987).

in Germany will starve as a consequence of 100% adoption of organic farming. At the same time he estimated the creation of about 90,000 new jobs in agriculture.

Bechmann's results correspond in many respects with those of Bossel *et al.* (1987). The most obvious difference lies in the development

Table 18.6. Relative changes in resource use and incomes, western Germany, projected to 2025 (1985 = 100).

	Scenario A (intensification)	Scenario B (business as usual)	Scenario C (ecologization)
Number of farms	38	45	61
Labour force (full time)	32	46	65
Farm profit	163	250	88
Nitrogen fertilizer use	141	115	75
% of which surplus (1985 = 58)	40	40	17
Pesticide use	259	216	57
Concentrate use	52	79	53
% of which home-produced (1985 = 37)	44	52	86
Soil erosion	77	83	56

Source: Adapted from Bossel *et al.* (1987).

of oilseed production, but the time scale of the two studies is also different. In a later study, Bechmann *et al.* (1993) were able to confirm most of these results, although the main question was whether 100% conversion to organic farming could be financed. The study concluded that conversion to organic farming in Germany was possible on a financially neutral basis given current levels of agricultural support, although conventional prices for all products were assumed. The precondition for this is that the conversion-related income reductions are compensated for by a redistribution of a proportion of current agricultural subsidies (about DM 10 billion). This support would be paid in the form of a direct income payment for reduced yields. Thus it is possible to make use of the current market regulation/structural funds.

Summary and Conclusions

The main results of the Bossel study are that food demand is likely to fall by 20% by 2025 and a reduced demand for meat would have a relatively greater impact on agricultural production than a reduction in demand for crop products. Ecologization would reduce cereal and sugarbeet output, while increasing the output of potatoes and grain legumes. Some reduction in livestock production is predicted and farm income, without premium prices, would remain at the level of 1985.

In comparison to the other scenarios, ecologization shows the clearest advantages for the environment and health, while the financial consequences are difficult to estimate. However, this path can only be implemented if farmers can be motivated to convert by risk reduction measures and an appropriate policy environment (Bossel *et al.*, 1989).

The changed initial situation resulting from German reunification led to an examination of the future scenarios for agriculture in the new federal states for the European Parliament. Kramer *et al.* (1993) confirmed the basic validity of the four scenarios under the conditions in the new states as well, although reunification resulted in completely new initial positions for agricultural development. The development of further environmental and nature conservation measures as well as specific support for extensive production methods can result in a significant increase in the proportion of organically managed farms in a short time. The authors found that organic farming had the greatest potential for development in the less favoured areas and in the vicinity of centres of population.

References

Bechmann, A. (1987) *Landbau-Wende, Gesunde Landwirtschaft – Gesunde Ernährung.* S. Fischer Verlag, Frankfurt am Main.

Bechmann, A., Meier-Schaidnagel, R. and Rühling, I. (1993) *Landwirtschaft 2000 – Die Zukunft gehört dem ökologischen Landbau.* Institut für ökologische Zukunftsperspektiven, Barsinghausen.

Bossel, H., Kramer, P., Schaffner, J., Weismantel, H. and Zerger, U. (1987) *Alternativen landwirtschaftlicher Produktionsweisen – Technikfolgenabschätzung für die landwirtschaftliche Produktion.* Materialenband V zur BT-Drucksache 10/6801, Bonn.

Bossel, H., Kramer, P., Müller-Reißmann, K.F., Schaffner, J., Weismantel, H. and Zerger, U. (1989) *Perspektiven zur Zukunft der Landwirtschaft (Agrarpfade) – Studie im Auftrag der Enquete-Kommission 'Technikfolgen-Abschätzung und -Bewertung'.* Universities of Hanover and Kassel.

EC (1991) Council Regulation (EEC) No 2092/91 of 24 June 1991 on organic production of agricultural products and indications referring thereto on agricultural products and foodstuffs. *Official Journal of the European Communities* 91(L198), 1–15.

Enquete-Kommission (1990) *'Technikfolgenabschätzung und -Bewertung': Landwirtschaftliche Entwicklungspfade.* Drucksache 11/7991, Deutscher Bundestag, Bonn.

IFOAM (1992) *Basic Standards of Organic Agriculture.* International Federation of Organic Agriculture Movements, Tholey-Theley, Germany.

Kramer, P., Müller-Reißmann, K.F. and Schaffner, J. (1993) *Perspektiven zur Zukunft der Landwirtschaft in den neuen Bundesländern – Studie im Auftrag der STOA des Europäischen Parlaments.* Systemforschung Stadt und Land, Sachsenhagen.

19 Impacts of Widespread Conversion to Organic Agriculture in the State of Baden-Württemberg, Germany

JÜRGEN BRAUN

Introduction

The current situation of agriculture in the European Union is characterized by decreasing product prices and decreasing incomes, by surpluses on the most important markets, which is a burden on public expenditure and world markets, and by increasing pollution of the environment due to agriculture. Against this background organic agriculture and extensification are becoming more important. Therefore it is necessary to get some indication of the implications of a region-wide transition to organic agriculture in terms of land use, output and income.

In this chapter this will be investigated using preliminary results of a model calculation for the state of Baden-Württemberg. Therefore the optimized current situation (1990) of agriculture in Baden-Württemberg is described. In a further step the scenario of state-wide conversion to organic agriculture is developed. The implications of a complete conversion of Baden-Württemberg's agriculture to organic agriculture are derived using a comparative-static analysis. First of all the development of the model is explained.

Methods

Requirements

The requirements of a model are determined primarily by the study's goal and the availability of data. The main goal of the study is to find out the impact of converting a region's agriculture on production (quantity and structure) and farm income.

Census data, regional data collections and estimated data are available as a data base. An empirical study of farm-level data is not possible considering the high input of time and work which would be necessary. Therefore the following requirements can be formulated:

- It must be possible to include the farm-level interrelationships between production enterprises in order to investigate the farm's optimal adjustment to changed input–output relationships and price levels.
- It must be possible to apply the farm-level results on a regional level without an intolerable level of aggregation bias.

Type of model

Process analysis models, especially linear programming models, are well suited to the analysis of the impacts of changing production circumstances on farm management. Linear programming models provide an optimal structure of production while keeping to specified secondary conditions. By varying different parameters, changes in farm organization and income implications can be analysed. As a result, the optimal adaptation of farm organization to changed circumstances can be obtained. Linear programming also seems to be well suited to the study's goal, which is the conversion of farms to organic management. By formulating a linear programming model, the complementary and competitive effects existing between production activities can be investigated and a valuation of the complementary effects which have to be emphasized in organically managed farms is possible. Linear programming therefore helps to illustrate Brinkmann's theory of competitive and complementary effects on which organic farming is based (Dabbert, 1990).

Planning period

A comparative-static comparison is used in order to compute the optimal quantity and structure of production of the region's farms and the adaptation of farms to changed circumstances between two time periods. Although a dynamic model would be desirable against the background of changes in marketing strategies due to conversion, and their implications for cash flow on farms, this has not been constructed for the following reasons:

- Adequate data, relating for example to changing rotational practices and marketing during the transition process, are not available or are insufficient.
- The development of a dynamic model would involve a significant increase in time and computational resources.

Delineation of model regions

Many macroeconomic problems are too complex to be completely described in a single model. Therefore suitable aggregates need to be constructed from the various model variables. However, aggregation problems can occur when investigating changes in aggregated variables without referring directly to the initial variable. Possible methods for partly solving the aggregation problem are described elsewhere (see Day, 1963; Lee, 1966; Weinschenck, 1967).

In this study, the state of Baden-Württemberg is divided into regions to reduce the aggregation bias. The regions represent combinations of administrative districts, which are the smallest units for which data are available. The yield index, a parameter which characterizes the cropping capacity of soil, is used as a criteria for the combination of the districts. On this basis, the state of Baden-Württemberg is divided in three regions with differing soil quality, for which similar production functions for individual farms can be assumed.

Selection of model farms

For the aggregation of farms, the factor which is most restrictive for the production of a specific commodity is used as an aggregation criterion. In the context of the existing problem, which is the conversion of the whole farm, this should be the extent of farmland. Therefore, the farms are first combined using farm-size category as a criterion. All farms in the state are allocated to the following farm-size categories in each region: less than 10 ha, 10–30 ha; 30–50 ha; and more than 50 ha.

In order to consider other factor relations determining production, farms in each farm-size category are combined according to type of farming. The following farm types are considered: cropping; forage-based livestock; intensive livestock; and mixed farms. Farms with permanent cropping and farms smaller than 1 ha in size are ignored because of insufficient data. For each region, there are therefore 16 different combinations of farm type and size. These models are further differentiated according to their main production enterprise so that altogether there are 52 farm models for each region. The results of these 52 farm models are then aggregated to provide results for the whole state using factors derived from statistical data.

Results of delineation

The number of farms and the area represented by the farm models are shown in Table 19.1. About 82,000 farms with a total agricultural land area of 1,300,000 ha are represented by the models. This area consists of 60% cultivated area and 40% permanent pasture. The region with medium cropping capacity (region 2) is the largest, occupying about 50% of all farms and total farmland in Baden-Württemberg. Seventeen per cent of the farms are located in the region with high cropping capacity (region 1). Significantly, these farms use 22.5% of arable area and only 8% of permanent pasture. In the region with low cropping capacity (region 3) this relation is inverse: 32% of all Baden-Württemberg farms occupy 38% of permanent grassland and only 27% of the arable area.

Data sources

Farm, cropping and output statistics are used to derive the capacities of the representative farm models and to determine the different regional yield relationships. Livestock statistics are used to derive the livestock housing capacity. Standard data form the basis for the definition of production activities. Product prices and input costs are derived from price statistics. Nitrogen-yield functions are used to estimate the financially optimal quantity of fertilizer used in the conventional farm models.

Model validation

In order to validate the preliminary model, the optimized 1990 situation is compared against the actual 1990 situation with respect to land use. As Table 19.2 shows, there are some differences between the model and the real situation. In the model, the share of cereals is lower than in the actual situation whereas the shares of grain legumes and forage crops are higher. Oilseed and root crop production are described well by the model. The differences can be explained, on the one hand, by the previously mentioned omission of farms smaller than 1 ha and farms with permanent cropping from the model, and the omission of some cereals such as durum and spelt wheat. On the other hand, there

Table 19.1. Number of farms and agricultural area in Baden-Württemberg represented by farm models.

	Region 1 (high)[a]		Region 2 (medium)[a]		Region 3 (low)[a]		
	Number/area	% of state total	Number/area	% of state total	Number/area	% of state total	State total: number/area
Farms (number)	14,070	17.1	42,337	51.3	26,054	31.6	82,461
Farmland (ha)	218,832	16.8	673,719	51.7	410,828	31.5	1,303,379
Arable (ha)	175,994	22.5	392,411	50.1	214,845	27.4	783,250
Permanent grass (ha)	42,838	8.2	281,308	54.1	195,983	37.7	520,129

[a] Cropping capacity based on yield index (see text).
Sources: Statistisches Landesamt (1990) and own calculations.

Table 19.2. Optimal and actual structure of land use (percentage of total arable area) in Baden-Württemberg, 1990.

Crop	Optimal situation	Actual situation
Cereals	58.54	64.30
Oilseeds	8.75	8.99
Grain legumes	6.19	1.00
Forage crops	22.30	16.70
Root crops	4.23	4.10

Sources: Ministerium (1992) and own calculations.

are differences in relative profitability of cereals and forage production. The higher share of forage crops and grain legumes in the model is also a consequence of taking into account the nitrogen fixation benefits of legume crops which are underestimated in the real situation. Furthermore, set-aside is not considered in the model, so that there must be a different structure of land use. Overall it can be concluded that the preliminary results describe the actual situation well.

Assumptions for organic management

The impacts of conversion to organic production of the whole agriculture of Baden-Württemberg are evaluated under the following assumptions:

- All farms in Baden-Württemberg must convert to organic agriculture.
- Previous foreign trade protection measures are maintained.
- The demand for organically produced commodities will not increase to the same extent as supply, which means that the price level will be the level currently paid for conventional products.
- Crop yields under organic management are estimated by time-series analysis of yields in the past for each combined region. Therefore it is supposed that the yield increase consists only of breeding progress. The share of breeding progress is taken from the investigations of Schuster *et al.* (1982). As a result grain yields are 30–40% lower than under conventional management and the difference in root crop yields is 20–40%.
- Performance of dairy cows decreases by 10% on farms with large herds and intensive management.
- Bull beef will have a daily weight gain which is about 200 g lower

than in the conventional situation due to lower feed inputs for finishing.

- Daily weight gain of finishing pigs will be 10% lower.
- Laying hens will have a performance 15% lower than conventional.
- The number of farms and their distribution by farm size and type will not change due to conversion.
- The German standards for organic agriculture (AGÖL, 1991) are followed when formulating production activities.
- Conversion involves costs for rebuilding livestock housing on farms which do not currently keep animals on straw litter. Costs are derived by applying the methods of Gartung *et al.* (1982a, b) and Hammer (1981).
- Fixed costs do not change due to conversion.
- The maximum and minimum percentages of Dabbert (1990) are used to determine crop rotation practices.

Impact of 100% Conversion on Land Use, Livestock Numbers, Output and Farm Incomes

Land-use changes

As shown in Table 19.3, the structure of land use changes considerably due to conversion to organic farming. In the optimized current situation (1990), about 58% of Baden-Württemberg's arable area is cultivated with cereals. The most important cereal crops are winter wheat (with a share of about 25% of the arable area) and barley (20% of arable area). After conversion the cereal cropping area decreases by 36% to 290,000 ha or 37% of the arable area. The production of lower-yielding cereals such as rye and oats gains importance, so that they are cropped on 6% and 10% of arable area, respectively.

Under conventional management oilseed production requires nearly 9% of the arable area, the majority of which is oilseed rape. After transition to organic farming, oilseed production loses importance and is reduced by 46% to 5% of the arable area. The production of grain legumes is increased by 116% to 105,000 ha, 13.3% of total arable area.

Whereas in the optimal current situation forage crops occupy 22.3% of total arable area and consist mainly of silage maize production, this figure changes considerably due to conversion: forage cropping is extended by 63% to an area of 280,000 ha and forage maize is substituted by temporary leys. Their production increases by 370% to 33% of the total arable area, due to their role in nitrogen fixation,

Table 19.3. Optimal structure of land use in Baden-Württemberg before (1990) and after 100% conversion to organic farming.

	Before conversion		After conversion		
Crop	ha	% of total arable area	ha	% of total arable area	Change as % of 1990 level
Arable area, of which:	783,250	100.0	783,250	100.0	0
Cereals	458,482	58.5	292,950	37.4	−36
Oilseeds	68,556	8.8	37,097	4.7	−46
Grain legumes	48,450	6.2	104,551	13.4	116
Forage legumes	174,651	22.3	284,827	36.4	63
Root crops[a]	33,111	4.2	63,825	8.2	93
Utilized permanent grass	422,800	n/a	294,600	n/a	−30

[a] Including field vegetables.
n/a, Not applicable.

whereas silage maize production decreases by 80% to 3% of total arable area.

The main root crop under conventional management is sugarbeet, production of which is reduced markedly by 38% after conversion, whereas potato cropping is extended by 200%. The share of field vegetable cropping is about 2.5% of the total arable area under organic management.

In the current optimized situation (1990) 423,000 ha of permanent pasture are used. After conversion the permanent pasture in use is reduced by 30% to 295,000 ha as a consequence of the increase in fodder production (e.g. temporary clover/grass leys) which is necessary for maintaining soil fertility.

Changes in livestock numbers

Table 19.4 shows the numbers of livestock before and after conversion. Due to lower performance per cow, dairy cow numbers increase by 12% to 555,000 animals. Lower growth rates and therefore lower productivity leads to a reduction in bull beef numbers by 21% to 210,000 animals. In the optimized current situation (1990), extensive beef production including suckler cows and beef heifers can be ignored because of its low importance. Under organic management these enterprises are substituted for bull beef.

Table 19.4. Optimal livestock numbers in Baden-Württemberg before (1990) and after 100% conversion to organic farming.

	Before conversion		After conversion		
Livestock type	Total	per 100 ha farmland	Total	per 100 ha farmland	Change as % of 1990 level
Dairy cows	496,800	38.1	555,073	42.6	12
Bull beef	266,200	20.4	209,710	16.1	−21
Suckler cows	–	–	15,108	1.2	∞
Beef heifers	–	–	18,372	1.4	∞
Breeding sows	287,000	22.0	285,833	21.4	0
Finishing pigs	2,429,500	186.4	447,773	34.4	−82
Laying hens	2,845,600	218.3	926,684	71.1	−67
Table poultry	2,234,200	171.4	0	0	−100

The number of finishing pigs is reduced by 81% due to conversion, from 2.4 million to 450,000 animals. The number of laying hens is reduced by 67% to 930,000 animals due to high investment costs for rebuilding animal housing as well as reduced performance (lower weight gain through feeding cereals, higher bedding costs with straw litter, and higher labour input). Table poultry are no longer produced. The number of breeding sows does not change because there is no reduction in performance due to conversion and most of the sows are currently kept in small straw litter groups so that there are no, or only minor, investments required for rebuilding animal housing.

Under organic management and under assumed price relationships the profitability of intensive livestock production is low. As a consequence, intensive livestock production is substituted by forage-based livestock. The latter are better able to utilize forage crops *in situ*, particularly those which contribute to nitrogen fixation.

Changes in physical output levels

Physical output levels before and after conversion are shown in Table 19.5. As a consequence of lower yields and the changed structure of land use there are marked changes in output. The output of cereals decreases by 56% to 742,000 t due to lower yields, reduced area and increased intra-farm use. Against that, the quantity of grain legumes, which are important on organic farms for nitrogen supply and as a protein source

Table 19.5. Optimal physical output in Baden-Württemberg before (1990) and after 100% conversion to organic farming.

Product	Quantity before conversion (kt)	Quantity after conversion (kt)	Change as % of 1990 level
Cereals	1701.2	742.4	−56
Grain legumes	136.3	214.9	58
Oilseeds	209.9	69.9	−67
Root crops[a]	1390.8	1653.1	19
Milk	2175.1	2175.1	0
Beef	214.7	193.2	−10
Pig meat	237.3	71.2	−70
Poultry meat	8.4	1.8	−78

[a] Including field vegetables.

for livestock, increases by 58%. The relative profitability of oilseeds is reduced due to conversion and they are substituted by temporary grass/clover leys. The output of oilseeds decreases therefore by 67% to 70,000 t. The quantity of root crops increases by 19% due to an extension of potato and field vegetable production greater than the reduction in sugarbeet production.

The quantity of milk does not change because of unchanged milk quotas and high relative profitability of milk production under organic management. The output of beef decreases by 10% to 193,000 t and the supply of pig meat is reduced by 70% to 71,000 t. Poultry meat production is reduced by 78%. These reductions are the consequence of a lower availability of feed grains which are fed more profitably to dairy cows.

Changes in farm incomes

The optimized income situation in 1990, before conversion, is presented in Table 19.6, which shows average incomes for the different types of farm included in the model. The income measure used is gross income including interest charges for working capital. The only state subsidy considered is a payment given to farms in disadvantaged regions. In the whole state, the average farm income is about DM 21,000 per farm. The income per farm varies depending on farm size and type, ranging from DM 1450 for a cropping farm with less than 10 ha to more than DM 121,000 for an intensive livestock farm larger than 50 ha.

The impacts on farm incomes due to conversion are shown in Table 19.7. Incomes are reduced for every combination of farm type and size

Table 19.6. Optimal farm income levels (DM per farm) in Baden-Württemberg before conversion (1990).

Farm size (ha)	Cropping	Intensive livestock	Mixed	Forage-based livestock	All farms
<10	1,452	9,363	5,914	8,227	5,081
10–30	6,619	37,348	27,288	29,856	25,401
30–50	30,536	62,112	51,733	63,159	55,244
>50	65,105	121,565	59,291	67,899	67,419
Total	7,808	31,667	23,068	27,208	20,703

Table 19.7. Change in optimal farm income levels in Baden-Württemberg after 100% conversion to organic farming.

	Change as % of conventional (1990) situation				
Farm size (ha)	Cropping	Intensive livestock	Mixed	Forage-based livestock	All farms
<10	−37.30	−78.08	−30.62	−29.53	−33.63
10–30	−46.95	−72.53	−22.18	−13.27	−20.04
30–50	−38.81	−76.44	−26.54	−25.78	−30.37
>50	−78.11	−104.07	−42.02	−28.84	−49.34
Total	−56.79	−77.46	−27.17	−20.69	−29.22

in Baden-Württemberg, falling by about 29% on average. The intensive livestock farms are most severely affected. Their income decreases on average by 77% to DM 7100 per farm. Forage-based livestock farms face the lowest fall in income of about 20% to an average of DM 21,600 per farm after conversion. In total, the highest income loss is experienced by farms larger than 50 ha, due to a very high percentage of intensively managed farms. These farms have to undertake high investments in rebuilding animal housing, must reduce livestock numbers considerably, and their land to labour ratio is high, so that they cannot introduce root crop production which is productive but needs a high input of labour.

Comparing the changes of income in the different regions, it becomes clear that the decrease in income is highest in region 1 (high cropping capacity) at 37%. In region 2 (medium cropping capacity), farms face the lowest income decrease of 24.3% on average. This difference can be explained, on the one hand, by the relatively high yield decrease in region 1 and, on the other hand, by differences in farm structure. Whereas in region 1 the proportion of cropping farms, which are the ones most severely affected by conversion, is 62%, this share is much lower (29%) in region 2, and the proportion of forage-based livestock farms in region 2 is higher at 60%. Although the proportion of forage-based livestock farms is highest at 65% in region 3 (low cropping capacity), the average income decreases more than in region 2. This can be explained by the smaller milk quota capacity and a higher percentage of intensive livestock farms.

Summary and Conclusions

Regional modelling of the conversion of all farms in Baden-Württemberg to organic farming gives the following implications:

- The structure of land use changes considerably: cereal cropping decreases by 36%, while oilseed production is reduced to 5% of the arable area, grain legume production is extended to 13%, and 33% of the arable area is used for temporary grass.
- The major changes in livestock numbers include a fall in bull beef numbers by 21%, pigs by 82% and poultry by 67%.
- The physical output of cereals and oilseeds decreases by 56% and 66%, respectively, whereas the supply of grain legumes and root crops (including field vegetables) is extended markedly.
- Farm income decreases by 29% on average, with big differences between types of farm. Forage-based livestock farms face the smallest loss of income, especially farms with large milk quotas. The intensive livestock farms are the most severely affected, particularly if they have large numbers of intensively kept finishing pigs and poultry: stock numbers have to be reduced and animal housing rebuilt in order to keep within the guidelines of organic production, with the result that they often cannot achieve a positive income. The income decrease is lowest overall in regions with a high percentage of forage-based livestock and mixed farms, which under conventional management already have a structure similar to that required under organic management.

These *preliminary* results showed that a conversion of all farms in Baden-Württemberg under the pessimistic assumption of the same price level as for conventional products would involve a considerable loss of agricultural income. If the current price differences between organic and conventional products were to continue, farm incomes under organic management would not be much lower than under conventional management. On the other hand, state-wide conversion to organic agriculture would make it possible to use foreign trade protection to raise prices in response to the reduced quantities of production. Another possibility to support an extension of organic agriculture would be direct payments for conversion of farms. If this type of support for organic agriculture leads to a temporary fall in prices on the current small markets for organic products, because supply increases faster than demand, existing organic farms should be supported at least for the period of time during which the administration strives for an extension of organic agriculture. Over and above that, demand should be supported simultaneously, for example by strict certification and labelling of organically produced products, by informing consumers, or by supporting farm organizations' marketing efforts.

It is obvious, however, that in the near future, and in the context of the current reform of the European Union's agricultural policy, Baden-Württemberg's agriculture will not convert to organic production in its entirety. But the scenario calculated shows the effects under pessimistic assumptions and the possibilities which organic agriculture raises with regard to the current situation. There is a big need for research and further investigation of support for organic agriculture (direct and indirect) and the impacts of conversion of all or only particular types of farm. This is valid, on the one hand, with regard to the current reform of agricultural policy which has markedly changed the circumstances of agricultural production, leading to a considerable increase in the attractiveness of organic agriculture, and, on the other hand, with regard to an anticipated reduction of environmental pollution due to agriculture (Rat von Sachverständigen, 1985).

References

AGÖL (1991) *Rahmenrichtlinien zum ökologischen Landbau* (13th edn). Arbeitsgemeinschaft Ökologischer Landbau, Bad Dürkheim.

Dabbert, S. (1990) *Zur optimalen Organisation alternativer landwirtschaftlicher Betriebe. Untersucht am Beispiel organisch-biologischer Haupterwerbsbetriebe in Baden-Württemberg.* Agrarwirtschaft Sonderheft 124. Verlag Alfred Strothe, Frankfurt am Main.

Day, R.H. (1963) On aggregating linear programming models of production. *Journal of Farm Economics* 45(4), 797–813.

Gartung, J., Krentler, J.-G. and Sievers, H.-G. (1982a) *Investitionsbedarf für den Bau von Milchvieh- und Mastbullenställen.* Sonderheft 63. Bundesforschungsanstalt für Landwirtschaft (FAL), Braunschweig-Völkenrode.

Gartung, J., Krentler, J.-G. and Sievers, H.-G. (1982b) *Investitionsbedarf für den*

Bau von Mastschweine- und Zuchtschweineställen. Sonderheft 64. Bundesforschungsanstalt für Landwirtschaft (FAL), Braunschweig-Völkenrode.

Hammer, W. (1981) Betriebstechnische Bewertung verschiedener Produktionsverfahren der Legehennenhaltung. In: *Legehennenhaltung.* Sonderheft 60. Bundesforschungsanstalt für Landwirtschaft (FAL), Braunschweig-Völkenrode.

Lee, J.E. (1966) Exact aggregation – a discussion of Miller's theorem. *Agricultural Economic Research* 18, 58–61.

Ministerium (1992) *Betriebsverhältnisse und Betriebsergebnisse von Buchführungsbetrieben 1990/91.* Ministerium für Ernährung, Ländlichen Raum, Landwirtschaft und Forsten Baden-Württemberg, Stuttgart.

Rat von Sachverständigen für Umweltfragen (1985) *Umweltprobleme der Landwirtschaft.* Sondergutachten, W. Kohlhammer Verlag, Stuttgart and Mainz.

Schuster, W., Schreiner, W., Leonhäuser, H. and Zschoche, K.-H. (1982) Über die Ertragssteigerung bei einigen Kulturpflanzen in den letzten 30 Jahren in der Bundesrepublik Deutschland. *Acker- und Pflanzenbau* 151, 368–387.

Statistisches Landesamt (1990) Statistischer Bericht CIV 7. *Agrarwirtschaft* 71/90, Statistisches Landesamt Baden-Württemberg, Stuttgart.

Weinschenck, G. (1967) Marktwirtschaft und Betriebswirtschaft – Möglichkeiten und Grenzen der Verknüpfung von Makro- und Mikroanalyse in der quantitativen Forschung. In: Schmitt, G (ed.) *Landwirtschaftliche Marktforschung in Deutschland, Arthur Hanau zum 65. Geburtstag.* Bayerischer Landwirtschaftsverlag, Munich.

20 Estimating the Impact of Widespread Conversion to Organic Farming on Land Use and Physical Output in the United Kingdom

NICOLAS H. LAMPKIN

Introduction

From a societal or policy perspective, the likely impacts of widespread conversion to organic farming on land use and physical output are as important as the financial implications are to the individual farmer. Any significant change in farming intensity will have knock-on effects in terms of dealing with overproduction, or conversely food security, as well as potentially positive or negative environmental impacts, balance of trade impacts, and employment impacts in the food and agricultural supply industries.

This chapter describes an attempt to develop a regional model of the impact of large-scale conversion to organic farming practices. This relies primarily on the aggregation of farm- (micro) level data to model a macrolevel situation. The dangers of this approach have rightly been raised elsewhere (Lockeretz, 1989), because it is not yet possible, on the basis of existing data, to predict accurately the feedback effects which could take place – for example the impacts on prices, input costs and consumer demand, both in the organic and non-organic sector. A general fall in output caused by widespread adoption of organic farming practices might lead to conventional prices increasing as well as organic prices falling, while a shift in consumer preferences to less processed, lower-meat diets could have significant effects on production patterns.

The approach taken here is based on a simple model of UK agriculture that treats the major statistical regions as regional farms. Assumptions are made as to the likely changes in land use and yields which would result from conversion of the regional farms to organic management. No attempt is made to incorporate feedback mechanisms, such as changes in price or consumer demand. The approach is based on work carried out for Elm Farm Research Centre (EFRC, 1988; Lampkin, 1989), updated in the light of more recent data.

Methodology

The United Kingdom is divided into regional 'farms' based on the standard Ministry of Agriculture, Fisheries and Food statistical regions, using June 1988 agricultural census data and 1988 regional yield data (MAFF, 1990) for current land use and output of cereals, turnips and swedes, and forage brassicas. Nineteen eighty-eight was chosen as the base year, because it was the most recent one for which published statistics were available, although yields generally were somewhat lower than in previous or subsequent years. Land-use data for Scotland and Northern Ireland were obtained from the 1988 UK *Agricultural Statistics* (MAFF *et al.*, 1990).

English national average data for potatoes and sugarbeet (MAFF, 1990) have been used for the English regions, while United Kingdom (UK) averages (MAFF, 1991) have been used for grain legumes and oilseed rape. Total output for each region has been estimated by multiplying the crop area by the average yield. These estimates may not therefore coincide with other published total output data for 1988. For example, estimated UK output for wheat in 1988 is 11.711 Mt, compared with the published figure of 11.723 Mt (MAFF, 1991).

Livestock numbers by statistical region were also obtained from the June 1988 agricultural census data for England and Wales, and from the 1988 UK *Agricultural Statistics* (MAFF *et al.*, 1990) for Scotland and Northern Ireland. Average milk output data by region have been estimated by using 1986 regional data adjusted to 1988 total output levels (MAFF, 1991). Livestock numbers for different classes of livestock were converted to livestock units (LU) using appropriate conversion factors (MAFF, 1977).

The projected changes in land use are based on subjective assumptions concerning the effect of adopting characteristic organic rotations in each of the regions concerned, in the absence of more reliable predictive methods or data. As such, the projections are intended to be more illustrative than predictive.

The main assumptions about land use which have been made are as follows:

- The area of cereals will fall by a maximum of 25% depending on the current levels. In the most intensive cereal-growing areas, a larger than average reduction has been assumed to allow for the reintroduction of fertility-building temporary leys. In areas with relatively few cereals, no reduction or possibly an increase has been assumed, due to increased reliance on home-grown cereals for livestock feed.
- Within the cereals area, the area of oats will expand at the expense of wheat and barley (a characteristic of current organic farms), and in certain regions the overall area of wheat will fall to match more closely the combined areas of other cereals (due to rotational constraints on wheat production in organic systems).
- The area of potatoes, vegetables and fodder crops will increase to compensate, at least in part, for reduced output of these crops.
- The sugarbeet and oilseed rape area will decline due to difficulties associated with organic production of these crops, but recognizing that a market will still exist for their products.
- The area of grain legumes will increase, particularly in arable areas, because of their role in nitrogen fixation and their potential for use as domestic protein sources for livestock feed and human consumption.
- The area of grass leys of less than five years' duration will increase, particularly in arable areas, to provide the fertility required for the arable part of the rotation.
- The area of permanent pasture and rough grazing will fall slightly, partly as long-term grass leys are brought back into cropping rotations, and partly due to increased forestry and environmental conservation activities which would lead to a general increase in the woodland and 'other land' areas.

The assumptions relating to crop and livestock output have been more closely specified, due to the greater availability of comparative data. The output assumptions, expressed as a percentage of the conventional yield in each region, are:

Wheat	60%
Barley	70%
Oats	80%
Oilseed rape	60%
Potatoes	60%
Sugarbeet	60%

Fodder crops	70%
Grain legumes	80%

Livestock numbers have been estimated on the basis of the following assumptions:

- Total numbers of dairy cows will not change, but milk yields per cow (and hence total output) will fall by 10% as a result of the change to a lower concentrate regime.
- Stocking rates will fall by 20% in the case of grazing livestock and 40% in the case of pigs and poultry, reflecting the latter's current dependence on cereal feeds and intensive housing, both aspects which would change under conversion to organic farming.

It must be emphasized that these assumptions, although as well founded as they can be, are necessarily speculative. In particular, the lack of data represents a significant impediment to the construction of more sophisticated models, although whether these would be more accurate is arguable. In addition, the current developmental state of organic farming means that in many cases the full potential of the system is not being met, so that even such data as do exist may not be of much help for predictive purposes.

The output assumptions made are conservative, and relate to an already low base year (1988). In case these assumptions are wrong, a sensitivity analysis has been included to evaluate the impact of crop and milk yields either 10% higher relative to the conventional values, or 10% lower. For example, the sensitivity analysis examines the effect of organic wheat yields at 70% and 50%, as well as 60%, of conventional yields.

The calculations were carried out using a budgeting approach, where the changes due to predetermined assumptions are evaluated. An optimizing approach would have been of limited value without including a number of other variables, the data for which are not readily available.

Results

The main projections of changes in land use are summarized in Table 20.1, giving an indication of a possible pattern of land use if all UK land were to be converted to organic farming. Tables 20.2 and 20.3 show the effect on cropping areas and output of a more realistic conversion of 10% of land area to organic farming for each of the UK countries.

The main crop output changes estimated are:

- a 4.5% fall in UK cereal production, arising from a combination of a 5% reduction in wheat and barley output and a 25% increase in the output of oats;
- a 2.5% reduction in potato output, while sugarbeet output falls by over 6%;
- a 7.4% reduction in the output of oilseed rape, which is offset partially by a 3% increase in field bean output.

These estimates are sensitive to the yield assumptions made. The results of the sensitivity analysis indicate that the change in cereal output might vary by ± 0.9 percentage points if the assumed organic yields vary by 10% relative to conventional yields (Table 20.4). In the case of oats, the difference is much greater, but the changes estimated in this case are influenced by the very low level of production in the base year. Any changes relative to this base are likely to appear large in percentage terms, but are small in absolute terms.

An earlier version of this model, using 1986 as a base year and assuming more generous organic yields, indicated a decline in UK cereal output of 3% following 10% conversion of UK agriculture to organic production (EFRC, 1988; Lampkin, 1989). Potato output was predicted to fall by 1.3%, sugarbeet by 5.8%, and oilseed rape by 6.1%. The closeness of these results is another indication that the model is relatively stable under modified assumptions. The major difference is the predicted increase in the output of grain legumes, which was estimated in the earlier study at 17.4%. Like oats, grain legumes have a small current production base, and the model appears to be very sensitive to assumptions made about changes in production area for such crops.

Livestock output is also predicted to fall by between 2% and 5% (Table 20.5). The changes in dairy LUs and milk yields arise directly from the assumptions stated above. The numbers of beef and sheep LUs are projected to fall by 2.3–2.5%, although these numbers will be similar because of an additional assumption in the model that the relative proportions of beef and sheep (and of pigs and poultry) remain constant. The effect of maintaining the number of dairy LUs constant is that a larger land area has to be allocated to this livestock class, reducing the land area available for other grazing livestock. This is compensated for to some extent by the overall increase in the area of leys less than five years old (Table 20.2).

The number of pig and poultry LUs is estimated to fall by over 5%, reflecting the 40% decrease in stocking rates assumed. Although pigs and poultry are not usually considered as land-using enterprises, they would become land-based under organic management requirements, and they rely on crops produced elsewhere under conventional

Table 20.1. Regional breakdown of the distribution of UK land use before (1988) and projected after 100% conversion to organic production (percentage of total, rounded to nearest 0.5).

Region	Status	Cereals	Wheat	Barley	Oats	Potatoes & veg.	Sugarbeet	Oilseed rape	Grain legumes	Leys <5 years
North	Before	15	5	9.5	1	0.5	0	1.5	0.5	9
	After	15	5	6	3.5	0.5	0	0.5	1	15
Yorks. & Humberside	Before	40	19.5	18.5	0.5	3.5	2	4.5	2.5	6
	After	30	15	10	5	4.5	1.5	1.5	4	20
East Midlands	Before	46	30	15.5	1	5	3.5	6.5	3.5	5.5
	After	35	20	10	5	6	2.5	3	7	17
East Anglia	Before	54	31	22	0.5	6.5	10.5	4	5	2
	After	40	20	15	5	7	7	2	8	14
South-East	Before	44	26.5	16	1	2	0.5	4.5	4.5	8.5
	After	35	18	12	5	2.5	0	2	7.5	20

South-West	Before	23.5	11	11.5	0.5	1	0	1	1	14.5
	After	20	10	7	3	1	0	0.5	2	20
West Midlands	Before	30	14.5	14	1.5	2.5	2	2.5	2	12
	After	25	12	10	3	3.5	1	1	3	17
North-West	Before	14.5	3.5	10	0.5	3.5	0	1	0.5	11
	After	15	5	7	3	4.5	0	0.5	1.5	15
Wales	Before	4.5	0.5	3.5	0.5	0.5	0	0	0	10.5
	After	5	1	2.5	1.5	0.5	0	0	0.5	15
Scotland	Before	10	2	7.5	0.5	0.5	0	1	0	8
	After	10	2.5	5.5	2	1	0	0.5	0.5	9
N. Ireland	Before	4.5	0.5	4	0.5	1	–	0	–	18
	After	5	0.5	3	1.5	1	–	0	–	20

Table 20.2. Organically managed area by crop type after 10% conversion, and percentage change in area occupied by crop relative to 1988, UK.

Land use	Organic area ('000 ha) after 10% conversion					Change in area of crop relative to 1988 after 10% conversion (%)				
	England	Wales	Scotland	N. Ireland	UK	England	Wales	Scotland	N. Ireland	UK
Cereals	261.3	7.5	52.8	5.3	326.9	−2.0	1.3	0.1	0.6	−1.6
Wheat	130.4	1.5	13.2	0.5	145.6	−2.6	5.3	3.4	0.8	−2.3
Barley	91.5	3.8	29.0	3.2	127.5	−3.4	−2.7	−2.5	−2.5	−3.2
Oats	38.9	2.3	10.6	1.6	53.3	40.4	37.4	19.8	51.0	34.4
Potatoes	16.5	0.8	3.7	1.4	22.4	2.3	5.9	3.0	1.5	2.5
Sugarbeet	12.8	0.0	0.0	0.0	12.8	−3.6	−10.0	0.0	n/a	−3.6
Vegetables	15.7	0.2	1.3	0.2	17.4	2.3	6.1	4.5	3.9	2.4
Other horticulture	6.6	0.1	0.5	0.2	7.4	1.0	2.0	3.1	0.3	1.1
Grain legumes	42.5	0.8	1.3	0.0	44.6	7.0	58.6	5.3	n/a	7.2
Turnips/swedes	2.5	0.5	15.8	0.1	19.0	2.7	7.5	41.3	6.2	25.0
Kale, rape, etc.	2.0	0.4	5.3	0.0	7.7	3.1	5.0	43.2	6.0	17.6
Maize, beets, etc.	4.5	0.9	0.0	0.6	6.0	1.7	0.0	−10.0	0.1	0.7
Oilseed rape	13.2	0.0	1.6	0.0	14.8	−5.6	−10.0	−6.2	−10.0	−5.7
Other crops	2.3	0.0	0.3	0.0	2.6	0.7	−2.1	2.1	n/a	0.8
Bare fallow	6.2	0.5	0.8	0.0	7.4	3.2	3.4	0.5	n/a	2.8
Leys <5 years	167.9	22.5	47.5	21.3	259.2	10.1	4.1	1.5	1.0	6.2
Leys >5 years	270.6	79.6	52.8	54.2	457.2	−1.0	−0.9	−1.7	−0.4	−1.0
Rough grazing	69.4	30.0	327.3	18.1	444.8	−0.5	−1.0	−0.5	−0.3	−0.5
Woodland	31.2	5.3	13.2	1.8	51.5	5.2	5.5	6.5	5.4	5.6
Other land	15.3	1.2	3.7	3.1	23.3	0.5	−0.7	−0.3	0.3	0.3
Total area	940.7	150.1	528.0	106.3	1725.0	0.0	0.0	0.0	0.0	0.0

Table 20.3. Estimated changes in crop output resulting from 10% conversion of UK agriculture to organic production.

	Organic output (kt) after 10% conversion					Percentage difference after 10% conversion				
Land use	England	Wales	Scotland	N. Ireland	UK	England	Wales	Scotland	N. Ireland	UK
Cereals	924.0	24.3	192.1	17.5	1157.9	−4.8	−2.0	−2.8	0.3	−4.5
Wheat	481.3	5.2	58.4	2.1	547.0	−5.6	−0.8	−2.0	−3.5	−5.3
Barley	299.3	11.4	92.5	10.2	413.3	−5.4	−4.9	−4.8	−4.7	−5.2
Oats	141.5	7.7	41.2	5.3	195.7	29.7	28.0	13.8	38.8	24.9
Potatoes	396.0	14.5	78.5	22.5	511.6	−2.6	−0.5	−2.2	−3.1	−2.5
Sugarbeet	317.7	0.0	n/a	n/a	317.7	−6.2	−10.0	n/a	n/a	−6.2
Grain legumes	125.8	2.2	3.2	n/a	131.2	3.6	44.9	2.2	n/a	3.7
Turnips/swedes	71.2	19.1	n/a	n/a	90.3	−3.1	2.2	n/a	n/a	−2.4
Kale, rape, etc.	43.2	8.7	n/a	n/a	51.9	−3.5	0.5	n/a	n/a	−3.0
Oilseed rape	23.8	0.0	2.9	0.0	26.7	−7.4	−10.0	−7.7	−10.0	−7.4

Table 20.4. Sensitivity analysis for relative changes in crop output (%) after 10% conversion of UK agriculture to organic production if organic yields are either 10% higher or lower relative to conventional than assumed.

Land use	Organic yield + 10% of conventional (%)					Organic yield − 10% of conventional (%)				
	England.	Wales	Scotland	N. Ireland	UK	England	Wales	Scotland	N. Ireland	UK
Cereals	−4.0	−0.9	−1.8	1.3	−3.6	−5.6	−3.1	−3.8	−0.8	−5.3
Wheat	−4.8	0.7	−0.6	−2.4	−4.6	−6.3	−2.4	−3.3	−4.6	−6.1
Barley	−4.7	−4.2	−4.0	−4.0	−4.6	−6.1	−5.6	−5.5	−5.5	−5.9
Oats	34.7	32.7	16.8	44.9	29.3	24.8	23.2	10.8	32.7	20.6
Potatoes	−1.4	1.1	−0.9	−1.9	−1.3	−3.8	−2.0	−3.5	−4.2	−3.8
Sugarbeet	−5.5	−10.0	n/a	n/a	−5.5	−6.8	−10.0	n/a	n/a	−6.8
Grain legumes	5.3	51.7	3.8		5.4	1.9	38.0	0.7	n/a	2.0
Turnips/swedes	−1.9	4.0	n/a	n/a	−1.1	−4.4	0.5	n/a	n/a	−3.7
Kale, rape, etc.	−2.2	2.0	n/a	n/a	−1.7	−4.8	−1.0	n/a	n/a	−4.4
Oilseed rape	−6.9	−10.0	−7.3	−10.0	−7.0	−7.8	−10.0	−8.1	−10.0	−7.9

Table 20.5. Estimated changes in livestock output resulting from 10% conversion of UK agriculture to organic production.

Output	Absolute output change					Relative output change (%)				
	England	Wales	Scotland	N. Ireland	UK	England	Wales	Scotland	N. Ireland	UK
Dairy (LU)	0	0	0	0	0	0.0	0.0	0.0	0.0	0.0
Milk (Ml)	−104.4	−15.5	−12.3	−12.3	−144.4	−1.0	−1.0	−1.0	−1.0	−1.0
Beef (LU)	−36,089	−14,247	−18,910	−27,231	−96,476	−1.7	−3.0	−3.0	−2.9	−2.3
Sheep (LU)	−23,247	−18,490	−3,761	−16,268	−61,766	−2.0	−3.0	−3.0	−2.9	−2.5
Pigs (LU)	−85,990	203	−4,596	3,073	−87,309	−6.4	0.8	−3.7	3.3	−5.5
Poultry (LU)	−45,646	311	−2,877	3,074	−45,137	−6.4	0.8	−3.7	3.3	−4.9

Table 20.6. Sensitivity analysis for relative changes in livestock output (%) after 10% conversion of UK agriculture to organic production if organic yields and numbers are either 10% higher or lower relative to conventional than assumed.

	Milk yields and stocking rate + 10% of conventional (%)					Milk yields and stocking rate – 10% of conventional (%)				
Output	England	Wales	Scotland	N. Ireland	UK	England	Wales	Scotland	N. Ireland	UK
Milk	0.0	0.0	0.0	0.0	0.0	−2.0	−2.0	−2.0	−2.0	−2.0
Beef	0.4	−1.6	−1.6	−1.8	−0.6	−3.7	−4.4	−4.4	−4.1	−4.0
Sheep	−0.1	−1.6	−1.6	−1.8	−0.9	−4.0	−4.4	−4.4	−4.1	−4.1
Pigs	−5.9	2.6	−2.7	5.5	−4.8	−7.0	−1.0	−4.8	1.1	−6.3
Poultry	−5.8	2.6	−2.7	5.5	−4.0	−7.0	−1.0	−4.8	1.1	−5.7

management, so it is appropriate to consider them as such in this model.

Modifications to the assumptions concerning stocking rates and milk yields result in larger percentage differences than with crop outputs (Table 20.6). The change in UK beef output, for example, varies by ±1.7 percentage points if the stocking rates are assumed to be either 10% or 30% less than conventional. The corresponding changes in the earlier study were −1.9%, −1.3% and −1.9% for milk, beef and sheep respectively (Lampkin, 1989).

The Wider Impacts of a 10% Conversion of UK Agriculture to Organic Farming

The changes in output predicted using this model can be interpreted in a number of different ways. Current performance of organic farms is determined by a small number of producers, some with limited farming experience, solving problems individually often without easy access to research and extension facilities. This situation is now changing, and it is likely that widespread conversion to organic farming would allow more resources to be invested in research and extension, thus improving the performance of the group as a whole. There is a big difference between the potential of organic farming as demonstrated by some producers and the average results achieved by organic farmers generally. It could be argued, therefore, that the more optimistic scenario portrayed in the sensitivity analysis may be closer to the real situation if 10% of UK farms actually did convert to organic farming, but evidence for this is clearly lacking at present.

Surplus production

Reduced output, at least up to a certain point, is an advantage rather than a disadvantage in the context of current overproduction (Table 20.7). This is particularly so in the context of the current CAP reforms (EC, 1992) and the introduction of set-aside schemes covering 15% of arable land. It has been argued that continuing intensification on the remaining arable land, and the benefits deriving from the set-aside 'break', will eliminate the surplus reduction benefits arising from this level of set-aside within ten years (Agra-Europe, 1992). At the same time, on the basis of the central yield scenario outlined above, 30% of UK agriculture would need to be converted to organic production to achieve the 15% reduction in arable output which is the aim of the set-aside programme. Set-aside has the advantage that it can be implemented quickly. Thirty per cent conversion to organic production

Table 20.7. Production of crops and livestock by type as a percentage of total new supply for use in the UK, 1986–1990.

Crop	1986	1987	1988	1989	1990
Wheat	120	125	103	121	127
Barley	162	145	140	158	128
Oats	97	100	99	101	99
Cereals	123	121	107	121	115
Oilseed rape	133	108	107	101	98
Sugarbeet	55	59	56	57	56
Potatoes	91	90	90	88	89
Milk powder	246	192	223	196	291
Beef	96	97	87	93	93
Sheep	81	83	89	93	94
Pigs	103	100	100	95	98
Poultry	95	97	98	98	95

Source: MAFF (1991).

would take much longer to achieve, but the benefits might also be longer lasting. The change in output from a less dramatic switch to organic farming is likely to be quite small in this context, but there will still be some impact *at the margin* on attempts to reduce surplus production.

Food security

The issue of food security, a prime objective of postwar agricultural policy (Sturgess, 1992), becomes relevant if rates of conversion to organic farming exceed 30% and domestic production of basic commodities such as wheat and barley falls below the levels of domestic consumption. How important this is depends on a number of factors: our ability to rely on less intensively produced imports from third countries, as was the case until the early 1970s; changing consumer preferences, including the reduced demand for meat that might accompany the very widespread adoption of organic farming; and the level of reliance on cereals and other livestock feeds potentially suitable for direct human consumption.

As an illustration of this, some 8.1 Mt of domestically produced cereals were used for livestock feeds in 1987 (MAFF, 1988). This was equivalent to 37% of new cereals production, 39% of domestic consumption and 45% of total new supply in that year. Total imports of all materials for concentrated livestock feed production were equivalent to 50% of domestic production used for this purpose. These figures illustrate the extent to which livestock production in the UK has become dependent on 'bought-in' crop land, either elsewhere in the country or abroad. The return to predominantly grass-fed, land-based livestock production of all types, and the substitution of the increased domestic production of grain legumes for imported protein, could release a significant proportion of both domestically produced and imported cereals and other foodstuffs for human consumption without materially altering the current level of food security.

Estimating the potential impact of changes in consumer demand for meat, and the reliance on cereals for livestock production, were beyond the scope of this study. However, the difference in cereal production of 21 Mt in 1988 and 12 Mt estimated if all UK land converted to organic farming, is equivalent to the combination of surplus production and the use of domestic cereals for livestock production. This does not necessarily imply a serious problem with respect to human food supply, with a level of self-sufficiency (57–59%) comparable to the situation in the UK before joining the European Community (EC, 1972). However, the environmental and other impacts on food-exporting countries, in particular in the Third World, need more detailed consideration.

Impact on the supply of organic produce

The impact of 10% conversion on existing levels of organic production will be highly significant. The projected changes represent an increase in organic production of more than 100 times the estimated 1989 production levels (Table 20.8). In the face of such a large increase in the output of organically produced food, the question of whether the market can cope becomes important, particularly given the significance of premium prices if organic producers are to maintain their relative competitiveness. A number of factors are involved in this issue which are not quantifiable at this stage. These include the potential economies of scale and reduced marketing costs resulting from an increase in supply; current demand elasticities for organic food for which few published estimates exist (OFC, 1992); potential increases in demand for organic food resulting from improvements in availability; and increased prices for conventionally produced food due to the general reduction in supply. An indication of the likely impact on the market and prices may be obtained from surveys of consumer willingness to pay premium prices for organic food (e.g. Mintel, 1989). These indicate that premium prices would need to fall from current levels of 50–100% to 10–20% if the market for organic food is to move from being a niche to a mass market.

Table 20.8. Estimated area, quantity and value of crop production from organic farms, UK, 1989.

Crop	Area (ha)	Quantity (t)	Value (£ thousands)
Cereals	2,866	10,105	1,834
Wheat	1,285	4,727	1,063
Barley	272	874	76
Oats	1,159	4,166	636
Potatoes	175	3,319	189
Horticulture	411	3,500	1,731
Livestock	14,956	–	4,756

Source: Adapted from Murphy (1992).

Summary and Conclusions

The impacts of widespread adoption of organic farming in the UK on land use and physical output have been estimated using a model based on the concept of regional farms and how they might be different under an organic scenario. Projections of crop and livestock output after 10% conversion of UK land to organic farming include: a 4.5% fall in cereals output arising from a combination of a 5% reduction in wheat and barley output and a 25% increase in the output of oats; a 2.5% reduction in potato output; while sugarbeet output falls by over 6% and oilseed rape by 7.4%. This is offset by a 3% increase in field bean output. Livestock output is projected to fall by between 2% and 5%. The projected reductions in output indicate the contribution which widespread adoption of organic farming might make to surplus reduction objectives, but do not necessarily represent a threat to food security, due to the likelihood that widespread adoption of organic farming would be accompanied by changes in consumption patterns, and due to the fact that the implicit levels of production reflect levels of self-sufficiency experienced in the UK in the early 1970s.

References

Agra-Europe (1992) Historic agreement on CAP reform. *Agra Europe* 1492 (22 May), 1–15.

EC (1972) *Yearbook of Agricultural Statistics*. European Commission, Brussels.

EC (1992) Regulation (EEC) No. 1765/92 establishing a support scheme for producers of certain arable crops. *Official Journal of the European Communities* L181 (2 July), 12.

EFRC (1988) *The Potential for Developing Organic Agriculture as a Mainstream Policy Option for Reducing Surpluses and Protecting the Environment.* Elm Farm Research Centre, Newbury.

Lampkin, N.H. (1989) Organic farming – a policy option for UK agriculture? Paper presented at the UK Agricultural Economics Society conference, Aberystwyth, 1989. In: Lampkin, N.H. (ed.) *Collected Papers on Organic Farming.* Paper 3. Department of Economics and Agricultural Economics, University College of Wales, Aberystwyth, pp. 1–12.

Lockeretz, W. (1989) Problems in evaluating the economics of ecological agriculture. *Agriculture, Ecosystems and Environment* 27, 67–75.

MAFF (1977) *Definitions of Terms Used in Agricultural Business Management.* Ministry of Agriculture, Fisheries and Food, Alnwick.

MAFF (1988) *Output and Utilisation of Farm Produce in the United Kingdom, 1981–1987.* HMSO, London.

MAFF (1990) *Annual Estimates of Crop Production – 1988 Harvest.* Ministry of Agriculture, Fisheries and Food, Guildford.

MAFF (1991) *Agriculture in the United Kingdom: 1990.* HMSO, London.

MAFF, DAFS, DANI and WOAD (1990) *Agricultural Statistics United Kingdom 1988.* HMSO, London.

Mintel (1989) *The Green Consumer.* Mintel, London.

Murphy, M. (1992) *Organic Farming as a Business in Great Britain.* Agricultural Economics Unit, University of Cambridge.

OFC (1992) *Marketing.* Final Report II. Organic Farming Centre, University of Edinburgh.

Sturgess, I.M. (1992) Self-sufficiency and food security in the UK and EC. *Journal of Agricultural Economics* 43(3), 311–326.

21 Input–Output Modelling of Organic Farming and the Rural Economy of England and Wales

Peter Midmore

Introduction

Much of the focus of academic interest in organic farming has concerned its potential to resolve acute short-term economic and political problems. These are described in detail elsewhere in this volume (Chapters 23–26), and include exploration of the impact of organic agriculture on rapidly rising government expenditure on support mechanisms, overproduction, agri-environmental issues such as soil erosion and pesticide residues, and the whole question of appropriate scale of land utilization by different types of economic activity.

A less well-explored, but equally important, set of issues relates to how such changes might affect the rural economy in general. In principle, organic farming systems are more self-reliant than conventional methods of production. As the latter have become more specialized and capital- and energy-intensive, so they have become more reliant on a complex of other industrial processes which, in a number of instances, have created employment within rural areas. Reduced demand for inputs, and a reduction in the quantity of output available for processing, which are potential consequences of widespread conversion of farming to organic methods, may reduce demand for off-farm labour in rural areas.

Will the conversion of farms require a higher level of labour input or more specifically, will it allow a structure of holdings within which greater numbers of farm units survive, thus minimizing the rate of decline of farm labour demand? Does the aim of producing a higher quality of organic product require more labour in the processing of output, and can higher margins be earned as a consequence? These are questions which, even in the context of the general dynamic of rural economic decline and depopulation, are not adequately dealt with by existing methods of policy analysis.

In this chapter some elements of the steps which might point towards answers to these outstanding questions are taken, showing how the method of input–output analysis has been used, and suggesting further steps to extend its informative capacity.

The Input–Output Method

Input–output analysis can be used to estimate the wider economic impact of agriculture. If there is a change – say a reduction – in the sales of a particular industrial sector, then that sector will need to purchase less inputs from other industries. As these purchases are sales of the other industries, they in turn will need to purchase less, and so on until the magnitude of the ripple effect diminishes to an imperceptible level. Agriculture is an industry which has both one of the highest proportions of purchased inputs to gross output, and one of the highest proportions of gross output sold to other intermediate industries for processing, and as such has a high level of such economic linkage (Midmore, 1991a).

The basis of the evaluation of these linkages begins with the identification of transactions between economic sectors, which are organized in an input–output *matrix*; then each round of this process can be summed by a simple manipulation, used as the basis for identifying relationships between direct effects of output changes in one sector with the total indirect effects on the economy as a whole (known as *multipliers*). This can be taken a step further by identifying the relationships between gross output and income generated, and thereby produce income multipliers; or by identifying the relationships between gross output and numbers of jobs, and thereby produce employment multipliers. These multipliers relate to backward linkages only, though forward linkages can also be analysed: in some circumstances it can be argued that forward linkages (i.e. a supply-driven relationship) do not *necessarily* exist, though for many rural industries there are considerable incentives for industries based on their outputs to be located within production areas.

The method relies on a number of simplifying assumptions which sacrifice accuracy for comprehension; in particular, it uses a production function which does not allow for economies of scale or substitution of one type of input for another, and it also assumes that the output of each industry is homogenous, and (normally) that prices are fixed. These assumptions are very much approximations, but resulting bias may be offset by using data which is as up to date as possible, and by examining the industrial sectors in question in as fine a scale of disaggregation as possible. On a regional level the scale and stability of regional leakages from the production flows are also important in determining the accuracy of the approach.

Recent Applications

Earlier input–output-based studies of the effects of a substantial change in agricultural systems to organic methods include studies by Ennis (1985) and Lampkin *et al.* (1987). Ennis's application of the method in the United States concluded that increases in farm and household income (due to higher output prices and reduction in farm costs) outweighed the decline in incomes borne by other sectors in the economy, although the model did not include livestock and it seems likely that the crop yield assumptions made were over-optimistic. Lampkin *et al.*'s examination of the local effects of a pilot scheme in the Teifi Valley, Wales, concluded that the effects on farm employment would be static, though jobs could possibly be safeguarded in the long run. The decline in input-supply industries as a result of reduced purchases could be more than outweighed by gains in marketing and processing activities.

In Britain, there have been two further recent attempts to analyse such changes; these are described in more detail here.

Effects of a proposed organic aid scheme in England and Wales

Jenkins and Midmore (1992) investigated the comparative implications of a number of proposed nature conservation and environmental protection schemes advocated by the national environmental conservation authority (the Nature Conservancy Council, now English Nature), including a scheme to aid conversion to organic agriculture. Each proposed scheme was evaluated in terms of environmental benefits, farm-level effects, direct budgetary costs, indirect budgetary costs and savings, and forward and backward linkage effects. The last-mentioned utilized an input–output analysis, based on economic changes at farm level which were identified in the prior investigation. The organic scheme's farm-level effect was based on earlier work by Midmore and Lampkin (1988); aggregate reductions in commodity

Table 21.1. Comparative total linkage effects of different environmental protection schemes, including conversion to organic farming, England and Wales.

	Income effect (£ millions)		Employment effect (thousands)	
Scheme	Total	Backward	Total	Backward
Habitat creation[a]	45.237	25.176	3.1	1.6
Organic conversion	36.323	19.648	2.5	1.3
Cereals management[b]	13.755	4.230	0.9	0.2
Moorland conservation[c]	1.767	0.753	0.1	0.1
Stock reduction[d]	0.650	0.256	0.1	0.0

[a] Restoration of hedges and other traditional field boundaries, creation of wetland, heathland areas, etc.
[b] Either sowing spring cereals or management of winter stubble for over-wintering birds.
[c] Traditional management of semi-natural grassland.
[d] Adherence to maximum livestock stocking rates in hills and uplands.
Source: Jenkins and Midmore (1992).

purchases by farms were applied to the agriculture and horticulture industry's position, as described by the 1984 United Kingdom input–output tables (CSO, 1988). This, in common with the rest of the analysis, relied on estimates of regional take-up and impact on representative farming systems. A summary of the results is provided in Table 21.1, together with those relating to the other environmental management schemes.

It is interesting that conversion of 5% of farmland to organic production has a lower cost, in terms of reduction in national incomes (just over £35 million) and lost employment (2500 jobs) than the habitat creation scheme. In fact, in terms of budgetary savings only the cereals management scheme also makes a net contribution to an overall saving in (British) government expenditure; all other schemes in this package were found to require additional expenditure, even though the direct cost of most of these would be mitigated by some budgetary saving. The savings identified in the study are currently much less important as a result of the recent changes in European agricultural policy, and as a consequence arguments for aid to promote agricultural practices which promote environmental conservation need to be based more on the (environmental) costs of *not* supporting them, and the potential for improved rural economic conditions which they might assist.

Two further comments are in order before detailed examination of the most recent study. First, the national input–output tables on which the estimates are based are considerably out of date and also aggregate all types of agricultural production into a single sector, both of which, as noted in the preceding section, tend to reduce the efficiency of the estimates of income and employment losses. Second, these estimates are based simply on the output-reducing effects of conversion to organic agriculture, and pay no attention either to the fact that conversion may lead to a more robust holding structure, and slow down the rate of farm amalgamation as individual businesses are strengthened, or to the creation of value-added income and employment through production of market-driven output which is moving from profitable niche markets to potential mainstream products. Both of these factors suggest that the income and employment losses identified by Jenkins and Midmore might be conceived as the maximum losses and that, depending on circumstances, they may well be more than offset by regenerative activity based on organic conversions.

Detailed comparisons of multipliers: Organic farming in the Welsh rural economy

The financial performance of organic farming was the subject of a detailed survey in 1989, reported extensively, if controversially, by Murphy (1992) (see Chapter 5). The Welsh regional results of this study were used to calculate an aggregate organic farming income, with details of gross outputs and input usage which allowed modification of an existing Welsh Input–Output Disaggregated Agriculture Model (christened WIODAM by Bateman *et al.*, 1993). The original model described the agricultural and food complex in Wales in terms of eight agricultural enterprise categories, together with four input/processing industrial sectors. Organic agriculture was incorporated by calculating enterprise input–output coefficients using regression techniques proposed by Errington (1987) and elaborated by Midmore (1991b). Essentially, levels of output of each enterprise group are used as predictors of total usage of each input category in a regression equation, and the resulting coefficients attached to each enterprise 'predictor' taken as input costs per unit of output, or input–output coefficients. There are considerable drawbacks to the use of this method, not least the appearance of negative coefficients due to a considerable degree of multi-collinearity between the predictor variables, though these are partially evaded by the use of ridge regression methods in estimation (Judge *et al.*, 1988).

The coefficients derived by these means were then multiplied by the gross organic outputs from each enterprise group, to form columns of input–output flows. These were inserted into the original WIODAM

Table 21.2. Input–output multipliers for Welsh organic and conventional agriculture, 1991.

	Output		Income		Employment	
	Organic	Conventional	Organic	Conventional	Organic	Conventional
Cereals	1.02	1.06	1.02	1.11	1.12	1.06
Pasture and forage crops	1.22	1.47	4.26	1.88	1.16	2.12
Other crops	1.13	1.03	1.29	1.04	1.17	1.04
Milk	1.80	1.79	2.46	2.12	2.96	2.29
Cattle	1.66	1.52	2.64	2.03	1.90	2.86
Sheep	1.60	2.07	1.51	2.21	2.44	2.57
Other livestock	1.35	1.34	4.13	3.29	2.36	2.69

Source: Bateman *et al.* (1993).

model by subtracting them from the original enterprise group columns, producing conventional and organic sectors for each enterprise group. Finally, the destination of enterprise outputs from the organic sector (identified from other material in the survey) was included and subtracted from conventional output flows. This allowed calculation of both conventional and organic multipliers from a fully disaggregated regional model of the Welsh agricultural sector, and description of the backward linkage effects of each with the wider economy. The resulting multipliers are shown in Table 21.2.

The relationship between organic and conventional multipliers is somewhat complex. In some instances their values are more or less exactly the same, whereas in others they diverge quite markedly; output, income and employment multipliers in some instances deviate in opposite directions. For Welsh agriculture, the most important agricultural products are milk, beef and sheep: for *milk*, all of the organic multipliers are higher; for *cattle*, organic income multipliers are higher, though conventional activity appears, per unit of output, to create more employment; but for *sheep*, all conventional multipliers are higher than their organic counterparts. The importance of these relationships lies in the opportunity to determine an effective mix of activities to maximize income and employment.

This model was also used to investigate the impact of changes on incomes and employment in Wales if the organic share of overall production were to increase. A number of assumptions were employed. For example, a high proportion of existing organic production in Wales is based on smallholding horticulture, though considerable scope for expansion appears to exist in dairying and other livestock production; thus the expansion was forecast entirely in terms of the latter. Also, there is considerable scope and interest in domestic processing of organic outputs, and forward linkage effects were calculated on the basis of the proportion of organic processing which currently takes place. The forecast balance of effect, in terms of incomes and employment, was small but positive; an increase in livestock output from the current level of 0.1% to 1% would in net terms create an extra £0.28 million income, and an extra 20 jobs. This would, however, involve creation of new income and employment in some sectors, particularly processing, compensating for reductions in agricultural requisites, feed production and energy.

Conclusions

Conversion to organic farming does have the potential to generate considerably wider social and economic impacts than simply on the farms involved. However, the input–output method merely describes a portion of these; and other, possibly more important, linkages are ignored. There are methods which have been proposed to incorporate ecological–economic interaction within a similar, social accounting matrix approach, though these have never been developed beyond the illustrative stage (e.g. Victor, 1973, Ch. 4). What is perhaps especially important to note with regard to input–output forecasts is that, first, they are only (almost) as good as the forecast of the initial change so that if this is flawed there is no hope of the forecast of linkage effects being accurate; and second, changes involving a marginal change in output may involve overestimates of impact, though this depends on whether changes involve an increase or a decrease (for further details, see Midmore, 1993).

References

Bateman, D.I., Hughes, G.O., Lampkin, N.H., Midmore, P. and Ray, C.S. (1993) *Pluriactivity and the Rural Economy in the Less Favoured Areas of Wales.* Report for the ESRC. Department of Economics and Agricultural Economics, University of Wales, Aberystwyth.

CSO (1988) *Input–output Tables for the United Kingdom 1984.* Central Statistical Office report. HMSO, London.

Ennis, J.L. (1985) The likely inter-industry effects of organic farming adoption in the United States. MSc thesis, Department of Agricultural Economics and Sociology, Ohio State University.

Errington, A.J. (1987) Labour use and labour requirements in UK agriculture. *Journal of Agricultural Economics* 38(2), 271–280.

Jenkins, T.N. and Midmore, P. (1992) The economic implications of selected nature conservation proposals for agriculture. Paper presented at the 30th European Association of Agricultural Economists Seminar, Château-D'Oex, Switzerland, November.

Judge, G.G., Carter-Hill, R., Griffiths, W.E., Lütkepohl, H. and Lee, T.-C. (1988) *Introduction to the Theory and Practice of Econometrics* (2nd edn). John Wiley, New York.

Lampkin, N.H., Jones, W., Mann, M. and Midmore, P. (1987) *Proposal for an Integrated Rural Development Programme Focusing on Organic Agricultural Production in the Teifi Valley, Dyfed.* Vol. I. Department of Agricultural Economics, University of Wales, Aberystwyth.

Midmore, P. (1991a) Input–output and agriculture. In: Midmore, P. (ed.) *Input–output Models in the Agricultural Sector.* Gower, Aldershot.

Midmore, P. (1991b) Problems in estimating agricultural input–output coefficients by regression methods. *Aberystwyth Rural Economic Research Papers* No. 91–02, (November).

Midmore, P. (1993) Input–output forecasting of regional agricultural policy impacts. *Journal of Agricultural Economics* 44(2), 284–300.

Midmore, P. and Lampkin, N.H. (1989) Organic farming as an alternative to set-aside and an option for extensification. In: Dubgaard, A. and Hjortshoj-Nielson, A. (eds) *Economic Aspects of Environmental Regulation in Agriculture.* Wissenschaftsverlag Vauk, Kiel, pp. 267–278.

Murphy, M.C. (1992) *Organic Farming as a Business in Great Britain.* Agricultural Economics Unit, University of Cambridge, Cambridge.

Victor, P.A. (1973) *The Economics of Pollution.* Macmillan, London.

22 Modelling the Impact of Widespread Conversion to Organic Farming: An Overview

PETER MIDMORE AND NICOLAS H. LAMPKIN

Introduction

The effect of conversion on a major scale has been the main focus of chapters in this section, especially with reference to a number of key policy variables. Here, the evidence that these and other studies provide is reviewed, particularly in terms of the effectiveness of the arguments which can be based on them. The issues which are of particular interest include: changes in output and land use; food security and international trade; employment and income impacts in agriculture and related industries; and environmental/social benefits and public expenditure.

Changes in Output and Land Use

The conclusions reached by Zerger and Bossel (Chapter 18), Bechmann (1987, reported in Chapter 18), Braun (Chapter 19) and Lampkin (Chapter 20) concerning changes in output levels and land use following 100% conversion to organic farming are remarkably similar, given the different starting positions and methodologies. In general they indicate a 40–50% decline in output of cereals as a consequence of reduced intensity and area cropped, with an increase in rye and oats at the expense of barley and wheat. All the studies indicate a substantial increase in forage crops and particularly grain legume production, accompanied by a reduction in oilseeds and sugarbeet in favour of potatoes and field vegetables. Estimates of the changes in livestock numbers and output are also similar: 10–25% for grazing livestock and 50–80% for pigs and poultry. These estimates are supported to some extent by survey evidence on the impacts of large-scale conversion as part of the European Community's extensification programme in Germany (Schulze Pals, 1993; Chapter 25, this volume).

There have also been similar studies from the United States. Oelhaf (1978, 1983) attempted to model the implications of US conversion to organic production relying primarily on neoclassical economic models and historical data for conventional cropping, as virtually no data for organic crops existed at that time. Oelhaf concluded that 17% more land would be required to produce the same quantity of crops, and that a 10% increase in costs and land area was likely if the United States were to convert to organic production. The US Department of Agriculture study team on organic farming (USDA, 1980) assumed that 30% of the US area currently harvested for maize and soyabeans would be converted to a seven-year rotation of alfalfa, alfalfa/maize, soyabeans, maize, soyabeans, and oats. Maize yields would fall by 10% and soyabean yields would remain the same. On this basis, maize production would fall by 14% and soyabeans by 12%, while oat production would increase by 40% and alfalfa hay by 41%.

One of the weaknesses of the European scenario models reported in this book is the absence of any direct assessment of the impact of changing levels of output on prices for agricultural produce, and the influence which these changes in turn might have on the structure of land use. In most cases, the models rely on current organic farming land-use patterns which reflect not only husbandry constraints but also current market prices, in particular the high demand for cereals and vegetables, and the lack of demand for oilseeds and sugarbeet (i.e. crops which require significant additional processing) and livestock products.

In part, the lack of consideration of price changes recognizes the fact that policy measures are often a more important determinant of prices than supply, particularly in the European context. This situation is now changing following reform of the Common Agricultural Policy and the GATT agreement (Chapter 26). But even if price changes are taken into account, there may be problems determining appropriate initial price levels. It is probable that premiums for organic produce will fall or disappear in the face of widespread conversion, but little information exists on price elasticities in the organic market. The relative changes in output levels of certified organic produce will in any case be so large that elasticities derived from the current market (e.g. McGregor *et al.*, 1991) are not of much help. Thus it may be that any attempt to account for the influence of changing price levels should be based on prices and price elasticities for conventional products, but possible shifts in demand from final consumers and intermediaries (particularly for livestock feeds) will also need to be taken into account.

The USDA (1980) report estimated price changes based on (then) current demand elasticities: maize prices were forecast to increase by 28%, soyabeans by 53%, and oat prices to fall by 80%. There would also be a fall in hay prices, which with higher maize prices would lead to a shift in the balance between grain and fodder. Total receipts would fall due to inelastic demand for livestock products (on the assumption that livestock numbers would increase). Exports of maize and soyabeans would also fall, but income from exports would increase due to export demand inelasticities. Changes in prices would lead to the remaining conventional farmers altering their production practices to compensate.

Another US study used linear programming to estimate the effects on production, supply prices, land use, farm income and export potential of 100% conversion to organic farming in the United States (Olson *et al.*, 1982; Langley *et al.*, 1983). Crop yields and production costs were estimated for 150 producing regions for seven crops under conventional and organic management. Crop and livestock demand were exogenously defined on the basis of 31 consuming regions, with interregional transfers possible. It was assumed that no supply control programmes were in force, so that the land base included set-aside areas. Organic yields were based on 1944–1946 yield levels adjusted for crop variety improvement and irrigation development. Five different scenarios were investigated, the three main ones being: (i) continuation of current conventional management; (ii) a 'high-export' strategy bringing back into production most set-aside crop land; and (iii) continuation of current organic management practices and yield assumptions. Widespread adoption of organic farming was estimated to lead to double the acreage required to produce 12% less wheat, 52% extra land for 25% less maize, and 38% extra land for 34% less soyabeans when compared with the continuation of current conventional management. Alternatively, 16% extra land would be required to produce 45% less wheat compared with the high-export scenario. Some interregional shifts in crop production would also occur.

The decrease in total production under organic management estimated by Olson *et al.* would lead to lower export potential, higher supply prices, higher value of production (16% above the high export scenario), 14% lower costs of production and 114% higher net farm income. However, the implied yield reductions under organic management are substantially greater than in other studies. Part of this is due to the need to bring lower fertility land into production, thus further reducing average per hectare yields. It is not clear to what extent the method of estimating yield levels actually succeeds in deriving results which correspond to the reality of organic management – no attempt was made to contrast the results with empirical surveys such as

Klepper *et al.* (1977). There was also no specific inclusion of livestock or rotational constraints, so that the fertility benefits of a preceding legume crop or manure applications were ignored.

Besides the highly aggregative, national-level approaches described above, there have been attempts to examine the implications of converting villages or districts to organic production. These are based on a detailed analysis of the current land use, farm structure, geographical, climatic, market, social and other constraints in the particular area under study. Estimates of the financial and other implications of conversion have then been made based on budgeting a number of farms of different types, in some cases raising the data to produce an aggregate result (Deerberg, 1985; Bokermann *et al.*, 1986; Lampkin *et al.*, 1987; Rist *et al.*, 1989; Breuninger, 1990). Typically, these studies include assessments of changes in land use, physical output, labour requirements, nutrient budgets, animal feed budgets, marketing options, farm income and capital investment. Their specificity to one or other particular locality, however, tends to limit their application to consideration of macrolevel issues.

Food Security and International Trade

A frequently expressed concern is whether, if organic farming were adopted on a large scale, domestic food requirements could still be met. This concern is understandable in the context of postwar agricultural policy, with its original emphasis on securing adequate food supplies in the light of wartime experiences. It seems less relevant in the current context of policy-driven overproduction in developed countries, and the knock-on effects of low world food prices on agriculturally-based developing countries.

The studies reported in this book indicate that domestic food demand could be met with organic methods, but food would be more expensive. Zerger and Bossel (Chapter 18) raise the issue of declining food demand in Germany due to demographic change, and Bechmann (1987) uses these results, and an assumption of reduced meat demand, to conclude that no one in Germany need starve given 100% adoption of organic farming. Lampkin (Chapter 20) suggests that the UK would need to increase reliance on food imports again, but that the level of self-sufficiency under 100% conversion to organic farming would be similar to that in the early 1970s (although current levels of food self-sufficiency may be overstated due to the reliance on livestock feed imports). Similar conclusions are reached in the US studies. Olson *et al.* (1982), however, argue that the US would not have large reserves of land which could be used to feed the world, because 99% of the available crop land would be utilized (including less productive land) compared with 66% under the continuing conventional management scenario.

There are a number of open questions, some raised by Zerger and Bossel, as well as Bechmann (1987), in Chapter 18 and Lampkin in Chapter 20, which need to be addressed in more detail in future studies. At present, livestock, particularly pigs and poultry, are competitors with humans for cereals and protein-rich feeds. To what extent might widespread adoption of organic farming be accompanied by a shift in consumer tastes to low-meat diets, thus making more efficient use of the crops which are produced? To what extent might demographic change in developed countries also contribute to reduced demand and offset reductions in supply? Large quantities of crops currently exported by the United States and developing countries such as Brazil are imported into Europe, not for human consumption, but for livestock production. To what extent would an increase in the area of domestically produced forage and grain legumes under organic management substitute for these imported livestock feeds? If, as a consequence, there were a net *decrease* in imports, would this benefit developing countries by removing a source of competition for food, or would the loss of foreign exchange earnings be more harmful? Conversely, if there were an overall net *increase* in imports into Europe as a consequence of widespread conversion, would this lead to greater competition for food and the potential for even more environmental damage abroad as new land is brought into production and production methods are intensified? Can ecologically-based, low-input systems offer the potential for *increasing* production in resource-poor countries where conventional inputs are not affordable and consume precious foreign exchange reserves?

Employment and Incomes in Agriculture and Related Industries

To the extent that employment and incomes are issues of social as well as individual concern, they should form an important part of any assessment of the *economic* performance of organic farming (see Chapter 4). Increased employment and similar or increased incomes on individual farms are a feature of many of the farm-level survey and case studies reported in this volume. However, the modelling studies present a mixed picture of the possible industry-level impacts of widespread conversion to organic farming. Zerger and Bossel (Chapter 18) suggest that the number of farms and the numbers employed in agriculture will continue to decline, but at a slower rate if an ecological approach was adopted, and that farm incomes would fall slightly if farmers were not able to obtain premium prices. Braun (Chapter 19) suggests that farm incomes would fall significantly, assuming no premium prices and maintenance of the existing number of farms and distribution by size and type. Midmore (Chapter 21), on the other hand, estimates that there might be a small increase in overall employment and incomes as a result of conversion of 5% of Welsh agriculture to organic production. The US studies (USDA, 1980; Olson *et al.*, 1982) indicate that incomes might increase due to the increase in *conventional* product prices resulting from reduced supply. The contrasting conclusions reached, compared with the farm-level studies, clearly relate to the different price assumptions adopted. They are also linked to the problem of extrapolating from current organic farming, which is only a very small part of the whole agricultural industry, to a situation where organic farming has much greater significance.

Linkages with other industries, both upstream and downstream, are also an important part of assessing the employment and incomes impact of widespread adoption of organic farming. The greatly reduced use of agrochemical and energy inputs indicated in the literature and in the farm-level studies is likely to have significant knock-on impacts in the agricultural supply industry. One approach to analysing this issue is the use of input–output models, e.g. Midmore (Chapter 21), and Ennis (1985, reported in Chapter 21). Midmore concludes that although jobs would be lost in the agrochemical industry, these would be compensated for by a limited increase in on-farm employment and in local added-value activities such as processing and distribution. Lockeretz (1989a) concluded that although high-input systems had greater local economic benefits on a per hectare basis, the benefits per unit value of production were lower, which meant that a greater proportion of the value of production left the local economy to pay for purchased inputs. The implication of this is that organic systems which are less reliant on purchased inputs may contribute more to local economic activity in the long term, particularly in rural areas which are highly dependent on agriculture.

Environmental/Social Benefits and Public Expenditure

The environmental benefits arising from the changes in land use and reduced intensity of production provide further examples of the positive externalities which might be derived from the widespread adoption of organic farming (Chapter 4). Zerger and Bossel (Chapter 18) estimated from their model that soil erosion might be reduced by 50% under the ecological scenario. Jenkins and Midmore (1992, reported in

Chapter 21) examined the promotion of organic farming as one of a range of environmental policy options, arguing that any negative income and employment impacts might be outweighed by the environmental and public expenditure benefits, at least when compared with other policy options such as habitat creation and cereals management schemes. Rist *et al.* (1989) focused specifically on environmental externalities in their more localized study of the potential of organic farming in the canton of Zug in Switzerland. Others have argued that the widespread adoption of organic farming could be environmentally damaging because much of the extra land which would be required to maintain current output levels would be in environmentally sensitive, marginal areas.

Berenschot (1989) attempted to place values on the environmental impact of conventional and organic (bio-dynamic) agriculture in the Netherlands. They estimated 40% less output from Dutch agriculture, but lower input costs, so that the added value on the national farms under bio-dynamic management would only be 18% less. The national farm income, after taking labour and capital costs into account, would be negative for both conventional and bio-dynamic scenarios (at Fl −3 and −5.8 billion, respectively). However, Berenschot argued that the environmental costs of conventional production, estimated at Fl 6.1 billion (against a questionable zero for the bio-dynamic farms), more than made up for the difference in farm incomes and that bio-dynamic agriculture would therefore be worth pursuing from a social perspective.

Positive environmental and social externalities can be used to provide a justification for government intervention to support organic farming, and their quantification can help to determine the appropriate level of support, but this need not necessarily lead to increased government expenditure. In the European context, there is some evidence that financial support for conversion to organic farming, at least in the short run, can be more than offset by the reductions in public expenditure on storing and disposing of surplus products on world markets (Chapters 21 and 26).

Conclusions

The models reviewed in this section share many of the characteristics and deficiencies of more general attempts to analyse the impact of changes in agricultural policy. Indeed, some of the more comprehensive attempts have formed part of a range of scenarios to provide a basis for policy choices. This is not surprising, as in order to convince policy-makers of the cost–benefit ratio involved in assisting organic farming, there is a need to share a common set of tools. It is, however, surprising that not more of the critical economic perspective of environmentalism has been incorporated into these analyses, given the development of the organic movement as part of the environmentalist critique of agriculture. In improving the quality of these analyses, the incorporation of a number of factors not normally considered relevant to economic modelling may prove to be important.

One common failing of analyses of the impact of both conventional policies and those which are 'ecological' in some respect is the use of the farm as the basis of analysis. Most models treat the whole system as if it were a single farm, and amalgamate responses of abstract or prototypical farms as the response of the system. The problem with this approach is that the individual business units are not independent of the systemic parameters that govern changes, which, if adopted by considerable numbers, would cause a change in the system itself. Moreover, such changes are unlikely to be well behaved and thus mathematically tractable. Most of the models discussed so far also suffer to some extent from being 'comparative-static' type models. That is, they do not incorporate positive and negative feedback mechanisms which would allow a more 'dynamic' analysis, although Zerger and Bossel (Chapter 18) have made some attempts in this direction. Vester *et al.* (1983–1989) have attempted to define the variables and the feedback links which a dynamic simulation model of widespread adoption of organic farming might incorporate. Their approach is based on ecological systems analysis, but also incorporates economic and policy variables and indicators. The model is defined at three levels: the national-level system and community-level and farm-level subsystems. The detailed theoretical treatment of the subject by Vester *et al.* goes further than any of the other models described above, although no attempt is made to quantify any of the relationships identified.

The tenor of models thus far developed remains neoclassical in orientation, preferring to demonstrate policy efficiency as a major goal (especially in terms of minimizing budgetary cost) rather than objectives which are associated with sustainability. Objectives related to the emerging paradigm of ecological economics (Ekins and Max-Neef, 1992), for example, are associated with the ideas of diversity, robustness and continuity. In this regard, the evolution of the number and type of farms is crucial, yet none of the studies, except perhaps Zerger and Bossel (Chapter 18), satisfactorily deals with this issue. Neither do they deal sufficiently with the institutional context, ignoring the hegemonic influence of the economically powerful agribusiness interest. Finally, none pays much attention to the global issues of rapidly growing populations, especially the need for densely populated industrial countries to command energy and other resources in order to meet their nutritional needs.

These criticisms are perhaps unfair, given the slender resources so far devoted to the study of the long-term impact of radical changes in farming practice (especially in comparison with those used to study more conventional policy options). In particular, although more sophisticated approaches are available which might improve the reliability of the models, appropriate data, for example on price elasticities or regional yield and farm income variations, are still not readily available. It is this lack of appropriate data which leads Lockeretz (1989b) to question any attempt at national or regional modelling. The data availability situation is improving gradually, to the extent that in some countries, such as Germany, it may now be possible to develop more realistic models. Given the key policy importance of the impact of widespread conversion to organic farming, efforts to improve both data availability and the quality of the models clearly need to be maintained and encouraged in the future.

References

Bechmann, A. (1987) *Landbau-Wende, Gesunde Landwirtschaft – Gesunde Ernährung.* S. Fischer Verlag, Frankfurt am Main.

Berenschot (1989) *Op zoek naar een duurzame landbouw. Een schets van de kosten en baten van de omschakeling van Nederland op biologisch-dynamische landbouw.* Berenschot Consultants, Utrecht.

Bokermann, R., Alps, H., Dersch, H. and Dimmer, W. (1986) *Betriebswirtschaftlicher Fachplan zur Dorfentwicklungsplanung Rambach.* Fachbereich Landwirtschaft, Gesamthochschule Kassel, Witzenhausen.

Breuninger, E. (1990) Die Umstellung eines Dorfes auf alternativen Landbau. Diplomarbeit, Institut für landwirtschaftliche Betriebslehre, Universität Hohenheim, Stuttgart.

Deerberg, F. (1985) *Gutachten zur Einschätzung der Möglichkeit einer Stabilisierung des ländlichen Raumes durch ökologischen Landbau im Rahmen des Dorfentwicklungsplanes Rambach.* Fachbereich Landwirtschaft, Gesamthochschule Kassel, Witzenhausen.

Ekins, P. and Max-Neef, M. (eds) (1992) *Real Life Economics: Understanding Wealth Creation.* Routledge, London.

Ennis, J.L. (1985) The likely inter-industry effects of organic farming adoption in the United States. MSc thesis, Department of Agricultural Economics and Sociology, Ohio State University.

Klepper, R., Lockeretz, W., Commoner, B., Gertler, M., Fast, S., O'Leary, D. and Blobaum, R. (1977) Economic performance and energy intensiveness on organic and conventional farms in the Corn Belt: a preliminary comparison. *American Journal of Agricultural Economics* 59, 1–12.

Lampkin, N., Jones, W., Mann, M. and Midmore, P. (1987) *Proposal for an*

Integrated Rural Development Programme Focusing on Organic Agricultural Production in the Teifi Valley, Dyfed. Department of Economics and Agricultural Economics, University College of Wales, Aberystwyth.

Langley, J.A., Heady, E.O. and Olson, K.D. (1983) The macro implications of a complete transformation of US agricultural production to organic farming practices. *Agriculture, Ecosystems and Environment* 10, 323–333.

Lockeretz, W. (1989a) Comparative local economic benefits of conventional and alternative cropping systems. *American Journal of Alternative Agriculture* 4, 75–83.

Lockeretz, W. (1989b) Problems in evaluating the economics of ecological agriculture. *Agriculture, Ecosystems and Environment* 27, 67–75.

McGregor, M., Hunter, T. and Blackholly, H. (1991) An evaluation of the response of consumers to the reduction in price/premiums and re-merchandising of organic produce. In: *Marketing.* Final report volume II. Organic Farming Centre, School of Agriculture, Edinburgh, pp. 44–61.

Oelhaf, R.C. (1978) *Organic Agriculture – Economic and Ecological Comparisons with Conventional Methods.* Allanheld, Osmun & Co., Montclair, New Jersey.

Oelhaf, R.C. (1983) The economic feasibility of widespread adoption of organic farming. In: Knorr, D. (ed.) *Sustainable Food Systems.* AVI Publishing, Westport, Connecticut, pp. 156–171.

Olson, K.D., Langley, J. and Heady, E.O. (1982) Widespread adoption of organic farming practices: estimated impacts on US agriculture. *Journal of Soil and Water Conservation* 37, 41–45.

Rist, S., Stoeckli, B. and Suter, H. (1989) *Möglichkeiten und Grenzen des biologischen Landbaus im Kanton Zug.* Forschungsinstitut für biologischen Landbau, Oberwil/BL.

Schulze Pals, L. (1993) *Regional und betrieblich differenzierte Untersuchung über Akzeptanz, Motivation und wirtschaftliche Folgen einer Umstellung auf ökologischen Landbau im Rahmen des Extensivierungsprogrammes.* Arbeitsbericht 2/93 (November). Institut für Betriebswirtschaft, Bundesforschungsanstalt für Landwirtschaft, Braunschweig-Völkenrode.

USDA (1980) *USDA Study Team on Organic Farming: Report and Recommendations on Organic Farming.* United States Department of Agriculture, Washington, DC.

Vester, F., Hesler, A v., Schwarz, R. and Storp-Benecke, M. (1983–1989) *Systemstudie Ökoland I-VIII.* Arbeitsberichte 1, 4, 5, 7, 9, 11, 12, 13. Institut für Interdependenz von Technik und Gesellschaft, Universität der Bundeswehr, Munich.

Part 5

Organic Farming and Agricultural Policy

23 Policy Issues and Impacts of Government Assistance for Conversion to Organic Farming: The Danish Experience

ALEX DUBGAARD AND HANNE HOLST

Introduction

Since the mid-1980s support for organic farming has been an issue high on the public and political agenda in Denmark. Conversion to organic practices is generally seen as a means of alleviating the environmental problems of conventional agriculture. In 1987 legislation was passed to support organic farming in Denmark (Act on Organic Farm Production). The programmes implemented since then aim at promoting organic agriculture in several ways, the most important being:

- financial support for conversion and for infrastructure development projects;
- an official labelling system for organic products and a legal system for certification and control of organic production.

This chapter presents the basic features of this legislation. It also examines the impact of the support on the development of organic agriculture in Denmark.

Conversion and Development Support Schemes

The financial support system, introduced in 1987, provides grants to farmers converting from conventional to organic practices as well as grants supporting the development of a research, extension and

Table 23.1. Organic farming development and conversion grants (DK millions) Denmark, 1987–1992.

	1987	1988	1989	1990	1991	1992	Total
Conversion grants	0 (0%)	4 (33%)	8 (17%)	8 (31%)	15 (75%)	4 (12%)	39 (28%)
Development grants	0.1 (100%)	8 (67%)	39 (83%)	18 (69%)	5 (25%)	28 (88%)	98 (72%)
Total	0.1	12	47	26	20	32	137

Sources: Danish Ministry of Agriculture (1992) and Directorate for Agricultural Development (pers. comm., May 1993).

Table 23.2. Differentiation of organic farming conversion grants, Denmark, 1987–1992.

Livestock units[a] per hectare	Conversion grant (DK ha^{-1})
Less than 0.3	2600
0.3–0.7	3400
More than 0.7	3800

[a]One livestock unit is the equivalent (measured in feed consumption) of one dairy cow.
Source: Danish Ministry of Agriculture (1988).

marketing infrastructure for organic agriculture. The amounts of support granted under the two schemes are shown in Table 23.1. By the end of 1992 a total of DK 39 million had been given to conversion, while development support amounted to DK 98 million.

The conversion grant scheme

Conversion support is a one-off grant payment ranging from DK 2600 to 3800 per hectare of the area in rotation. As shown in Table 23.2, the level of the subsidy is based on the livestock density on farms entering into a conversion agreement. Model calculations indicate that total transition costs (under optimal conditions) amount to about DK 4200 ha^{-1} (Danish Ministry of Agriculture, 1992). These costs are incurred mainly through yield reductions and extra labour requirements which cannot be balanced by price premiums since products can be classified as organic only after a two-year transition period.

To be eligible for conversion support, farmers must comply with the official standards for organic practices specified in the Act on Organic Farm Production, 1987 and later amendments. This defines organic farming as an agricultural production system which avoids the use of synthetic fertilizers, pesticides, growth regulators and livestock feed additives. In addition, Danish regulations demand a high degree of nutrient self-sufficiency within the organic sector (cf. Dubgaard, Chapter 7, this volume). Farmers receiving the conversion subsidy are bound to convert the whole farm to organic methods within a four-year period. At least one hectare must be converted per year and the farmer commits himself to run the farm as organic for at least two years after completion of the conversion.

The development support scheme

Development support can be given to projects enhancing the scientific basis of and the marketing system for organic agriculture. Projects

Table 23.3. Sectoral distribution of organic farming development support, Denmark, 1987–1992.

	DK millions	%
Collection and processing of organic produce	18	18
Information and PR activities	20	21
Extension	14	14
Research and experiments	46	47
Total	98	100

Source: Danish Ministry of Agriculture (1993).

improving collection, processing and marketing of organic produce may receive development grants covering up to 40% of 'extraordinary' implementation costs. Training, information and research programmes can be given support covering up to 100% of the programme costs. Grants can be extended to private individuals, companies, associations and public institutions.

The sectoral distribution of development support is shown in Table 23.3. Nearly half the support has been given to research and experiments (field trials, market analyses, etc.). Processing and collection systems for organic produce have received 18% of the grants, while one fifth of the development support has been used to increase the awareness of organic farming and organic products, *inter alia* in the form of promotion campaigns with much emphasis on the beneficial environmental aspects of organic production practices.

An official labelling system for organic produce

In connection with the implementation of the Act on Organic Farm Production, the government and the organic farmers' organizations agreed that there was a need for a legal framework which could give guarantees to consumers of produce bearing organic labels. Subsequently an official labelling system was established for organic products together with public control of production and marketing. Only certified farms, companies and shops are permitted to market produce under an organic label. Control is in the form of on-farm checks by government inspectors. Similar control takes place in companies, organizations and shops dealing with processing, labelling, marketing and retail sale of organically produced food products.

Consumers' confidence in organic labelling is fairly high. In an investigation by Bjerke (1992), three out of four consumers stated they felt confident that products bearing the organic label were in fact organically produced. Still, official control may not have played a decisive role in this context since only 15% of the respondents knew that organic labelling is controlled by the state (Bjerke, 1992).

Assessment of the Support Programmes

The conversion grant scheme

Since the conversion support scheme was introduced, the number of certified organic farms has increased from 219 in 1988 to 672 in 1992 (Table 23.4). In the same period the area converted or under conversion rose from 5880 ha to 18,500 ha. However, the net increase in the organic farm sector is somewhat lower than indicated by the figures in Table 23.4. In 1988 there were about 350 organic and bio-dynamic farms in Denmark with a total agricultural area of 7000 ha (Dubgaard and Sørensen, 1988). There are probably still some uncertified organic and bio-dynamic farms in Denmark, but since the conversion support scheme was introduced, the number of organic farms in Denmark has approximately doubled, and the agricultural area converted or under conversion has increased by 150%.

In 1992, the Danish Ministry of Agriculture made an assessment of the support measures introduced by the Act on Organic Farm Production (Danish Ministry of Agriculture, 1992). An interview survey was conducted (by the private consultants PLS Consult) to assess the impact of the conversion grant on farmers' decisions to convert. The results of this survey are reported in Table 23.5. Only one out of four respondents indicated that the conversion grant had exerted a decisive influence on the decision to convert. Close to 50% of the respondents stated that the conversion grant (they had received) had been of no influence at all to their decision. Altogether economic considerations ranked very low among the reasons given for converting. The motives stated by the converting farmers themselves were mainly 'idealistic'. More than 80% of the respondents specified concern about the environment as a motive of particular importance.

Nevertheless, economic factors may still have exerted a decisive influence on the rate of conversion in recent years. In the peak period of conversion (1989–1991) the leading dairy cooperatives in Denmark offered price premiums on organic milk in the range 25–35%. However, processors overestimated consumers' willingness to pay premium prices for organic dairy products. At present, dairies have to dispose of about half the amount of organic milk which they have contractual commitments to buy from organic producers (at premium prices) through conventional market outlets (National Ecological Information

Table 23.4. Increase in number of certified organic farms, Denmark, 1988–1992.

	1988	1989	1990	1991	1992
Farms certified at start	–	219	401	523	672
+ applications submitted	–	194	166	213	62
– applications withdrawn	–	8	23	20	1
– certifications suspended	–	4	21	44	61
= farms certified at end	219	401	523	672	672
Farm area (ha)	5,881	9,553	11,581	17,963	18,500
Conversion grants (DK millions)	4	8	8	15	4

Source: Danish Ministry of Agriculture (1992).

Table 23.5. Reasons stated by farmers for converting to organic agriculture, Denmark, 1992.

Reasons for conversion[a]	Not at all important (%)	Of some importance (%)	Of particular importance (%)
Environmental concerns	3	14	84
Better working conditions on the farm	28	20	50
Animal welfare	24	27	47
Improvement of food quality	30	31	37
Conversion grants	48	27	24
Recommendations from colleagues	60	22	17
Better economic returns	65	20	15
Other	64	2	30

[a] Based on interviews with 100 farmers who had received conversion grants.
Source: Danish Ministry of Agriculture (1992).

Centre, 1992). Consequently, most dairies no longer offer price premiums to new producers of organic milk – and the prospects for dairy farmers who have already converted are gloomy.

Conversion to organic farming came to a standstill in 1992 (cf. Table 23.4). The number of support applications submitted fell from 213 in 1991 to 62 in 1992 and the amount of support granted dropped from DK 15 million in 1991 to DK 4 million in 1992. The most recent statistical information indicates that by mid-1993 the number of certified organic farms had declined by 17 compared to the year before. This is the first time a reduction in the number of organic farms has been registered in Denmark.

The development support scheme

Retail distribution of organically produced food was previously confined to specialist shops. Organic products are now available on a daily basis in nation-wide supermarket chains. For example, the largest Danish supermarket group, the consumer cooperative FDB, has been marketing organic food products since the beginning of the 1980s. The whole country is covered by FDB's distribution network and its market share for convenience goods is 25–30%. In addition, a large private supermarket group (with 15% market share) has been distributing organic/bio-dynamic products for several years. Therefore, it seems safe to conclude that a lack of appropriate marketing channels no longer represents a substantial obstacle to the expansion of organic farming in Denmark.

Consumers' awareness of organic production is relatively high. In a representative group of consumers three out of four knew about the existence of organic agriculture and 45% of the respondents were familiar with the basic principles for this type of farming – such as the non-use of chemical fertilizers and pesticides (Bjerke, 1992).

Bjerke's (1992) investigations indicate that a majority of consumers believe that organic products are more wholesome than conventional products. Yet, two out of three respondents stated they did not believe that a switch to organic food would be of *significant* importance to healthy living. Only 9% of the respondents were fully convinced that it would. On the other hand, a majority of the respondents in Bjerke's investigations stated that they considered conventional agriculture more polluting than organic farming.

The evaluation made for the Danish Ministry of Agriculture by PLS Consult concludes that development support has enhanced knowledge and awareness of organic farming and organic products – in addition to improving processing and marketing channels for such products (Danish Ministry of Agriculture, 1992). However, several development projects receiving support have been unsuccessful. For example, a supported initiative to establish an independent processing and marketing network for organic milk ('Dansk Naturmælk') failed after a couple of years. Since then, the two largest 'conventional' dairy cooperatives in Denmark (MD and Kloever Mælk) have taken over processing and marketing of milk from most organic producers. Only liquid milk products have acquired a market share of some significance (2–3%). Attempts to create demand for processed dairy products, such as butter and cheese, have largely failed.

In 1988 and 1989, organic pork and beef were introduced in most FDB stores, followed by a massive promotion campaign. Nevertheless, sales did not reach an economically sustainable level and retail marketing of organic meat was discontinued by FDB in 1990 (Obel, 1993). Several other attempts to establish organized market outlets for organically produced pork and beef have also been unsuccessful.

In general, the promotion campaigns conducted in mass media and supermarkets do not seem to have had a measurable effect on demand. Follow-up interviews indicate that promotion campaigns in the late 1980s had no significant impact on purchasing behaviour by the vast majority of consumers (Bjerke, 1990). Likewise, a promotion campaign for organic milk, running for several weeks in 1991, had no visible effect on weekly sales (National Ecological Information Centre, 1992).

The lack of consumer response to such promotion efforts is not surprising in fact. For more than a decade organic farming and organic food products have received massive and benevolent media attention in Denmark. Therefore, consumer awareness of organic products was probably high even before the commercial promotion campaigns were initiated.

Bill on Amendment of the Support Scheme for Organic Agriculture

In a bill to repeal the Act on Organic Farm Production of 1987, the Danish Minister of Agriculture has proposed a change to the present support system from one-off grant payments to annual maintenance payments providing ongoing support to farmers who adopt environmentally beneficial practices (Danish Ministry of Agriculture, 1993). The new scheme will enhance the level of support considerably. The proposed annual payments to organic farming amount to DK 1700 ha^{-1} during the first two years after conversion and DK 1400 ha^{-1} thereafter. The Danish Organic Farmers' Association (pers. comm., May 1993) claims that an annual subsidy of DK 3800 ha^{-1} is required to compensate for the losses incurred under organic practices.

In the absence of price premiums, substantial support is required to secure factor remuneration in organic farming at the level obtained in conventional agriculture (Dubgaard, Chapter 7). According to estimates made by Dubgaard *et al.* (1990) organic practices increase (annual) production costs (per unit of output) by about DK 2500 ha^{-1} on the 'average' organic farm. Thus, in most cases price premiums will still be required to make organic farming economically viable – even if the proposed support scheme is implemented.

Conclusions

Insufficient demand for organic products – at the price premiums required to cover extra production costs – has often been attributed to inadequate distribution channels and a lack of consumer awareness of organic products. In Denmark, such barriers have been removed to a considerable extent since the 1980s. Still, the expected take-off in the sales of organic food products has not occurred.

Consumers in general do not seem to place much emphasis on the perceived nutritional superiority and healthiness of organic food. The benefits associated with a conversion to organic practices are perceived as mainly environmental. However, an improvement of the environment constitutes a public good. It is not surprising, therefore, if most consumers are unwilling to pay for the provision of such benefits on an individual/voluntary basis. Some sort of government intervention is usually required to ensure an adequate supply of a public good.

This indicates that consumers' inclination to pay price premiums for organic products will not suffice to bring organic farming beyond its present niche position. Likewise, the support scheme introduced in 1987 has not given conversion the expected momentum. In other words, if organic farming is to play more than a symbolic role in Danish agriculture, public support will have to be increased substantially.

The new support scheme proposed by the Minister of Agriculture will considerably enhance support for organic agriculture in Denmark, but not to such an extent that (significant) price premiums will no longer be required to ensure the economic viability of organic farming.

References

Act on Organic Farm Production (Act No. 363; 10 June, 1987; Amended May 1990). Folketinget, Copenhagen.

Bjerke, F. (1990) *Organic Produce in Danish Supermarkets.* Working Paper, Roskilde University Centre, Roskilde.

Bjerke, F. (1992) *Forbrugernes interesse for økologiske produkter* (Consumers' preferences for organic products). Institut for Samfundsøkonomi og Planlægning, RUC, Roskilde.

Danish Ministry of Agriculture (1988) *Executive Order (No. 4, January 1988) on Grants for the Promotion of Organic Farm Production.* Copenhagen.

Danish Ministry of Agriculture (1992) *Evaluering af lov nr. 363 af 10. juni 1987 om økologisk jordbrugsproduktion* (Evaluation of Act No. 363 of 10 June, 1987 on organic farming). Jordbrugsdirektoratet, Copenhagen.

Danish Ministry of Agriculture (1993) *Forslag til lov om ændring af lov om støtte til forbedring af strukturen og effektiviteten i jordbruget og om ophævelse af lov om økologisk jordbrugsproduktion og lov om tilskud til skovforbedrende foranstaltninger* (Bill on the amendment of the Act on support to improvement of structures and efficiency in agriculture and on the repeal of the Act on organic farm production). Copenhagen.

Dubgaard, A. and Sørensen, S.N. (1988) *Økologisk og biodynamisk jordbrug – en statistisk undersøgelse* (Organic and bio-dynamic farming – a statistical survey). Rapport 43. Statens Jordbrugsøkonomiske Institut, Copenhagen.

Dubgaard, A., Olsen, P. and Sørensen, S.N. (1990) *Økonomien i økologisk jordbrug* (Economics of organic farming). Rapport 54. Statens Jordbrugsøkonomiske Institut, Copenhagen.

National Ecological Information Centre (1992) *HV Guide.* No. 4. Økologisk Landscenter, Aarhus.

Obel, C. (1993) Forbrugernes efterspørgsel vurderet ud fra etiske og miljømæssige krav (Assessment of ethical and environmental determinants of consumer demand). *Tidsskrift for Landøkonomi* 180(2), 84–89.

24 Agricultural Policy and Impacts of National and Regional Government Assistance for Conversion to Organic Farming in Switzerland

OTTO SCHMID

Introduction

Situation of Swiss agriculture

In Switzerland in 1990, there were about 108,000 farms with an average size of 9.9 ha, of which about 62,000 were mainly living from agriculture, with an average size of 15.2 ha (Bundesamt für Landwirtschaft (BLW), 1992). The average size of farms is, compared with other countries, very small, which is due to several factors, but mainly to an agricultural policy which until 1992 was aiming to preserve traditional family farm structures, using a number of protective measures (price guarantees for milk and cereals, taxes, import restrictions, etc.). In 1993, a reorientation to a more liberal agricultural policy was decided (mainly because of GATT pressures), but which must take into consideration the achievement of important non-economic functions of agriculture (landscape diversity, tourism, etc.) (EVD, 1992).

Situation of organic farming in Switzerland

The number of organic farms in Switzerland rose rapidly from 803 in 1990 and 940 in 1991 to 1500 farms in 1993, which is more than

1% of the total number of farms in Switzerland. There are various reasons for this development: a more positive attitude among farmers towards organic farming; market developments; increasing consumer demand; more training courses and advisory services for organic farming; and the impact of government assistance for conversion to organic farming.

Different systems of government assistance for organic farming

In 1988, the canton of Bern, and a year later the canton of Basel-Land, decided to pay subsidies for organic farms in conversion, mainly because of political pressures in their parliaments. The Research Institute of Organic Agriculture (Forschungsinstitut für biologischen Landbau (FIBL)) in Oberwil made proposals on how to support these farms, based on economic studies (Schmid, 1988). A few more cantons subsequently introduced such subsidies, Zug in 1990, Zürich in 1991 (after a referendum) and Graubünden (Grisons) in 1992. The majority of the other 20 cantons and the federal government did not follow these first five cantons, mainly for financial reasons. The impact of the conversion subsidy system in the above-mentioned cantons will be discussed.

In December 1992, the parliament of the Swiss Federation decided to give annual direct payments to farmers, starting in 1993. This system and its possible impact will be compared with other systems of promoting the conversion to organic farming, such as training courses and advisory services, market development projects, consumer information, support for research on organic farming, etc. This chapter focuses mainly on the policy issues and their impact on organic farming, i.e. mainly on conversion subsidies.

Conversion Subsidies

Importance

The number of farmers interested in organic farming has been rising rapidly in Switzerland in recent years. But many farmers are still afraid of the economic risks during the conversion period. A survey conducted in Switzerland in 1987 (Roux, 1988), found that 25% of the farmers surveyed could imagine converting their farm to organic agriculture. The main argument against converting was the idea held by conventional farmers that the risks involved are too high. One of the problems of conversion is indeed that during the usual two-year conversion period, the farmer and his family have more work and lower yields without having access to a full 'organic' premium to cover the lower income. Products have to be sold as 'conventional' or as 'products from farms in conversion' at a lower price than products from fully-converted organic farms. Other problems during conversion include: the increased labour input needed to test new methods; the time needed to follow special courses; and new investments.

Definition of the conversion period

In Switzerland, the standards for organic agriculture of the Federation of Swiss Organic Farming Organizations (Vereinigung schweizerischer biologischer Landbau-Organisationen (VSBLO)) prescribe a conversion period of two years (VSBLO, 1992). During this time the entire holding should meet the standards. Exceptions are possible for intensive crops (vegetables, fruit orchards and vineyards): a farm can convert on a field-by-field basis, but within five years the full standards requirements have to be fulfilled. The two-year conversion period rule was adopted by the cantons.

Existing systems of conversion subsidies in the Swiss cantons

As already mentioned, five cantons pay conversion subsidies. A short description of the amount of money given by the cantons of Bern, Basel-Land, and Zug is given (Graubünden and Zürich are not described, because the policies in these cantons are still in an introductory phase). All cantons pay a single premium (i.e. not an annual payment): one part per farm and another part per hectare (depending on the kind of crop – see Table 24.1). They specify inspection and certification by VSBLO as a condition for the payment of subsidies.

Other conditions for the payments include:

- 50% of the money will be paid to the farmer after the first year if requested;
- the farm must be operated organically for 12 years, otherwise the subsidies must be repaid;
- state farms cannot receive a subsidy;
- farmers with a very high income (because of other incomes) receive less money.

Economic justification

There has not been very much research into the problem of the conversion period in Switzerland. Most of the studies made are about

Table 24.1. Cantonal subsidies (SF) for conversion to organic farming, Switzerland, 1993.

	Basel-Land	Bern	Zug
One-off subsidies per farm			
Training courses and conversion-related investments	4,000	5,000 (3,000 for farms < 3 ha)	5,000 (3,000 for farms < 5 ha)
One-off subsidies per hectare			
Arable land	700	800	1,000
Leys	700	500	700
Meadows	500	500	700
Mountain meadows (extensive use)	300	500	–
Permanent pastures	100	200	300
Intensive crops (vegetables, fruit, vineyards)	3,000	3,000	3,000
Maximum one-off subsidy/area per farm	25,000 SF	20 ha	–
Farm examples			
20 ha dairy farm (with 4 ha extensive meadows and 4 ha permanent pasture)	11,600	13,800	17,400
20 ha arable farm (with 10 ha arable land)	15,000	18,000	22,000

Source: Schmid (1993).

farms which have already completed the conversion. The above-mentioned subsidies were based on a study conducted by FIBL on how farms in conversion to organic agriculture should be supported (Schmid, 1988). This was based mainly on an economic survey of organic farms by the Swiss Federal Research Station for Farm Management and Agricultural Engineering (Forschungsanstalt für Betriebswirtschaft und Landtechnik, Tänikon (FAT)) (Steinmann, 1983; Mühlebach and Näf, 1990). These results are described in more detail in Chapter 8 of this volume.

The existing cantonal conversion subsidies in Switzerland will be analysed and compared with the economic requirements indicated by the above-mentioned studies.

Compensation for the lack of an 'organic' premium

The FAT studies on the economic situation of organic farms showed that a special premium for milk and milk products from organic production is essential, because usually more than 50% of the income of organic farms is derived from the livestock sector (Steinmann, 1983; Mühlebach and Näf, 1990). But this market has first to be developed if premiums are to be obtained.

For *fodder areas,* one study estimated that compensation of SF 720–930 ha^{-1} $year^{-1}$ would be required and would have to be covered by a higher milk price (MIBA, 1990). In mountain areas the compensation required can be lower due to the lower productivity of the land (SF 300–500 ha^{-1} $year^{-1}$, depending on the region). Because farms in conversion are not able generally to obtain or benefit from an organic premium, it was proposed that the state pay a minimum of 50% of this amount. The canton of Bern decided to pay a one-off amount of SF 500 ha^{-1} for leys and SF 200–500 ha^{-1} for permanent pastures and meadows. This amount of money would cover less than half of the necessary compensation during two years of conversion.

For *arable crops,* the market for organic products is better developed. For example, the farmer receives for cereals an organic premium of SF 700–900 ha^{-1} $year^{-1}$. Calculations based on existing studies (Looser and Strasser, 1990) show that the existing premiums should be much higher (SF 1200–2000 ha^{-1} $year^{-1}$). In 1989, the canton of Bern decided to make a one-off payment of about SF 800 ha^{-1} for all arable crops, the other cantons between SF 600 and SF 1000, which is about 50% of the necessary compensation. For cereals the situation for farms in conversion has improved since 1991, because all farmers not using any insecticides and fungicides in cereals receive a premium of SF 800 from the Federation (a policy decision to reduce overproduction and oversupply of cereals on the market).

For special crops, such as *vegetables, fruit orchards and vineyards*, the cantons with conversion subsidies are paying SF 3000 ha^{-1} for farms in conversion, which represents almost a 50% compensation for the lower income during conversion.

A degree of differentiation between arable land, fodder areas and intensive crops, as is currently the case, makes sense as long as it is based on real cost calculations. For certain crops the risks and the labour requirements are higher. In this way it is easier to avoid certain crops only being chosen because of their conversion premium. On the other hand, a differentiated system is more complicated to administer.

If market development has to be carried out by the farmers who are in conversion themselves, the subsidies should be the highest in the first year. The existing system could be improved in this respect.

Costs for training courses, testing new methods and new investments

Farmers in conversion to organic farming have not only to follow special courses, but also need more time to practise and to test new methods. The time needed was estimated as 15–20 days per year during conversion. If the government were willing to cover half of the costs over two years, a one-off support payment of SF 2500 per farm would be necessary.

Most of the farms in conversion make new investments in weed control and manure management. The calculated additional costs are around SF 3000–5000 $year^{-1}$ (for annual amortization, interest, repair costs, etc.). Farmers who have to change their livestock housing (access to open air, etc.), or even to change their stocking rate, will also have very high investments, which differ greatly between farms. It was estimated that, on an annual basis, these supplementary costs during conversion would be as high as the investments for arable crops. These costs are not covered through a higher price and not compensated by lower variable costs.

The cantons decided to make a single payment of SF 4000–5000 towards these costs to each farm converting. This would meet about half of the costs for new investments and half of the remuneration of the extra time needed for training and testing new methods.

Farms selling direct to the consumers have even more investments, but these costs and the higher level of effort are expected to be compensated through higher prices (special margin).

Yield decrease during conversion and soil biological activity

Existing studies do not give a clear indication whether there is an additional, conversion-related decrease in yields in the first two to three years of conversion compared with the yields when the farm is fully converted. Until now this yield decrease problem has not been taken into account in the cantonal support schemes.

A Swiss study of the microbiological activity of soils on conventional and organic farms showed that farms which have been managed organically for more than ten years have significantly higher activity measured with the ATP biomass method (Maire *et al.*, 1990). It seems that the conversion period for the soil is longer than the usual two years which is needed for certification.

Regional cost–benefit calculations of conversion subsidies

A regional study made in the canton of Zug (Rist *et al.*, 1989) compared the costs of conversion subsidies with the costs of re-establishing an ecological balance for the lake of Zug and with the costs of water pollution created by agriculture, mainly due to phosphate use.

The calculated costs for the conversion of about 520 farmers (73% of all farmers) in this region were estimated at SF 14.4 million for conversion subsidies (based on the system in the canton of Bern) and SF 3.5 million for the advisory service, a total of SF 17.9 million. The costs of regaining an ecological balance for the lake by other technological means would be much higher, at SF 20–30 million (some experts estimated even SF 80 million).

General impact

Impact on the number of converting farms

It is interesting to compare the numbers of farms converting in those cantons which pay conversion premiums with the rest of Switzerland (Table 24.2). The number of farmers starting the conversion process in the cantons of Bern, Basel-Land, Zug and Zürich was initially very small. Different explanations can be given for each of the cantons.

- *Bern* (mainly dairy farms in the mountain area): Interest started to grow after an information campaign through the regional agricultural school.
- *Basel-Land:* For arable farms the amount of financial support was not high enough to start conversion. For dairy farms in this region a major pest control problem is the cherry moth (*Rhagoletis cerasi*) which is not easily controlled in organic farming. The situation is likely to change in 1993 following an information campaign.
- *Zug:* The situation is similar to that in Basel-Land, but because the conversion subsidies were proposed by the minority (green and red) political parties, the majority of the farmers did not convert.

Table 24.2. Number of organic farms starting conversion (first year) in cantons paying conversion subsidies, Switzerland, 1989–1993.

Cantons	1989	1990	1991	1992	Total 1992[a]	1993[b]
Bern	10	30	42	45	183	(40–50)
Basel-Land	2	2	2	6	39	(5–10)
Zug	1	1	2	9	17	(2–13)
Zürich	10	14	9	14	118	(15–20)
Graubünden (Grisons)[b]	(5)	(8)	13	131	182	(50–70)
Switzerland	65	111	145	291	1150	(300–320

[a] Total of certified organic and in-conversion farms.
[b] Approximate figures in parentheses.
Source: Schmid (1993).

- *Zürich:* Here there is a similar situation to that in Bern. Many arable farms in this area do not convert.
- *Graubünden:* For many traditional farms in the mountain areas, the step to organic farming is very small. The conversion premiums are therefore financially much more attractive for these farmers. The regional farmers' association and some cheese-makers have shown an interest in organic farming.

An explanation for this kind of impact might be that if farmers do not have an income guarantee for a period of years, they will not be willing to change to organic agriculture. The development of the organic milk market might play an important role. As soon as one or more of the important processors starts to market organic products successfully and is actively promoting organic farming, the interest in conversion to organic farming increases strongly.

The situations in the mountain and tourist canton of Graubünden and in the mountain area of the canton of Bern are different from those in areas with mainly mixed or arable farms. From the beginning, a very high number of farmers decided to convert after the decision of the cantons to pay conversion subsidies.

Influence of conversion subsidies on the organic market

An interesting question is whether, if the conversion payments are too high and many farmers convert, the limited market for organic products will be heavily distorted. In particular, some established organic farmers have complained that such conversion subsidies are creating a price decrease for organically grown products in the long run. So far, it has not been possible to reach a clear conclusion on this issue. An investigation in the canton of Graubünden (Canova, 1993), on 149 out of about 250 farms either fully organic or in conversion, showed that a part (about a third) of their products (mainly meat and milk) continued to be sold as conventional. On the other hand, the organic market in this region is not very well developed, even though a regional survey of consumers, hotels and butchers showed an interest in organic products which was higher than in other regions (Geissbühler and Sotoudeh, 1993).

Impact of conversion subsidies on the commitment of farmers and consumers

A strong commitment to organic farming is still necessary for farmers and consumers. A system which does not cover all supplementary costs during conversion, as in Switzerland, might help to avoid a situation where farmers only convert for financial reasons. The motivation to convert cannot be only financial, if the conversion is to succeed. Consumers should also contribute, e.g. by paying an extra premium for products from farms in conversion.

Impact of subsidies on the social situation of organic farmers

As long as a new organic farmer risks social isolation, only a very few convinced farmers will convert. The introduction of conversion subsidies has changed attitudes to organic farming in a positive way.

Regional (cantonal) or national (federal) conversion subsidies for organic farming?

While only a few regions of a country are paying such subsidies, in other areas some of the farmers are waiting until their regional parliament and administration also provides such subsidies before they convert. This is counterproductive. The organic farming organizations think that a basic national regulation is necessary. In addition, a degree of regionalization of the payment system might still be beneficial and allow a certain dynamism.

Subsidies for already converted farms?

In Switzerland subsidies are only paid for farms in conversion. Farms already converted cannot benefit from this regulation. This seems to be unjust. Farms in conversion also receive a 'lump sum' in compensation for their training and special investment costs. In regions where there are no markets established for milk and meat, an area payment should also be paid.

Direct Payments for Farmers

New agricultural policy in Switzerland

At the end of 1992 the Swiss national parliament decided to change the national law for agriculture (Article 31a and b). It was decided to implement two categories of direct payment for farmers (EVD, 1992):

- *Article 31a:* Complementary direct payments for all farmers, which should allow agriculture to fulfil all functions of general interest. These payments will be determined in relation to the type of farm, the surface area and the limitations due to the production location. All cantons pay a one-off premium (not an annual payment) – one part per farm and another part per hectare (depending on the type of crop).
- *Article 31b:* Contributions to those farmers who produce in a sustainable way, respecting the environment and the basic needs of animals, as in organic farming and integrated production. In addition to these payments contributions are given for areas which are used for ecological purposes (habitats). The payments must be high enough that the farmers can achieve an income similar to conventional farmers.

The budget in 1993 for these payments was SF 599 million for Article 31a and around SF 56 million for Article 31b, rising after a few years to the same level as for Art. 31a (Anon., 1993a, b).

General direct payments for all farmers

These are listed in Table 24.3. The Federation of Swiss Organic Farming Organizations has proposed that a higher amount of money should be paid per farm and that the payments per hectare should be raised also (VSBLO, 1993).

Payments for more sustainable farming methods

Such payments are set out in Table 24.4. All farmers should receive from the government compensation for landscape conservation from 1992 onwards. In addition, some cantons pay subsidies for hedges, extensive pastures, etc.

Table 24.3. Federal government annual payments (SF) for all farms, Switzerland, 1994.

Type of payment	Payments	Comments
Payment per farm	400–2000	Depending on the production area and the size of the farm
Additional payment to farms for farms with animals (> 5 livestock units)	2500	Only for farms > 9 ha
Basic payment	380 ha^{-1}	Farms 3–50 ha
Payment for meadows and pastures (without alps)	180–290 ha^{-1}	Farms 3–50 ha

Source: Anon. (1993a).

Table 24.4. Federal government annual payments (SF) for more sustainable farming, Switzerland, 1994.

Type of payment	Payments	Payments proposed by the organic sector for 1994
Payment per farm (> 3 ha)	1000 (IP) 1500 (O)	min. 2000 (O)
Payment per hectare for meadows	100 (IP) 150 (O)	300 (O)
Payment per hectare for arable land	400 (IP) 750 (O)	800–1000 (O)[e]
Free range livestock production (per livestock unit)	30–120[a]	150–200
Payments for nature conservation per hectare (extensive meadows, hedges, etc.)[b]	450–1000	600–1600
Payments for traditional fruit trees[c] (per tree)	10	30
Payment for costs of inspection (per farm)	No payment[d]	200

IP: Integrated production, based on standards recognized by the government.
O: Organic farming based on VSBLO standards recognized by the government.
[a] Depending on the type of animal (> 5 livestock units).
[b] Minimum 6 years contract.
[c] 300 trees, > 5 years old, height min. 1.6 m.
[d] Some cantons pay 100–200 per farm.
[e] For vegetables, orchards and vineyards, 400 per ha is proposed.
Sources: VSBLO (1993) and Anon. (1993b).

Aim and possible impact of the new system of direct payments

The political reasons for the change in agricultural policy are, on the one hand, compensation for starting the deregulation of the Swiss agricultural market (mainly due to GATT pressures) and, on the other, encouragement for farmers to produce in a more sustainable (ecological and ethological) way, which would lower the costs of overproduction and environmental problems.

As the figures show, the organic farmers receive a 20–50% higher surface area payment than their colleagues adopting integrated methods (although not double, as the organic farmers would have preferred). In the author's judgement, it seems likely that the proposed system of direct payments will mainly motivate traditional dairy farms in mountain and hill regions to convert to organic farming, but not arable farms in good production areas.

An economic study for the Social Democratic Party criticized the proposed payments (Eichenberger and Kaufmann, 1992). Their conclusions were as follows:

- The proposed system is not sufficiently motivating to bring about change to more sustainable methods.
- Most of the problems of agriculture will not be solved, e.g. the problems of overproduction, the environmental costs (estimated by these authors at SF 0.6–1.5 million per year), and the reduction in the number of farmers in Switzerland (2% annually), which will be mainly a problem for rural areas.

Therefore the report proposes:

- Direct payments should only be given to a more restrictive form of integrated production and to organic farmers, but not to all farmers.
- Taxes on fertilizers and pesticides of 50–100% should be introduced.
- Training courses, conversion premiums, and financial support for organic marketing projects, besides subsidies, are important in promoting a more ecological agriculture.

The government's proposed system was criticized similarly by other organizations such as the small farmers' association and the environmental organizations.

Other Governmental Assistance for Organic Farming

Support for research

The Swiss government paid the Research Institute of Organic Agriculture (FIBL) about SF 1.1 million in 1992 for research into organic farming. This constituted about 1% of the total research budget for agriculture (excluding private commercial research). Very little research is done at university level. In relation to the future importance of organic farming this amount of money would seem to be insufficient.

The impact of research to promote organic farming is difficult to judge, depending very much on the research priorities, in particular the kind of technical problems worked on and the proposed solutions to be adopted by farmers.

Marketing projects and consumer information

It can be argued, as some market specialists do, that it is mainly consumers and landowners who will benefit from conversion subsidies and direct payments and not primarily the organic farmers. From their point of view, these subsidies can be used as an argument to put pressure on farmers to pay more for agricultural land and to accept lower producer prices. Whether this will be the case is difficult to judge. Experience in other countries (such as Denmark) shows the importance of governmental support for good consumer information about organic farming and market development. In Switzerland only two or three regions have supported small-scale organic marketing initiatives. In view of the increasing number of organic farms these kinds of support would be a major help, particularly for new organic farmers.

The role of specialist advisory services and training courses

In many cases, good training courses on organic farming (for beginners and on specialist subjects), with convincing contributions from experienced organic farmers and experts, as well as additional individual and group advice, are very important for the farmer's decision-making process. Of particular importance seems to be good follow-up with small group meetings of farmers in conversion. Almost all of the cantons in Switzerland are financing organic farming advisory services (50–100% of the costs, depending on the canton) and courses for farmers (60–80% of the costs). In 1993, there were altogether about 15 full- or half-time organic farm advisers and about 20 farmers partly working as farm advisers or inspectors (five organic farm advisers working at agricultural schools, about seven advisers at FIBL, and three organic farm advisers paid by organic farmer organizations). These advisers are helping to reduce the risk during conversion due to a lack of information. But even with a good advisory service, some of the main problems of conversion, mainly the economic risk and the difficulty of selling these products at a higher price, still remain.

Private and local initiatives for farms in conversion

In one Swiss canton (Schwyz) a private firm (a milk processor) decided to pay for a detailed conversion plan for all of its farmers willing to convert to organic farming (15 farmers started the programme in 1990). In some towns, the local administration has paid for conversion plans for their farms (Zürich, Basel).

In 1988 a private foundation (Bio-Stiftung) collected money for a private fund to support farms in conversion. This was not very successful. Many private individuals believe that this support has to be made through the state. The experience shows that only for a very limited number of farms (mainly bio-dynamic farms) can consumers be found who are prepared to give some financial support.

Summary and Conclusions

Between 1989 and 1993, five cantons in Switzerland introduced conversion subsidies. The schemes are of two types:

- a one-off area payment, depending on the crop;
- a one-off payment for the farmer for attending special courses on organic farming, for testing new methods and for new investments (special equipment for organic farming).

In general these financial subsidies cover less then half of the supplementary costs which a farm in conversion has if the products cannot be sold at a higher price. In mountain and hill areas these conversion subsidies motivated a higher number of farmers to convert to organic farming than in other regions. For arable farms, the conversion subsidies do not seem to be high enough to encourage more farmers to convert to organic farming, unless there is a strong demand for organic products.

Important issues would appear to be how the organic market, especially for meat and milk (products), is developing, and how the federal agricultural policy, particularly the direct payment system

Table 24.5. Impact of different policies for promoting organic farming, Switzerland, 1993.

Service/concept to promote organic farming	Actual situation	Help for farms in conversion	Help for organic farms
Advisory services			
Individual advice	++	+++	++
Small group advice	+	+++	++
Regional groups	++	++	+++
Conversion plans	+	+++	–
Introductory courses	+++	+++	+
Advanced courses	++	+++	+++
Farmer publications	++	+++	+++
Visits to pilot farms	++	+++	+++
Inspection	+++	+++	++
Research for organic farming			
On-farm research	++	++	+++
General research	+	+	++
Consumer information	+	+++	+++
Label regulation			
Private regulation	+++	+++	+++
State regulation	+	+++	+++
Conversion subsidies			
Private funds	+	+	–
Regional support	++	+++	++
National support	–	+++	++
Direct payments (from 1993)	++	+++	+++
Market development			
Plant products	++	+++	+++
Animal products	+	+++	+++

–, Not existing or irrelevant; +, little importance; ++, medium importance; +++, very important.
Source: Schmid (1993).

starting in 1993, will support the development of organic farming.

Table 24.5 gives a personal judgement of the ways in which different concepts of promoting organic agriculture could help farms in conversion as well as fully-converted farms. The author's indications will need to be verified in different regions by means of long-term studies. Table 24.5 shows the relevance of different systems of support for organic agriculture. Conversion subsidies alone are not sufficient to motivate more farmers to take up organic farming. What is needed is a holistic approach on a national as well as regional level.

References

Anon. (1991) *Verordnung über Produktionslenkung und extensive Bewirtschaftung im Pflanzenbau.* Eidgenössische Drucksachen- und Materialzentrale, Bern.

Anon. (1993a) *Verordnung über ergänzende Direktzahlungen in der Landwirtschaft.* Eidgenössische Drucksachen- und Materialzentrale, Bern.

Anon. (1993b) *Verordnung über Beiträge für besondere ökologische Leistungen in der Landwirtschaft.* Eidgenössische Drucksachen- und Materialzentrale, Bern.

BLW (1992) *7. Bericht über die Lage der Schweizer Landwirtschaft und die Agrarpolitik des Bundes.* Eidgenössische Drucksachen- und Materialzentrale, Bern.

Canova, G. (1993) *Umfrage über das Angebot an Bio-Produkten im Kanton Graubünden.* Zentralstelle für landwirtschaftliche Betriebsberatung Graubünden, Chur.

Eichenberger, M. and Kaufmann, M. (1992) *Landwirtschaftspolitik am Scheideweg.* Agrarpolitische Szenarien und Vorschläge für die SP-Fraktion der Bundesversammlung, Bern.

EVD (1992) *Berichte zur Verordnung über ergänzende Direktzahlungen und über Beiträge für besondere ökologische Leistungen.* Vernehmlassungs-Entwürfe. Eidgenössisches Volkswirtschaftsdepartement Bern.

Geissbühler, F. and Sotoudeh, A. (1993) Nationale und regionale Markenzeichen für biologischen/ökologischen Landbau am Beispiel des Kantons Graubündens. Unpublished Diplom-Arbeit, Institut für Agrarwirtschaft, ETH, Zürich.

Looser, B. and Strasser, A. (1990) Produktionskosten von Bio-Weizen im Vergleich mit konventionellem Anbau. Unpublished report. Landwirtschaftliche Schule Strickhof, Lindau.

Maire, N. *et al.* (1990) *La conversion des domaines agricoles en mode biologique: Effect sur l'équilibre physico-chimique et biologique des sols.* Report No. 40. National Forschungsprogramm 'Boden', Schweizerischer Nationalfond, Liebefeld-Bern.

MIBA (1990) *Produzentenpreiszuschläge auf Verkehrsmilch aus biologisch-organischen Betrieben.* Internal Report. Milchverband Basel, Basel.

Mühlebach, J. and Näf, E. (1990) *Die Wettbewerbsfähigkeit des biologischen Landbaus.* Schriftenreihe der FAT 33. Eidg. Forschungsanstalt für Betriebswirtschaft und Landtechnik, Tänikon.

Rist, S., Stöckli, B. and Suter, H. (1989) *Möglichkeiten und Grenzen des biologischen Landbaus im Kanton Zug.* Forschungsinstitut für biologischen Landbau, Oberwil/BL.

Roux, M. (1988) *Umweltrelevantes Handeln von Landwirten.* Report No. 11.

Nationales Forschungsprogramm 'Boden', Schweizerischer Nationalfond, Liebefeld-Bern.

Schmid, O. (1988) *Umstellungsbeiträge für biologischen Landbau – Vorschläge zur Realisierung für die Landwirtschaftsdirektion des Kantons Bern.* Internal Report. Forschungsinstitut für biologischen Landbau, Oberwil/BL.

Schmid, O. (1990) *Beiträge für die Umstellung auf biologischen Landbau.* Forschungsinstitut für biologischen Landbau, Oberwil/BL.

Schmid (1993) *Umstellungsbeiträge sind wirksame Anreize.* Bio-Berater Bulletin, Nr. 6. Forschungsinstitut für biologischen Landbau, Oberwil/BL.

Steinmann, R. (1983) *Biologischer Landbau. Ein betriebswirtschaftlicher Vergleich.* Schriftenreihe der FAT 19. Eidg. Forschungsanstalt für Betriebswirtschaft und Landtechnik, Tänikon.

VSBLO (1992) *Richtlinien über Verkaufsprodukte aus biologischem Landbau.* Vereinigung schweizerischer biologischer Landbau-Organisationen, Basel.

VSBLO (1993) *Stellungnahme der Bio-Bauern zur Verordnung Artikel 31a und b LWG.* Vereinigung schweizerischer biologischer Landbau-Organisationen, Basel.

25 Financial Assistance for Conversion to Organic Farming in Germany under the European Community's Extensification Programme

LUDGER SCHULZE PALS, JÜRGEN BRAUN AND
STEPHAN DABBERT

Introduction

Since 1989, conversion to organic farming in Germany has received financial support under the European Community's extensification programme (EC, 1988). The central goal of the programme is to reduce the market supply of surplus products: cereals, oilseeds and leguminous crops as well as beef and lamb. This goal has not been realized because, since the introduction of the programme, only about 2.4% of the total agricultural area in the former Federal Republic of Germany (FRG) has been converted to organic farming. Even so, through the introduction of the programme, the framework for organic farming has changed greatly, so that the supply of organic products has more than quadrupled in the past four years.

This chapter is intended, based on results from the University of Hohenheim and the Federal Agricultural Research Centre Braunschweig-Völkenrode (FAL), as a first contribution to the evaluation of the extensification programme from the perspective of organic farming. The following questions will be addressed:

- How have the conversion requirements been carried out in practice?
- In which regions is conversion of particular interest?

- What motivated the farmers to convert?
- How have the production structures, yields, financial performance and marketing changed as a result of the conversion?
- How has the conversion affected income?

The Impact of the EC Extensification Programme on Organic Farming in Germany

The extensification programme as a mechanism for promoting organic farming

The EC extensification programme (EC, 1988) anticipated two different forms of extensification:

- a quantitative approach requiring an absolute reduction of 20% in either land devoted to specific crops or numbers of livestock;
- a production methods approach requiring the adoption of production practices which would be expected to result in the same 20% reduction in output.

For organic farming, only the production methods approach, which demands the conversion of the entire farm, is of interest. The guidelines of the extensification programme as implemented in Germany approximate to a great extent to the standards of the German umbrella organization for organic farming (Arbeitsgemeinschaft Ökologischer Landbau: AGÖL), which comprises the seven most important certification bodies in Germany. The central points are a ban on the use of synthetic plant protection chemicals and fertilizers, as well as limiting the stocking rate to 2.0 livestock units (LU) per hectare. Differences between the government guidelines and the AGÖL standards occur mainly with regard to the purchase of conventionally produced animal feed and in the determination of appropriate animal husbandry. Here, the AGÖL standards are significantly stricter than government regulations (König, 1989; AGÖL, 1991).

The amount of the extensification subsidy is based on average crop areas before conversion. Generally, for the land areas on which surplus products (grains, oilseeds and legume crops) were grown before conversion the farms receive DM 425–510 ha^{-1}, depending on the state. For the other farm areas (potatoes, beets, forage and grassland), participating farmers receive a subsidy of DM 300–360 ha^{-1}. Some federal states pay an additional subsidy of DM 1416 ha^{-1} for the conversion of orchards and vineyards.

The subsidies are authorized for a period of five years and are paid annually. They are intended to compensate for lower yields which can occur at the beginning of the conversion and cannot be made up through higher prices. This issue gains importance in the context of the EC regulation which gives a legal definition to organic farming and came into effect on 1 January 1993 (EC, 1991). The regulation specifies that, from 1994, products cannot be marketed as organic during the first two years of conversion, and must, therefore, to a large extent be sold at conventional prices, although the European Commission now plans to amend the regulation removing this restriction. The subsidy is intended to make up for the additional costs arising in the first years of conversion. Only farms that can prove expenditure on conventional practices with appropriate documentation (e.g. invoices for plant protection chemicals and fertilizers) can participate. Farms that have already converted are excluded from the programme.

Regional differences in uptake of the conversion aid scheme

The significance of organic farming has grown dramatically since the introduction of the extensification programme. Since 1989, more than 10,000 farms with an area of about 377,000 ha have converted in Germany. This is about 2.4% of the total agricultural area of the country (Table 25.1). The number of organically managed farms has increased almost fourfold in the past four years; until the extensification programme was introduced only about 2300 farms were registered with the AGÖL member organizations (as at 1 July 1989).

Strong regional differences emerge in the acceptance of the programme. The proportion of land area converted ranges from 7.7% in Mecklenburg-Vorpommern to 0.5% in Rheinland-Pfalz. Just over half of the total area converted within the framework of the extensification programme (but only 14% of the farms) lies in the former German Democratic Republic, even though the programme has only been offered there since 1991 (Table 25.1).

In western Germany, the largest uptake has been primarily in the grassland and sub-alpine mountain areas of southern Germany (for example in the pre-alpine region, the Black Forest, the Saar-Pfalz and the Hessian mountains). In some districts in these regions, more than 4% of agricultural land had been converted to organic farming by 1991/92 (Fig. 25.1). In eastern Germany, the programme has met with most acceptance in Mecklenburg-Vorpommern and Brandenburg. By far the greatest response has been in eastern Vorpommern, where just under 20% of the agricultural area was converted in 1991/92 alone. In the Mecklenburg-Vorpommern lake area and in Spreewald, nearly 10% of the area was converted in 1991/92. The low liquidity levels of many farms in eastern Germany may account for the high uptake. Farmers

Table 25.1. Numbers of farms and land area converted to organic farming with financial support under the European Community's extensification programme, Germany, 1989/90–1992/93.

State	Number of farms	as % of total	Land area (ha)	as % of total	Average size (ha)
Western Germany					
Baden-Württemberg	2,666	3.0	35,647	2.4	13.4
Bayern	2,982	1.5	66,394	2.0	22.3
Rheinland-Pfalz	220	0.6	3,032	0.5	13.7
Saarland	67	2.5	3,733	5.3	55.7
Hessen	1,427	3.5	31,257	4.1	21.9
Nordrhein-Westfalen	625	0.9	12,333	0.8	19.7
Niedersachsen	622	0.7	19,916	0.7	32.0
Schleswig-Holstein	177	0.7	9,291	0.9	52.5
City states	13	1.1	454	0.4	34.9
Sub-total	8,799	1.6	182,047	1.6	20.7
Eastern Germany[a]					
Mecklenburg-Vorpommern	497	15.1	99,901	7.7	201.0
Brandenburg	286	7.5	50,069	3.3	175.1
Sachsen	431	9.3	16,998	2.0	39.4
Sachsen-Anhalt	142	4.5	14,892	1.4	104.9
Thüringen	70	2.2	12,755	1.7	182.2
Sub-total	1,426	7.9	194,615	3.6	136.5
Total	10,225	1.8	376,662	2.4	36.8

[a] Extensification programme only available in eastern Germany from 1991/92.
Sources: Wolfgarten and Wimmers (1991), König (1993) and Kruse (1993).

see, in some circumstances, the possibility of saving on variable costs, and, at the same time, improving the cash flow of the farm business through participation in an extensification programme.

Structural characteristics of participating farms

The regional variation in the numbers of farms converting under the extensification programme indicates that participation is largely dependent on the current intensity of farming and stocking rates. The more extensively an area was farmed before conversion, the more attractive the option to convert. This is true, for example, for forage-based livestock farms in the less intensive grassland regions of the medium-altitude mountain areas. The introduction of milk quotas led

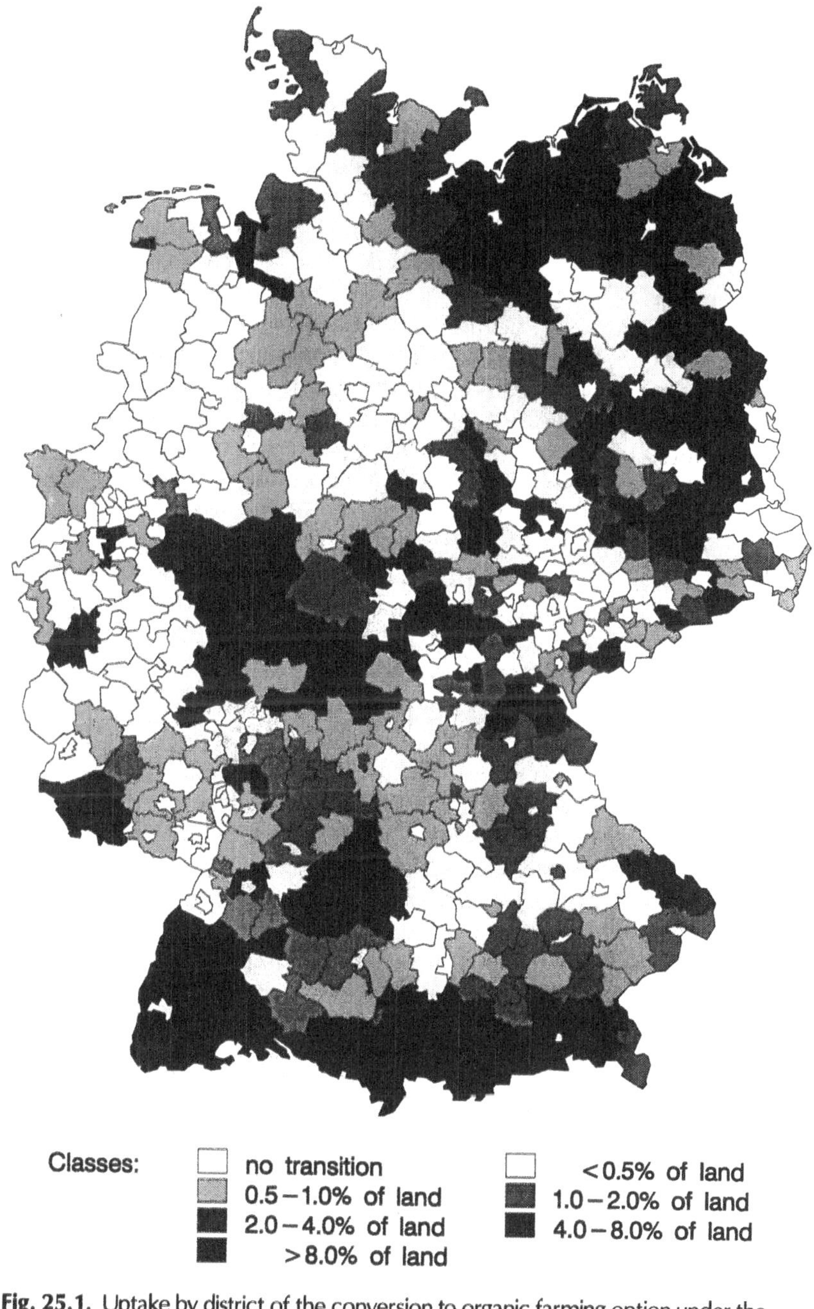

Fig. 25.1. Uptake by district of the conversion to organic farming option under the European Community's extensification programme, Germany, 1989/90–1992/93. Source: Schulze Pals (1993b).

to a general reduction in the use of fertilizers and concentrate feeds in these areas because of a lack of viable production alternatives. The conversion to organic farming under the extensification programme hardly entails any significant changes in production methods on most participating dairy farms. Similar observations may be made for farming in areas with predominantly light soils. Here, because of the minimal natural yield capacity, the step to organic farming is not difficult, because there is already a minimal intensity of input use. This particularly holds true for the extremely light soils in some regions of eastern Germany.

In contrast, cropping farms in better locations show very little interest in conversion, because of the high profitability of conventional crops such as sugarbeet. Conversion is also less attractive for farms specializing in pig or poultry production because, according to the guidelines, stocking rates must be reduced to 2.0 LU ha^{-1}, and the purchase of feed produced by conventional methods must be reduced to less than 20% of the total feed in terms of dry matter content. In general, participation in the extensification programme leads to a large reduction in livestock numbers on intensive livestock farms. The resulting income reductions must be compensated through higher prices or the implementation of new production processes which increase the financial risks of conversion.

The size of the converting farms is significantly larger than the average size of all farm and forestry enterprises in Germany. While the average size for all farms in western Germany was 18.8 ha, the farms converting were on average 10.1% larger (20.7 ha). Larger farms have, in general, fewer animals per hectare of land than smaller farms. Smaller farms are under more pressure to produce intensively in order to cover their fixed costs, and are less able than larger farmers to reduce their intensity because of this.

In addition to the larger than average size of converting farms, the age of the farm manager may also play a role in the decision to convert. In general, younger farm managers run the larger farms (see below), and they show themselves to be more open to new production alternatives.

Furthermore, the level of the payments accounts to a large extent for the wide acceptance of the extensification programme on farms with large land areas. On the larger farms, the extensification programme (which does not have a farm size limit) contributes a larger share of the necessary income and therefore reduces the risk of conversion.

In eastern Germany, several contrary trends can be seen. Here, the farms converting under the extensification programme have about 42% less land than the average farm (Table 25.1). This may be due to a greater willingness on the part of individuals and partnerships or limited companies (for example, GmbH) to undertake conversion to organic farming than is the case for the larger cooperatives and farming companies with share capital (for example e.G., AG).

National Survey of Farms Converting to Organic Farming under the EC Extensification Scheme

Since the start of the extensification programme, the Institute of Agricultural Economics at the Federal Agriculture Research Centre Braunschweig-Völkenrode (FAL) has been conducting an evaluation of the impact of the conversion aid scheme (Schulze Pals, 1992, 1993a, b). At the core of the FAL study is the analysis of the economic consequences of conversion to organic farming under the extensification programme. The goal is to document and suggest possible solutions to the problems of conversion by drawing in different natural and socioeconomic locational conditions which until now have received little attention in the literature. Furthermore, the results of horizontal and vertical farm comparisons of converting farms are to be compared with the findings from conventional farms in cases where the converting farms and the conventional farms have similar starting situations. Farm types are to be identified for which the change to organic farming in comparison to conventional reference systems under given circumstances is either more or less advantageous.

In order fully to consider the dynamics of the conversion process, detailed investigations of the converting farms over several years are needed to form the data base for the planned analysis. A total of 107 farms from all over western Germany are being monitored over the whole five-year period of the extensification programme. The farms in question began their conversion in the fiscal year 1990/91. The analysis time period includes the last conventional fiscal year 1989/90 as well as the first five years of conversion (1990/91–1994/95). The following preliminary results are based on the investigations of the last conventional fiscal year and the first two years of conversion.

Factors influencing the decision to convert

The FAL study shows that almost 80% of the managers of the 107 selected farms are less than 44 years of age (Table 25.2). In Germany as a whole, just under 47% of farmers are this young. Their willingness to convert can be explained by the fact that, in general, younger farm managers are more open to new business and farming approaches and are also more innovative in the implementation of these new ideas.

Table 25.2. Age structure (%) of surveyed farmers converting to organic farming with extensification subsidies compared with all farmers, Germany, 1991.

Age (years)	AGÖL registered	Not AGÖL registered	All survey farms	All German farms
< 35	45.1	27.8	39.3	15.1
35–44	36.6	41.7	38.3	21.4
45–54	12.7	19.4	15.0	29.9
> 55	5.6	11.1	7.5	33.6

Sources: BMELF (1992) and Schulze Pals (1992).

From the perspective of the farm manager, the conversion should also contribute to maintaining the farm's existence over the medium to long term. This also proved significant in the analysis of motives for conversion. In response to the question whether they would have converted their farms in the next few years without financial assistance, 82% of the 107 farm managers said that they would have converted their farms even without the extensification premiums. Only 18% said they would not have converted without the premiums. The introduction of the extensification programme meant that the financial risks of conversion were calculable and therefore fostered the readiness to convert earlier. The conversion aid scheme was, however, in no way a decisive factor in the basic decision to convert. The conviction that the farm will benefit financially over the long term, and the desire to do something personally for nature conservation and the environment, count among the most important conversion motives. Many farm managers feel the conversion is a challenge which demands more from them as managers.

Advantages and disadvantages of group membership

Most federal states allow the participating farms to join an AGÖL-affiliated organic certification or producer association. Of the 107 farms surveyed, only about two thirds joined an AGÖL association. The advantages and disadvantages of joining an association are evaluated differently according to farm type. While the large majority of cropping, intensive livestock and mixed farms are AGÖL association members, just under 40% of the forage-based livestock farms, particularly milk producers, are not members.

There are two main reasons, from the perspective of the farm manager, for not joining an association: the lack of marketing advantages and the level of the membership fees. In the eyes of the farm manager it only makes sense to join an association if the group in question can provide good marketing opportunities. This, for example, is not the case for the marketing of milk. Of 64 milk producers who were included in the sample, only 11 were in a position to sell their milk to a specialist organic dairy at the beginning of the fourth conversion year.

The high level of organization (almost 80%) in the cropping farms surveyed indicates that membership of an association makes particular sense in the marketing of crop products. Trade and processing are often linked contractually to an organic farming association so that the marketing is only possible with proof of appropriate membership.

The structure of the survey farms

The 107 survey farms consisted of 39 cropping farms, 61 forage-based livestock farms, five intensive livestock farms and two mixed farms. This makes clear that the two dominating farm types are forage-based livestock and cropping farms. The hypothesis that the conversion to organic farming under the extensification programme is of particular interest for these farm types was not rejected.

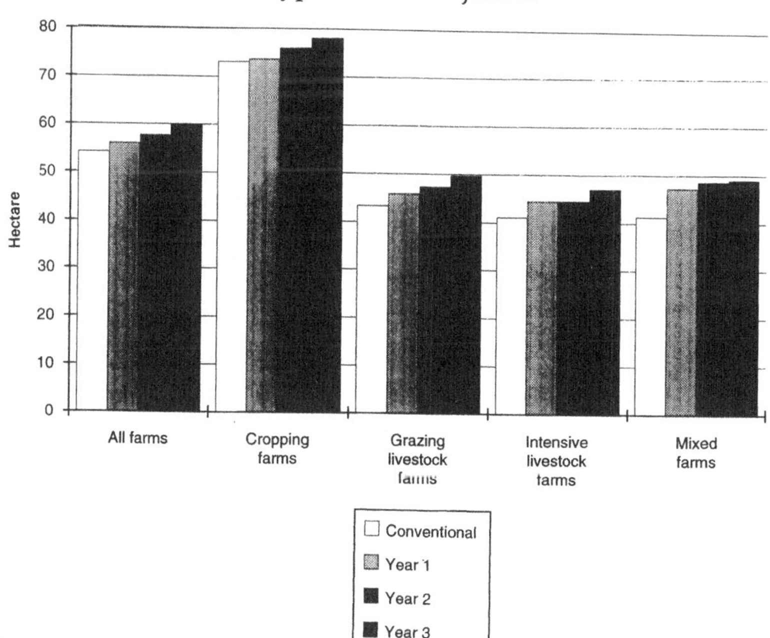

Fig. 25.2. Changes in average farm size on farms converting to organic farming with extensification subsidies, by farm type, western Germany, 1989/90–1992/93. Source: Schulze Pals (1993b).

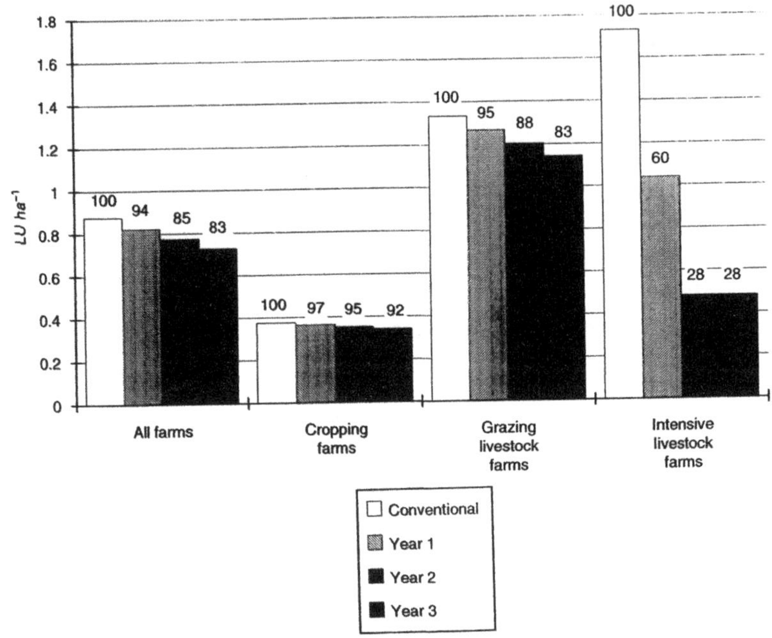

Fig. 25.3. Changes in stocking rate on farms converting to organic farming with extensification subsidies, western Germany, 1989/90–1992/93. Source: Schulze Pals (1993b).

The agricultural area of the survey farms is large (Fig. 25.2), nearly three times the national average size at the start of conversion. Farm size increased by 10.5% from the last conventional fiscal year (1989/90) to the third year of conversion. Particularly significant was the increase in area of the forage-based and intensive livestock farms. On the one hand, these farm types have the least land area of the four farm types studied; on the other hand, they have the largest livestock numbers. Reduced stocking rates as a result of conversion resulted in a degree of financial pressure to increase the amount of land farmed so as to maintain total livestock numbers. Forage-based livestock farms, faced with a decline in forage yields as a result of conversion, can only maintain milk production in line with milk quota availability by increasing the available land area. Intensive livestock farms are forced to draw more income from arable enterprises when the conversion results in a reduction of their livestock enterprises.

This is also evident from the changes in livestock numbers. On average, for all the farms surveyed, the stocking rate per hectare fell about 17% (Fig. 25.3). On intensive livestock farms, the stocking rate falls even further to only 28% of the level before conversion. The strong reduction in the stocking rate per hectare has two causes. First, the land area of the farm increases significantly. Second, especially on the forage-based and intensive livestock farms, the absolute livestock numbers are also reduced. The livestock reductions affect primarily pigs and beef cattle because these enterprises become less economically viable after conversion.

Because of the limited marketing opportunities for organically produced beef or pork, the intensive livestock enterprises seldom offer a viable alternative. Generally, only small amounts can be sold at a higher price through direct marketing. At the same time, the cost of feeding pigs in organic farming as opposed to conventional farming is much higher. First, the purchase of conventionally grown feeds (primarily protein sources) is limited to 20% of total feed use. Where farmers are certified as organic, the use of feedstuffs from Third World countries is often totally prohibited. The farmer is therefore forced to use local protein sources, with lower biological value, which ultimately has a negative effect on the finishing rates. Second, the farmer must calculate the value of his own organically grown grain for feeding at a much higher price than that in conventional agriculture.

At the beginning of conversion, many cattle producers faced great uncertainty concerning how the yields of permanent grassland and other forage crops will change. Mixed farms with dairy and beef cattle tend, therefore, more towards limiting the number of beef cattle in order to ensure enough forage for the dairy cattle.

Changes in land use

Figure 25.4 shows the changes in land use after the switch to organic farming. The cropping pattern is very strongly dependent on farm type. On all farm types, the proportion of cereals and oilseeds declined. Instead, larger areas of land were set aside or planted with forage crops. Above all, the proportion of nitrogen-fixing legume crops (grain legumes, leys and set-aside) rose from about 3% before conversion, to more than 30% of the farm land in the third year of conversion. The increased area of catch crops and increased use of undersown crops, from 16% to about 40%, indicate that the farmers placed great value on conserving soil fertility following conversion.

Within biological and technical restrictions, the crop rotation and the proportion of crops is primarily dependent on marketing opportunities. Due to the largely non-existent market for organically grown oilseed rape and sugarbeet, these products lose their ability to compete with wheat, rye and potatoes. But even among the cereal crops, the relative competitiveness of rye and oats has improved, due to increased

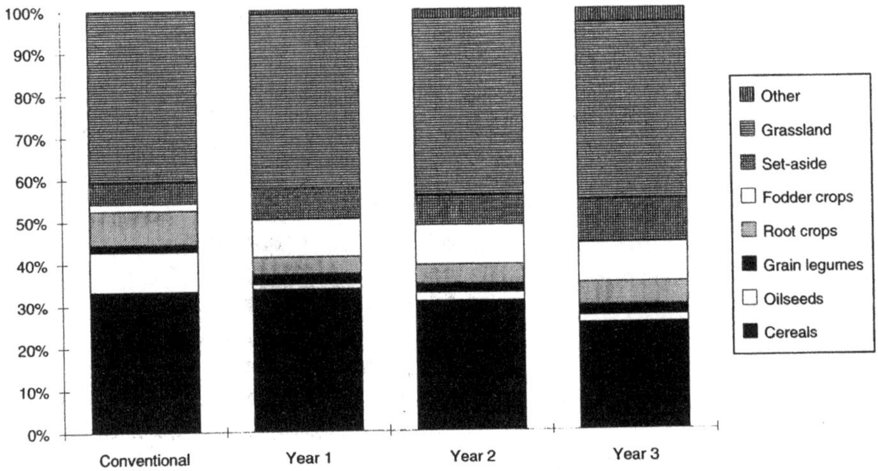

Fig. 25.4. Changes in land use on farms converting to organic farming with extensification subsidies, western Germany, 1989/90–1992/93. Source: Schulze Pals (1993b).

marketing opportunities, at the expense of wheat and, above all, barley. Vegetable production increased only minimally in the first three years of conversion: farmers are concerned about the large amount of work involved and their lack of experience in this area makes them wary of taking risks.

Economics of the conversion to organic farming

Gross margins

The whole-farm gross margin can be used as an initial criterion for assessing the viability of organic farming. Table 25.3 shows the changes in farm gross margin on the survey farms in the first two conversion years compared with the last conventional fiscal year.

Taking into account the extensification subsidies, the large cropping farms in particular were able to benefit from conversion. More than 70% of the cropping farms studied were able to improve their gross margins in both the first and the second years of conversion, in comparison to the conventional year. The forage-based and intensive livestock farms, on the other hand, benefited less from the conversion. While the forage-based livestock farms were able to maintain their gross margins at the level of previous conventional management, the intensive livestock farms faced significant gross margin losses.

The cropping farms benefited from the better marketing opportunities for crops, in particular cereals and potatoes. The farms were able to achieve price premiums of more than DM 500 t^{-1} above average market prices in the first two conversion years and thereby, in large measure, make up for the conversion-related yield reductions. The market for organically produced animal products is, in contrast, much more difficult. As a rule, higher prices can only be achieved through direct marketing. The clientele necessary for direct marketing can only be built up gradually. The worsened situation for intensive livestock farms, out of proportion to the other farm types, can be traced back to the large reduction in pig numbers without simultaneous development of other enterprises.

Without the extensification subsidy, only one-third of the forage-based and intensive livestock farms would be in a position to maintain the gross margins they achieved before conversion. About 60% of the cropping farms studied could have maintained their gross margin levels without the conversion subsidies, or, alternatively, with subsidies but selling produce at conventional price levels.

Labour use

The labour requirements for production and marketing increased on all farm types except the intensive livestock farms (Table 25.3). Due to the reduction in pig breeding and finishing, more time was made available on these farms. Increased labour requirements following conversion were most significant on the large cropping farms, due to the introduction and expansion of labour-intensive crops such as vegetables and potatoes. In addition, more time was required for marketing, since the preparation and delivery of the goods is largely carried out by the farmers themselves. Before conversion, collection of goods was largely the responsibility of the buyers.

Additional labour requirements also arose due to organic farming-related training needs, particularly in the areas of production and marketing. On almost all farms, this meant an average of two additional labour hours per hectare (Table 25.3). Important training sources include the local group of AGÖL-affiliated producers' associations, which meet regularly to exchange ideas and information.

Despite the increased labour requirements, the gross margins per labour hour obtained remained relatively constant across the farm types (Table 25.3).

Farm income

Farm income, including extensification subsidies, increased on average by 22% (Table 25.4). The increased labour requirements, however,

Table 25.3. Changes in land and labour productivity on various farm types converting to organic farming with extensification subsidies, western Germany, 1989/90–1991/92.

		All farms (*n* = 107)			Cropping (*n* = 39)		
	Unit	1989/90	1990/91	1991/92	1989/90	1990/91	1991/92
Land productivity							
Land area per farm	ha	54.1	55.9	57.6	73.1	73.6	75.9
Enterprise gross margin	DM ha^{-1}	1929	1920	1785	1541	1630	1590
Extensification subsidy	DM ha^{-1}	–	319	324	–	321	326
Total gross margin	DM ha^{-1}	1929	2239	2109	1541	1951	1916
Labour productivity							
Production and marketing	h ha^{-1}	55.3	58.0	62.2	27.2	32.3	38.4
In-service training	h ha^{-1}	–	1.9	1.9	–	1.6	1.6
Total labour input	h ha^{-1}	55.3	59.9	64.1	27.2	33.9	40
Gross margin per unit labour	DM h^{-1}	35	37	33	57	58	48
Proportion of farms with higher total gross margin after conversion	%	–	62	59	–	79	74

meant that profit per family labour unit increased by just over 4%. Without the extensification subsidies, income would have fallen significantly, by 15–20% in comparison to the conventional starting situation. The income reductions would have been lower on the cropping farms, while the intensive livestock farms suffered serious income decreases. These results also need to be seen in the context of the general decline in conventional farm incomes over the study period (see Chapter 6, Fig. 6.5).

The fixed costs of the surveyed cropping farms during the first year of conversion remain, in comparison to the last conventional fiscal year, largely constant. Although additional fixed costs arose due to the purchase of harrow-comb weeders, which were purchased by nearly every cropping farm, this was compensated by the reduction in other plant protection equipment. On farms which cultivated sugarbeet before conversion, costs were reduced through the sale of sugarbeet machinery.

Table 25.3. continued

	Unit	Forage-based livestock (*n* = 61)			Intensive livestock (*n* = 5)		
		1989/90	1990/91	1991/92	1989/90	1990/91	1991/92
Land productivity							
Land area per farm	ha	43.3	45.8	47.3	41.1	44.5	44.5
Enterprise gross margin	DM ha^{-1}	2281	2129	1976	3484	2883	2168
Extensification subsidy	DM ha^{-1}	–	314	319	–	345	345
Total gross margin	DM ha^{-1}	2281	2443	2295	3484	3228	2513
Labour productivity							
Production and marketing	h ha^{-1}	80.1	80.6	84.2	96.3	79.3	74.9
In-service training	h ha^{-1}	–	2.1	2.1	–	2.6	2.6
Total labour input	h ha^{-1}	80.1	82.7	86.3	96.3	81.9	77.5
Gross margin per unit labour	DM h^{-1}	28	30	27	36	39	32
Proportion of farms with higher total gross margin after conversion	%	–	56	53	–	40	40

Conversion status: 1989/90 = conventional farming; 1990/91 = year 1; 1991/92 = year 2.
Source: Schulze Pals (1993b).

Table 25.4. Changes in farm income (DM) on farms converting to organic farming with extensification subsidies, western Germany, 1989/90–1991/92.

	Conventional, 1989/90	Year 1, 1990/91	Year 2, 1991/92	Change (%), 1989/90–1991/92
Income including extensification subsidy				
Per farm	50,018	65,973	60,880	21.7
Per FTE[a] labour	30,980	37,878	32,266	4.2
Per hectare	933	1,190	1,065	14.1
Income excluding extensification subsidy				
Per farm	50,018	48,308	42,382	−15.3
Per FTE[a] labour	30,980	27,728	22,462	−27.5
Per hectare	933	871	741	−20.6

[a] Full-time equivalent (2200 h $year^{-1}$).
Source: Schulze Pals (1993b).

A conversion-related increase in fixed costs is, however, to be expected over the medium term on the farms studied. The introduction, or rather the expansion, of potato production makes investments in the areas of production, sorting and preparation necessary in order to reduce the labour intensity of this enterprise. Many farm managers are planning a series of investments in grain storage and dressing in order to meet market specifications (cleaning, quality, regular delivery).

Additional fixed costs arose on the forage-based livestock farms due to the conversion-related leasing of land for forage production. The dairy farms showed few conversion-related changes in fixed costs. In the next few years, it is to be expected that farms which have joined an AGÖL association and not yet met the AGÖL conditions for livestock husbandry will have to make investments in farm buildings. For example, within five years the conversion from slatted-floor housing will have to take place. Furthermore, the dairy cattle must have outside access, at least to pasture during the summer.

Conclusions

The results for the first two years of conversion show clearly that the extensification subsidies must be strengthened if they are to remain attractive to all types of farm in the future. Cropping farms do not, at this time, necessarily need the full extent of the extensification premiums paid up until now, since the markets for organically grown crop products are better developed and premium prices can be achieved. It would be sensible, therefore, in the future, to pay the cropping farms relatively high levels of subsidy during the first two years, when full organic prices might not be achieved, and then to reduce the subsidies over the successive years.

On the forage-based livestock farms, the extensification subsidies paid until now are adequate on average to maintain the same level of income as before conversion. However, intensive forage-based livestock farms, for example those with high milk production and minimal forage area per cow, will not be able to make up the conversion-related reductions in income (due to reduced milk and forage output) with the current payments. Furthermore, the majority of forage-based livestock farms will not be able to market a large portion of their products at higher prices within the subsidized conversion period, because organic meat processing and marketing are developing very slowly. At the end of the subsidy programme period, therefore, the majority of the forage-based livestock farms which continue farming organically may sustain large income losses.

The current organization of conversion support is of little interest to intensive livestock farms because these farms must make the largest structural changes in order to convert to organic farming. The small number of participants from this farm type illustrates this.

The prices for organically produced agricultural products are also declining because of the strong increase in supply within the framework of the extensification programme. Although the demand for food produced in an environmentally friendly manner is growing, the markets for organic food products are marked by structural deficiencies. If the government plans to take action in this area, it would be more advantageous over the medium and long term to promote the markets rather than the production.

Regional Impact of the Extensification Scheme on Conversion to Organic Farming in the State of Baden-Württemberg

Approach

This regional study is based on a sample survey of 148 farms participating in the EC extensification scheme in Baden-Württemberg in 1989/90. In Baden-Württemberg 603 farms participated in this scheme at that time (König, 1990) so that the sample includes 25% of the surveyed population.

The transition to organic farming is a dynamic process taking several years. Therefore, a correct evaluation of the conversion process requires the comparison of farm development under conversion to organic management with farm development under conventional management during a given period of time. The static comparison of a farm under conventional management in year t with the situation under organic management in year $t + n$ gives incorrect results: first, because the dynamics of the transition process are ignored; and second, because the situation under conventional management is compared with the wrong time period. Even if the comparison is restricted to a static analysis, this should include the situation under conventional management in the year $t + n$. Changes in agricultural policies and technical progress should be considered because they have an impact on the development of the farm even when keeping it under conventional management. However, in deciding whether keeping the farm under conventional management or conversion to organic farming is preferable from a financial viewpoint, both development paths are unknown and can only be derived hypothetically.

The results presented here are based essentially on three comparisons. In the first comparison, the sample farms under conventional management are compared with all farms in Baden-Württemberg to identify for which farm types the extensification scheme is attractive. The second, which compares the farms under conventional management in 1989 with the situation in the first year under organic management, shows the changes at the farm level due to the transition to organic farming. The third comparison determines the income effects within the programme by comparing the farm profits under organic management in 1990 with the profits under conventional management projected from 1989 to 1990.

In determining the income effects, the methodology for calculating gross margins of conventional farms (KTBL, 1992) is used. It has been modified, because the standardized values of this methodology are derived from conventional farms and do not reflect accurately the input–output relationships of organically managed farms. In particular, production activities have been modified by using actual survey-derived on-farm data for yields and prices. Variable costs are taken from Dabbert (1990). The fixed costs are derived from fixed-cost functions of conventional farms, to which are added the annual fixed costs due to investments which are directly related to conversion. The farm income projections for 1990 under conventional management are derived assuming an unchanged structure of production. Only crop yields, product prices and costs have been updated.

Characteristics of the converting farms

The participating farms are more likely to be located in disadvantaged regions. The average farm size of participants is 25.2 ha, implying that they are approximately twice as large as the average farm in Baden-Württemberg (12.4 ha). They produce less cereals, while the share of forage legumes is higher than the state average.

Fertilizer and plant protection expenditure of DM 236 ha^{-1} was about 45% lower than the state average. The average stocking rate at 1.09 LU ha^{-1} is slightly higher than the average for Baden-Württemberg. Similarly, crop yields and milk production per cow are close to the state average.

Farm-level changes due to conversion

The average farm size increased by 1.9 ha and the transition led to changes in land use and stocking rates (Table 25.5): cereal production fell slightly; root crops and oilseed production declined significantly,

Table 25.5. Land use and stocking rates on 148 farms before (1989) and during conversion (1990) to organic farming with extensification subsidies, Baden-Württemberg.

	1989	1990	Change as % of 1989 level
Crops (% of arable area)			
Cereals	57.2	53.3	−6.8
Root crops	3.7	2.2	−40.5
Oilseeds	6.8	2.1	−69.1
Grain legumes	2.3	4.2	+82.6
Other crops[a]	30.0	38.2	+27.3
Livestock (LU ha^{-1})			
Stocking rate	1.09	1.02	−6.0

[a]Including forage legumes.

by 40.5% and 69.1%, respectively; while grain legume production increased by over 80%. Forage legume production increased markedly by about 27%. The stocking rate decreased from 1.09 to 1.02 LU ha^{-1}. On some farms, dairy cows were replaced by suckler cows.

Crop yields and livestock performance fell considerably (Table 25.6). Cereal yields decreased in the first year after conversion by between 15% and 35%. Yields of potatoes fell by 12% from the already low level under conventional management. Milk production per cow fell considerably from 4530 kg to 2993 kg, a decrease greater than one would expect in a typical situation (see Chapters 6 and 14, this volume; BMELF, 1992).

Marketing also changed following conversion. Direct marketing to consumers was used by 23.2% of the farms while still under conventional management. This share doubled after conversion to 46.4% of farms. A further 21.2% of the farms intend to go into direct marketing. Product prices increased strongly. However, there were differences between the two thirds of farms that joined an AGÖL-affiliated organic farming association and those that did not. Certified farms received higher prices; on average, the difference was about 30% for winter wheat and about 160% for oats.

In the year of conversion, 50% of the sample farms undertook investment in machinery. These farms invested an average of DM 12,000 per farm in machines, especially in harrow-comb weeders and preparation sprayers. Almost 25% of the farms invested in buildings, especially slurry storage and direct marketing facilities. The average investment costs on these farms were DM 37,900 per farm.

Table 25.6. Crops yields (t ha^{-1}) on 148 farms before (1989) and during conversion (1990) to organic farming with extensification subsidies, Baden-Württemberg.

	AGÖL registered		Not AGÖL registered		All farms		All farms, relative yield (1989 = 100)
	1989	1990	1989	1990	1989	1990	
Winter wheat	6.0	4.1	5.9	3.9	5.9	4.1	69
Spring wheat	5.3	3.4	–	–	5.9	3.4	58
Rye	4.6	3.3	42.0	3.7	4.5	3.4	75
Winter barley	5.5	4.5	5.1	3.8	5.5	4.4	80
Oats	4.5	3.8	4.1	3.1	4.4	3.7	84
Potatoes	22.3	18.9	14.8	12.5	19.8	17.5	88

Income effects

Determination of the income effects makes it possible to ascertain the relative competitiveness of the organic and conventional farm management techniques. In the following, two scenarios are investigated: the first assumes that the extensification scheme exists, while the second scenario assumes that it does not. For both scenarios income effects were calculated.

For the scenario with the extensification scheme, standard farm income I is calculated by deducting from the standard farm income lease payments, annual fixed costs of investments due to conversion and the membership fee for the organic farming association. The state subsidy paid for programme participation is included. Standard farm income I reflects the actual income situation with the extensification scheme.

Table 25.7 shows the difference between standard farm income I under organic management and the projected income under conventional management. On average, the standard farm income increased to DM 1168 ha^{-1} under organic management. On farms that did not join an organic farming association the increase was higher than on farms that were members of an association. The difference of DM 421 ha^{-1} shows that the transition was profitable for the average farm. This figure, however, masks the fact that about 88% of the farms experienced positive income effects, while 12% of the farms had lower incomes after conversion.

In the second scenario, which assumes that the extensification scheme would not exist, standard farm income II is calculated. This differs from the standard farm income I only in the absence of the state subsidy. Table 25.7 shows that without the extensification scheme, the average farm income would have increased by DM 48 ha^{-1}. On farms without any membership of an organic farming association, the income would have increased more than on farms that became members. The proportion of farms that would have a higher income in this situation is 57%. Thus, it is about 30% lower than in the scenario with the extensification scheme. However, it is likely that in such a situation fewer farms would have converted to organic management. Thus the product prices which fell as a consequence of the extensification scheme would not have fallen or would have fallen less (Hamm, 1991). For the calculation, however, we use prices that were obtained through farm interviews. Standard farm income II marks the lowest farm income level.

The results only refer to the first year under organic management. In evaluating them, it should be considered that farms joining an organic farming association can expect higher prices once they achieve fully certified organic status. Therefore, it is incorrect to draw the

Table 25.7. Average adjusted standard farm income[a] (DM ha^{-1}) under conventional management projected from 1989 to 1990 and under organic management in 1990, with and without extensification subsidies, Baden-Württemberg.

	AGÖL registered	Not AGÖL registered	All farms
Number of farms	99	49	148
Conventional projection	812	601	747
Organic including subsidy	1199	1094	1168
Difference[b]	387	493	421
Organic excluding subsidy	829	745	795
Difference[b]	17	144	48

[a] Standard farm income minus lease payments, annual fixed costs of investments due to conversion, and membership fee for the organic farming association.
[b] Organic minus conventional projection.

conclusion that not joining an organic farming association is advantageous. The conclusion is also not valid because farms not joining an organic farming association were using their land less intensively already under conventional management, so that they had better conditions for conversion than farms joining an association.

Production volume effects

To determine effects on production volume, the quantity which was produced under conventional management in 1989 is compared with the quantity under organic management in 1990. The result of this comparison is the estimated change of production quantity due to conversion. It marks the lowest level because it is assumed that under conventional management there would not be any change in land area and in livestock numbers.

Cereal and oilseed production were reduced by 33% and 40%, respectively. The production of grain legumes was more than doubled. Altogether, the volume of crop production decreased by 23.5% as a consequence of the transition to organic management. Milk production fell by 35%. However, the volume of beef production actually increased slightly as a consequence of the suckler cow expansion.

Overproduction pressures on the markets for crop products are further reduced because of increased on-farm utilization by converting farms. As far as milk production is concerned, the data do not allow us to determine whether production decreased as a consequence of transition or whether other factors are responsible. Thus the reduction in milk output cannot be interpreted as a sole consequence of the extensification scheme.

Summary and Conclusions

In Germany, about 10,200 farms with an area of about 380,000 ha (2.4% of the total agricultural area) have converted to organic management since the extensification scheme was introduced. On the basis of two studies, reported in this chapter, of farms converting to organic farming under the extensification scheme, it was found that farms entering the programme were more likely to be found in disadvantaged regions and were larger and less intensive than average. The farm managers were younger than average and most would have converted even without the subsidies. Following conversion, the structure of land use changed considerably, stocking rates were reduced markedly, crop output fell by a quarter and milk production by a third, and about two-thirds of the farms joined an AGÖL association in the expectation of better marketing opportunities.

More than 70% of the farms in both studies experienced income increases as a result of conversion with subsidy support, but differences in the methods of calculating the programme's income effects between the two studies lead to contrasting results for the likely income impacts in the absence of the payments. In the FAL study, incomes would have fallen by 15–20%, while in the Baden-Württemberg study 57% of the farms would still have had higher incomes even if the subsidy had not been paid. The results of both studies are, however, preliminary. It is clear that the complete evaluation of a five-year programme is only possible after the programme is finished.

Although the two studies use different data and methods and have different aims, the results allow some preliminary conclusions:

- The objectives of the extensification programme with respect to a 20% reduction in crop output were achieved, whereas it failed in the case of beef production, which increased slightly due to an expansion in the number of suckler cows.
- As a consequence of the transition to organic management there will be a reduction of negative environmental effects, although these effects will be relatively small because the participating farms were already managed less intensively in the conventional situation (Rat von Sachverständigen, 1985).
- The output reduction and environmental impacts of the scheme would be much greater if intensive livestock farms and intensively farmed cropping farms were persuaded to convert.

The less intensive management of the farms is because the majority of them are located in disadvantaged regions. The results, however, invalidate the hypothesis often forwarded that organic farming is only possible on good soils. They indicate, moreover, that the opportunity costs of conversion in these regions are lower than in more advantaged regions. Certainly, this observation relates to the farm structure in these regions which is closer to the optimal farm structure for organic farming.

The decision to convert to organic farming represents a major turning point in the development of a farm. It is mainly the younger farm managers who are willing to accept such a change of farm management. Generally, the younger farmers are also managing larger farms; the farms participating in the extensification scheme are larger and appear to have a greater increase in farm size than the average farms.

If the extensification scheme is considered as a support programme for organic farming there are four main points of criticism:

- The majority of the participating farms, especially cropping farms, would not have needed the programme to make conversion profitable.
- The existing problems in the markets for organic products have become more severe.
- Existing organic farms that were not eligible for subsidies are disadvantaged by the negative effects of the programme on the markets for organic products.
- The enforcement of organic farming standards is unsatisfactory on those farms that do not join an organic farming association.

Although the output reduction effects at the European Community level are negligible, the impact on the small markets for organic products is very significant. The problems already existing, especially in the markets for crop products, have been exacerbated. At the same time, consumer uncertainty has increased because of the unsatisfactory procedures for guaranteeing that the standards of organic farming are maintained on those farms which do not join an organic farming association. These problems will probably be moderated by the EC Regulation (EC, 1991) for labelling products of organic farming, which came into force on 1 January 1993, although this does not yet cover livestock products. However, it would have been better to introduce clear labelling procedures and demand-side support before (or at least simultaneously with) the introduction of supply-side support, in particular with respect to livestock products for which markets are currently poorly developed.

The existing organic farms are disadvantaged because they are not allowed to benefit from the subsidy although they already contribute to the extensification programme's objectives of surplus reduction and environmental protection. However, the disadvantages of the negative effects of the extensification programme on the markets for organic products, especially crop products, constitute a bigger drawback for these farms than exclusion from directly paid public support.

References

AGÖL (1991) *Rahmenrichtlinien zum ökologischen Landbau* (13th edn). Arbeitsgemeinschaft Ökologischer Landbau, Bad Dürkheim.

BMELF (1992) *Agrarbericht der Bundesregierung.* Bundesministerium für Ernährung, Landwirtschaft und Forsten, Bonn.

Dabbert, S. (1990) *Zur optimalen Organisation alternativ wirtschaftender Betriebe. Untersucht am Beispiel organisch-biologischer Haupterwerbsbetriebe in Baden-Württemberg.* Agrarwirtschaft Sonderheft 124. Verlag Alfred Strothe, Frankfurt am Main.

EC (1988) Commission Regulation (EEC) No. 4115/88 of 21 December 1988 laying down detailed rules for applying the aid scheme to promote the extensification of production. *Official Journal of the European Communities* 88(L361), 13–17.

EC (1991) Council Regulation (EEC) No. 2092/91 of 24 June 1991 on organic production of agricultural products and indications referring thereto on agricultural products and foodstuffs. *Official Journal of the European Communities* 91(L198), 1–15.

Hamm, U. (1991) Starke Ausweitung des ökologischen Landbaus kann zu Absatzproblemen führen. *Agra-Europe,* Special Issue 16, 1–4.

König, M. (1989) Grundsätze für die Förderung der Extensivierung. *Agra-Europe,* Special Issue 25, 1–9.

König, M. (1990) Extensivierung – vorläufige Ergebnisse 1989/90. *Agra-Europe,* Special Issue 20, 1–14.

König, M. (1993) Extensivierung 1988/89 - 1992/93. Vorläufiges Endergebnis. *AID-Informationen, Arbeitsunterlagen für Berufsbildung und Beratung* 42(18), 1–25.

Kruse, S. (1993) Entwicklung der Betriebstruktur in den neuen Ländern. *AID-Informationen, Arbeitsunterlagen für Berufsbildung und Beratung* 42(20), 1–37.

KTBL (1992) *Standarddeckungsbeiträge 1990/91 und Rechenwerte zur Betriebssystematik für die Landwirtschaft.* Kuratorium für Technik und Bauwesen in der Landwirtschaft, Darmstadt.

Rat von Sachverständigen für Umweltfragen (1985) *Umweltprobleme der Landwirtschaft.* Special Report. Kohlhammer Verlag, Stuttgart and Mainz.

Schulze Pals, L. (1992) Extensivierungsprogramm und ökologischer Landbau. *AID-Informationen, Arbeitsunterlagen für Berufsbildung und Beratung* 41(5), 1–14.

Schulze Pals, L. (1993a) Ökonomische Wirkungen des Extensivierungsprogrammes auf die Umstellung landwirtschaftlicher Betriebe. In: Zerger, U. (ed.) *Forschung im ökologischen Landbau - Beiträge zur zweiten Wissenschaftstagung im ökologischen Landbau.* Stiftung Ökologie und Landbau, Bad Dürkheim, pp. 328–335.

Schulze Pals, L. (1993b) *Regional und betrieblich differenzierte Untersuchung über Akzeptanz, Motivation und wirtschaftliche Folgen einer Umstellung auf ökologischen Landbau in Rahmen des Extensivierungsprogrammes.* Arbeitsbericht 2/93 (November). Institut für Betriebswirtschaft, Bundesforschungsanstalt für Landwirtschaft, Braunschweig-Völkenrode.

Wolfgarten, H. and Wimmers, F. (1991) Betriebsgrossenstruktur 1990. *AID-Informationen, Arbeitsunterlagen für Berufsbildung und Beratung* 40(22), 1–17.

26 Organic Farming and Agricultural Policy in Western Europe: An Overview

NICOLAS H. LAMPKIN AND SUSANNE PADEL

Introduction

Since the early 1970s, significant problems have arisen from the implementation of the European Union's Common Agricultural Policy (CAP). These are reflected in the recent measures for the reform of the CAP (EC, 1992a, b) and the four current major objectives of agricultural policy: maintaining farm incomes; reducing surplus production; reorienting agriculture towards areas of market demand; and protecting the environment. Pressure for continued agricultural policy reform, further shifting the emphasis from agricultural production to environmental protection, will come from the continuing high budgetary costs of existing policies, the 1993 GATT agreement, and the proposed enlargement of the European Union to include, by 1995, Norway, Sweden, Finland and Austria.

The changing emphasis of agricultural policy is not unique to the European Union. In most OECD countries, policies are shifting away from commodity-specific price and income support towards policies which view rural countryside assets as a source not only of agricultural products, but also of environmental services, and encourage farmers to recognize that it is in their and society's interests to maintain and enhance the environmental asset base of farms (OECD, 1991). This shift has long been advocated by environmental groups and others, in recognition of the linkages between commodity support programmes, agricultural intensification, surplus production and environmental damage (e.g. Weinschenck and Dabbert, 1989; Jenkins, 1990; NCC, 1990; WWF, 1991; Greenpeace, 1992).

Most proposals in this direction focus on a combination of measures, including the ending of commodity price support to encourage deintensification of production and an expanded, long-term environmentally sensitive area scheme. This would include a baseline payment for avoiding damage to the environment, and higher rate payments for active environmental management (the provision of environmental goods and services), possibly financed by a levy on agricultural inputs. A common element in all these proposals is the distinction between payment in return for environmental benefits and compensation for loss of profits or direct income payments to continue farming (Jenkins, 1990).

Organic farming has the potential to make a significant contribution to these agricultural policy reforms, particularly in the areas of environmental protection, reduction in the output of surplus products and the reorientation of agriculture towards areas of market demand. A number of policy options exist to support the growth of the organic sector and to ensure that the potential environmental and surplus reduction benefits of organic farming are actually achieved. These include support for research and development, education, training and extension, market development and certification, as well as financial support for conversion and continued organic production.

The aim of this chapter is to review the contribution which organic farming can make to agricultural policy objectives and the impact of agricultural policy developments on organic farming in western Europe, based on the detailed analyses presented in the other chapters of this section, and brief reviews of the approaches adopted in other countries. From these experiences, conclusions are drawn concerning the factors which need to be considered in future policy developments.

Why Support Organic Farming?

Environmental protection

The contribution of organic farming to environmental protection has been raised in Chapter 4, while more detailed reviews such as Greenpeace (1992) and Redman (1992) provide some indication of the current evidence on its environmental impact. From these studies, it is clear that a variety of general environmental benefits can be expected, ranging from conservation of the soil resource, wildlife and non-renewable resources to reduced pollution. Organic farming offers the additional benefit of allowing the market to provide some of the recompense for these environmental goods, rather than relying just on policy support, and could therefore be a useful adjunct to other environmental schemes. The restructuring involved in establishing an organic farming system, and the consumer-led underpinning of organic production standards, can result in longer-term commitment and hence more durable benefits than might be the case where farmers are responding solely to temporary fiscal incentives such as payments for reduced fertilizer use in environmentally and nitrate-sensitive areas (EC, 1985, 1992b).

There are, however, limitations to the contribution which organic farming in its current form can make. First, organic farming may not be able to guarantee the achievement of specific environmental benefits such as heather regeneration, the recreation of chalk grasslands or the maintenance of broad-leaved woodlands. Although some additional conservation requirements could be incorporated into production standards, specific environmental benefits would need to continue to be the focus of other schemes such as those envisaged under the new agri-environment programme (EC, 1992b).

Second, organic producers are currently taking some of the external costs, which would otherwise be borne by society, onto themselves voluntarily. Unless all farmers are required to abandon potentially harmful practices, this process of 'internalizing' social and environmental costs can only be sustained in a competitive environment through the market (premium prices) or through agricultural policy support. In the absence of policy support, it is not clear that current premium prices are high enough to ensure incomes equivalent to or above conventional farming levels for *all* farm types (Chapter 12), and the financial costs of conversion may be acting as a significant barrier to the many farmers who are interested in changing to organic methods (Chapter 17). On the other hand, excessive support could substitute entirely for the market's contribution and thus be counterproductive.

In this context, policy support need not necessarily take the form of direct income payments. Quotas, levies or marketable pollution rights designed to restrict levels of input use (Dubgaard, 1987) and regulations to limit undesirable practices such as straw burning or excessive manure applications represent possible ways of achieving desired environmental objectives. These would also strengthen the competitive position of organic farming, particularly if any income generated is redistributed to encourage positive environmental management.

Output reduction and public expenditure savings

Organic farming's potential to reduce production of surplus commodities arises from a combination of lower yields per unit land area or per animal *and* reduced areas of crops and livestock numbers. The output reduction benefits are achieved as an *integral* part of the farming system and, as with the environmental benefits, are therefore likely to be more durable, thus overcoming some of the key criticisms of other output reduction policies such as set-aside (EC, 1987). These criticisms include the failure to address the underlying problem of market imbalance due to overpricing; the likelihood that the most marginal land with below average yields will be set aside; the likelihood that intensity on the remaining land, or on other farms as a response to improving prices, will be increased, counteracting any supply reduction benefits (so-called 'slippage'); and the lack of direct environmental benefits (Potter, 1986; Ervin, 1988; Koester, 1989; Anon., 1992a).

The extent of the output reduction benefits identified in Chapters 22, 25 and Schulze Pals (1993) is such that policy support for organic farming offers significant potential for net public expenditure *savings*, even where direct payments are made to farmers to assist conversion. Before the CAP reform in 1992, the costs of storing and subsidizing the export of surpluses took a significant proportion of the CAP budget. On the basis of several studies (Midmore and Lampkin, 1989; Russell and Power, 1989; Jenkins and Midmore, 1992; Lampkin, 1993a), a net reduction in public expenditure in the UK could be expected if payments of up to £150 ha^{-1} $year^{-1}$ were made to assist conversion.

The CAP reform (EC, 1992a) introduced a major shift from price support for cereals and beef to direct income payments for the production of specific commodities. These include arable area payments linked to 15% set-aside, and headage payments for certain types of livestock linked to progressively declining stocking rate limits and/or marketable quotas. Some of the direct public expenditure implications of conversion to organic farming following the CAP reform can be readily estimated. Significant reductions would be achieved where arable area payments are reduced due to a shift to grassland and livestock production, while livestock premium payments would not increase correspondingly because the necessary quota purchases or sales result in equivalent changes elsewhere. At current rates payable in the UK, these direct public expenditure savings might be as high as £60 ha^{-1} in the case of specialist arable farms, but would be lower for mixed arable/livestock farms and zero for livestock-only holdings (see Table 26.1, p. 447). Farms increasing their arable area would not be eligible for increased arable area payments.

The indirect budgetary impacts of output reduction due to organic farming are less easy to estimate following the CAP reform, as it is not yet known what effects the policy changes will have on the surplus problem and hence disposal costs. In principle, the surplus problem should disappear, so that further reductions in output would make no contribution to public expenditure savings. In practice, the original concerns about slippage and failure to achieve output reductions apply as much to the new set-aside scheme as to the old (Anon., 1992a). A 1993 cereals harvest only marginally below that in previous years, and a rapid increase in the number of beef cattle in the European Union with the possibility of increased intervention purchases, indicate that the CAP reform has not yet solved the surplus problem. In fact, the move to area payments linked to specific commodities and the introduction of quotas on payments for livestock subsidies may serve only to reinforce existing production structures (see below).

Market adjustment and development

The rapid expansion of the supply base to achieve a critical mass of product which would allow mainstream as well as niche market development is vital if the contribution of the market to supporting organic farming's environmental and output reduction benefits is to be maintained. The market also underpins the higher levels of employment and labour incomes on organic farms identified in some of the farm-level studies. The current relatively slow expansion of supply of organic livestock in the UK, for example, is leading retailers to withdraw from the market, despite continuing consumer demand, because of the high distribution and marketing costs associated with small quantities, as well as supply continuity problems. Greater availability of organic raw material would also allow food manufacturers and processors to substitute organic ingredients in existing mainstream brands, as in the case of the two largest German baby food brands, rather than face the very high costs of establishing new, niche-market brands.

At the same time, there is a risk that a rapid expansion in supply could, in the absence of market development, lead to downward pressure on price premiums, and financial problems for existing organic producers. The expansion of the market can be viewed as a series of discrete steps, rather than a continuous linear function, with a new critical mass required to get to each level. The difficulty lies in achieving a balance which allows the market to grow at an appropriate pace to the next level, while minimizing the pain of getting there.

Costs of conversion

A key factor inhibiting the adoption of organic farming, despite the premium market and the other benefits, is concern about the implications of conversion. The process of transition to organic systems involves the restructuring of the whole farm business, the extent of which is likely to be greater for specialized than for mixed farms. It is a complex process involving a high degree of innovation and learning on the part of the farmer, as well as some financial expense (Chapter 17). The costs include conversion-related investments and information-gathering expenses, as well as the loss of income and additional costs arising from additional yield reductions as biological processes become established; the trialling of new enterprises and techniques; errors resulting in higher than necessary yield reductions; and restricted access to premium prices during the statutory two-year conversion period. In many cases, it is difficult to see these costs as a form of investment by the farmer in future higher returns because there is no guarantee of improved incomes in the long run.

Financial Support Programmes

The costs of conversion and the fact that these costs are incurred by individual farmers in the pursuit of environmental and other objectives which are of social benefit has led several countries, in particular the Scandinavian and German-speaking nations, to introduce conversion aid schemes and other forms of financial support for organic farmers since 1987. In assessing the effects of these schemes, a careful evaluation of the objectives of the underlying policy programmes is necessary, in particular whether the emphasis has been on output reduction or environmental benefits. The aim here is to review previous and existing programmes to support conversion to organic farming, to assess the potential impact of the CAP reform on organic farmers and to draw some conclusions concerning the nature of future financial support programmes for organic farming.

Conversion support schemes, 1987–1992

Denmark was the first country to introduce, in 1987, a financial support scheme for organic farming on a significant scale (Chapter 23). The scheme covered the development of extension, information and marketing services as well as financial assistance during the conversion period, and led to a three-fold increase in the size of the organic sector. The other Scandinavian countries soon followed Denmark's lead.

Sweden started to integrate agricultural and environmental policies as early as 1985, with the introduction of policies aimed at reducing environmental pollution from manures and fertilizers. In 1986, a levy on all pesticides was introduced with the aim of reducing pesticide use, cutting surplus production, and raising funds for research and extension. Specific policies to support organic farming on environmental and surplus reduction grounds were introduced in 1989. These policies included support for a state advisory service for organic producers as well as a one-off financial support scheme for conversion to and, uniquely, continuation of organic production (Liden and Anderson, 1990). The conversion support ranged between SK 750 and 2900 ha^{-1} $year^{-1}$ for a maximum of three years, depending on land quality, yield potential and land use, but did not cover all the land under conversion and was not payable on horticultural crops. Grassland and green manures received funding for one year. Funding was only available for farms registered in 1989, although conversion could start up to 1992. One thousand farms started the conversion in 1989, representing more than a doubling of the size of the organic sector and an increase from 10,000 to 30,000 ha. The rate of growth of the organic sector subsequently, in the absence of financial support, has been much slower. Svensson (1991), in a survey of the 1781 converting and existing organic farmers receiving the payments in 1989, found that 97% of farmers would probably or definitely continue with organic management after the end of the support period, while 52% of those converting in 1989 stated that the availability of grants had greatly or fairly greatly influenced their decision to convert and 11% cited the subsidies as a major motive for conversion (compared with 67% citing environmental, 22% financial and 17% job satisfaction motives). Svensson also found, in contrast to the German results (Chapter 25), that there was no change in the cropping pattern on 60% of the farms converting, while 23% of farms increased the area of grass leys.

In *Norway*, a conversion support scheme was introduced in 1990. Each farm converting received a one-off payment of NK 15,000 for farms less than 5 ha, and NK 20,000 for farms larger than 5 ha (the average farm size in Norway is about 10 ha). This payment was supplemented by NK 2000 per hectare of cropable land (including pastures where machinery can be used) in the first year of conversion and NK 1000 ha^{-1} in each of the second and third years. In 1991, payments were limited to NK 60,000 per farm. From 1992, farmers were to receive NK 500 ha^{-1} for each of the three years following conversion. The whole farm must be converted within ten years. In addition, favourable credit terms may be available to farms converting to organic production. The total budget for 1990 was NK 16 million, including expenditure on advisory services and research. In 1990, 218 farms entered the first year of conversion out of a total of 450 farms registered with the scheme.

In *Finland*, a conversion aid scheme paying FM 2800 ha^{-1} for three years was launched in 1990 (Vaeisaenen, 1990). Six hundred and sixty farms converted, with an area of more than 9000 ha, representing a trebling of the existing number of organic farms. Organic production has to be maintained for three to five years after the end of subsidy.

Funding also covers ten full-time advisers. Aid for research and extension in organic farming increased from FM 50,000 in 1985 to FM 2.2 million in 1990.

In *Switzerland*, a number of cantons have introduced conversion support schemes since 1989 (Chapter 24). These have now been supplemented by a federal support scheme for organic and integrated production, details of which are also discussed in Chapter 24.

Austria provided AS 2 million and 6 million in 1989 and 1990, respectively, to help the organic farming organizations with advisory, certification, publicity and marketing activities (Posch, 1990). The emphasis was on building up the extension and marketing infrastructure before making conversion payments generally available. Some pilot projects for conversion payments were started in 1990 in some regions. Farmers received AS 3000 per hectare of arable land and AS 1500 per hectare of grassland, with a maximum area limit of 5 ha per farm (i.e. AS 15,000). In the Steiermark, payments were not restricted to the conversion period, unlike in other regions. In Oberösterreich, the organic farmers received a double payment of the generally available direct income payments under the 'Ökosozialen' (eco-social) programme for farmers in specially disadvantaged regions. Widespread financial support for the conversion period did not start until 1992 under the 'Biobauernzuschuss 1992' (organic farmer grant) programme (Anon., 1992b). Full details of payment levels were not available at the time of writing. The conversion payments have resulted in a dramatic increase in the number of organic farms (from under 1500 in 1989 to an estimated 9000 in 1993) and the area of organically managed land (from 20,000 ha in 1989 to an estimated 135,000 ha in 1993, nearly 3% of Austrian agricultural land) (Pircklhuber and Gründlinger, 1993).

Germany was the first EC member state to introduce, in 1989, a scheme to support conversion to organic farming in the context of the EC's extensification policy, leading to a more than seven-fold increase in the size of the organic sector (Chapter 25). The key objective of the scheme was output reduction rather than the achievement of environmental benefits. The EC's extensification policy was first linked to the original set-aside scheme and provided for a 'quantitative' 20% reduction in other surplus commodities, in particular the numbers of sheep and beef animals on individual farms (EC, 1988a). This proved unsuitable as a means to support conversion to organic farming. Revised legislation (EC, 1988b) introduced a 'production methods' option, the aim of which was to achieve a 20% reduction in output based on the adoption of less intensive practices, and it was this legislation which provided the basis for the German scheme to assist conversion. Germany's example was not followed by other EC countries except

Denmark, which used the extensification policy to help finance its existing scheme.

The German state of Saarland had introduced a small scheme two years earlier in 1987, paying individual producers a total of DM 10,000 per farm spread over three years (with 50% in the first year), as well as a single payment of DM 300 for training and DM 500 to cover the costs of replacement labour while attending a one-week training course. The payments were later raised to DM 20,000 per farm (Hoffmann, 1990). The conversion payments were related to the family and financial circumstances of the farmer, with the provisos that the payments should not account for more than 50% of the income losses resulting from conversion, and that farmers receiving the payments would not be eligible for other types of environmental grants available at that time. One consequence of the implementation of this programme ahead of other German states is that Saarland now has over 7% of its land area managed organically, including the 5% converted later under the extensification programme (see Chapter 25, Table 25.1). A number of other communities in Germany also introduced conversion aid schemes before the full extensification scheme was introduced, most of which had per hectare payments, some with maximum limits per farm (Volkgenannt, 1991). Some also introduced payments to employ trainees, and one, Ludwigshafen, included existing organic producers. Many of these schemes were superseded by the extensification payments.

The impact of the CAP reform

For the first time, specific measures to encourage conversion to and continued organic farming are included in European Union (EU) policy, as well as provision for related training needs as part of the additional measures to the CAP reform known as the agri-environment programme (EC, 1992b). About 1% of the funding for the reform package is allocated to these additional measures. The implementation of the agri-environment programme is likely to mean that the majority of western European countries will have financial support schemes for organic farming in place by the end of 1994. Full details of the schemes proposed in each EU member state were not available at the time of writing. In Germany, typical payments are DM 300 ha^{-1}, but there is considerable variation between states in relation to the payment levels for different crops, and in relation to payments for land in conversion and under continued organic management (Rupp, 1993). Other northern EU member states, including Belgium, Denmark, France, Luxembourg and the Netherlands, have proposed payments which are broadly in line with the payments originally envisaged (ECU 250 ha^{-1}), as has Spain. A scheme restricted to farms in conversion only is planned in the United Kingdom, but the level of payments is much lower, averaging £50 ha^{-1} $year^{-1}$ over five years and only £10 ha^{-1} $year^{-1}$ in the less favoured areas of England and Wales.

However, the profitability of organic farming, and the costs of conversion, will also be influenced by the mainstream measures of the CAP reform, with the result that the conclusions derived from many of the studies from EU member states reported in this book are likely to change significantly. Although existing organic producers might be expected to gain some benefit from the mainstream measures of the CAP reform, they look likely to discourage the adoption of organic farming. This is because the arable area payments and livestock aid eligibility quotas have the potential to freeze current production patterns and levels of intensity. The enterprise restructuring which conversion to organic farming entails will have implications for the balance between arable and livestock enterprises on the farm, as well as for the level of intensity of livestock production. Arable farms converting to organic production will tend to reduce the arable area and increase the area of grassland and numbers of livestock, while livestock farms will tend to reduce livestock numbers and may introduce or expand arable production.

Arable farmers converting now will, in addition to the other costs of conversion, lose eligibility for some arable area payments, without compensation, but can only get access to some livestock premiums through quota purchase. Livestock farmers converting will receive livestock payments on fewer animals, yet will not be entitled to arable area payments for any new arable land introduced, although this may be offset by quota sales. Existing organic livestock producers, who had reduced livestock numbers before 1992, have received lower livestock quota allocations than would have been the case had they remained under more intensive, conventional management, with a potentially adverse impact on asset values. Existing organic arable farmers will benefit from the arable area payments if the prices for organic crops remain at pre-1992 levels, and in the UK have been granted some useful exemptions, such as the inclusion of clover in set-aside land. However, their income from the arable area payments will still be reduced relative to what they would have obtained had they remained as conventional producers.

To illustrate the potential impacts of the CAP reform, models of six different farm types in the UK were developed (Lampkin, 1993b, updated), illustrating the extent of the loss (if any) of arable area payments, the gain or loss of livestock payments and the costs or revenue generated by related quota transactions (Table 26.1). The capital costs have been converted to an annual equivalent by using a 25% conversion factor which reflects the current relationship between

Table 26.1. Changes in arable and livestock aid payments and quota costs/revenue (£) per hectare resulting from conversion to organic production, Britain, 1993.

	Arable aid	Livestock aid	Quota (annual)[a]	Net position
Lowland arable	−58	+54	−17	−22
Dairy/arable	−14	+18	−8	−4
Specialist dairy	(+26)[b]	0	0	0
Lowland livestock	(+8)[b]	−23	+17	−6
Upland livestock	0	−36	+18	−18
Hill livestock	0	−23	+10	−13

[a] Additional allowance would have to be made for quota siphoned off for the national reserve when sold, and for allocations from the national reserve to organic producers.
[b] Payments which would be made were for land eligible for arable area payments, but not available under current regulations.
Source: Lampkin (1993b).

quota leasing and purchase prices. All farm types except specialist and mainly dairy farms experience a net loss in subsidy payments as a result of conversion to organic production. This is particularly significant on arable farms, and on hill and upland livestock farms where the revenue generated by quota sales would be insufficient to compensate for the loss of livestock payments.

At present, it seems likely, in the UK context at least, that the detrimental effects of the CAP reform will outweigh the limited benefits which might be obtained from the introduction of the agri-environment programme. As a consequence, the rate of adoption of organic farming is likely to remain slow, in marked contrast to the trends in some other northern European countries.

Lessons for future financial support programmes

It is clear that financial payments during the conversion period can have a significant impact on the rate of adoption of organic farming, as the experiences in Germany, Austria and Scandinavia have shown. However, these experiences also give rise to a number of issues which need to be addressed as part of any package of measures to support the organic sector:

- *Support services* for training, extension and certification may be inadequate or non-existent, increasing the risk of financially or environmentally damaging mistakes, resource wastage or even failure of the conversion process. Improved information, conversion planning and marketing can help to reduce the costs of conversion and to improve returns during the conversion period, thus reducing the need for direct financial support.

- Existing marketing channels may also be inadequate to deal with a large increase in supply, so that *market development* becomes a key issue – support programmes should not substitute entirely for the market as a means of recompensing organic farmers for internalizing external environmental costs.
- The *type and level of payment*, in particular the choice between lump-sum and per hectare payments, the rates of payment for different farm types and intensity levels (see below), and the timing of payments, should be related to the characteristics of the various costs of conversion (Chapters 17 and 24, this volume); for example, a flat rate payment per farm as well as per hectare may help to ensure that smaller farms are not disadvantaged.
- A *flat rate payment* per hectare, as implemented under most existing policies, may not be sufficiently flexible to encourage the most intensive producers to convert (where the greatest environmental benefits and financial costs of conversion might be expected) without at the same time providing unnecessarily generous support to the already extensive producers (where the environmental benefits and costs of conversion are more limited). The regional distribution, farm size and intensity characteristics of the western German farms converting, and the liquidity benefits identified as a major factor in the very high adoption rates among eastern German farmers, support this view, as does the fact that many farmers indicated they would have converted even without the benefit of subsidies (Chapter 25, this volume).
- A *staged conversion process* (Chapter 17) would allow experience to be gained and minimize the risks of financially and environmentally damaging mistakes in the early stages of conversion – many schemes do not permit this as an option.
- *Existing organic producers* may be severely disadvantaged by schemes which limit support to farmers converting to organic farming only, particularly if the increase in supply leads to a reduction in market prices, but also on equity grounds (see below).
- Although the adoption/diffusion model (Chapter 17, this volume; Padel, 1994) suggests that once a *critical level* of adoption has been reached, the diffusion of an innovation will progress under its own momentum, this might not always be true in the case of induced changes through legislation or subsidies. If the original value of the innovation (independent of the payment) has not been made very clear, disillusionment and withdrawal at the end of the payment period might be the outcome.

Of these, two issues deserve more detailed consideration: the inclusion of existing organic producers in any financial support scheme, and the factors determining appropriate levels of payment.

Existing organic producers

If payments are to be made to support organic farming, then both existing organic producers and farmers converting to organic farming should be entitled to support for the following reasons:

- Existing organic producers are *adhering to a detailed code of good environmental practice* and should therefore be achieving and maintaining the desired environmental and surplus reduction benefits. Other schemes, such as nitrate and environmentally sensitive areas, do not preclude farmers from receiving payments because they are already meeting the conditions the schemes impose.
- Existing organic producers are an important *source of information and expertise* for farmers converting, often without any financial reward for their efforts. Some payment for existing organic producers could be seen as acknowledgement of their contribution and might help to maintain their willingness to share their knowledge and experience with others.
- There is a case for *equity between existing organic producers and converters* since existing producers have had to finance the conversion unaided, and converters may be receiving payments for land with similar organic certification status.
- The potential exists for significant *market distortion* and other adverse impacts on existing organic producers due to the reductions in price premiums caused by expansion of supply ahead of demand.
- Existing organic producers have been disadvantaged by *reduced eligibility for arable area payments and the introduction of quotas on livestock subsidy payments* (as previously with milk quotas), due to the voluntary changes in land use and reductions in stocking rates already undertaken, with a potential adverse impact on incomes and asset values.
- Their *competitive position* with regard to existing organic producers in other countries who may be receiving payments also needs to be considered.

Level of payment

The level of payments to farmers converting, and to farmers continuing with organic management, should be determined with regard to:

- *the environmental and other conditions* which farmers are expected to meet and the costs of meeting them;

- *the environmental and surplus reduction benefits* which are likely to be obtained;
- *the costs of conversion,* in particular any restrictions on access to premium markets;
- *the costs of continued organic farming,* since despite the availability of premium prices, some farm types, particularly specialist arable farms, may still face reduced incomes;
- *payment levels and eligibility for related environmental schemes* which also involve adherence to codes of good environmental practice;
- *levels of payments in other countries* so that producers are not put at a competitive disadvantage;
- *the reduction in eligibility for other support schemes,* in particular arable and livestock aid payments due to changes in land use and the reduction in intensity of livestock production;
- *the potential for other savings in public expenditure* due to the reduction in output of surplus crops and livestock.

Technical and Other Support Programmes

Financial incentives are not the only approach to supporting organic farming, and should in any case be backed up by technical and other support programmes to avoid some of the potential problems identified above.

Research and development

Research and technological development conducted *within* functioning organic systems is essential to overcome some of the technical problems which still exist, to improve further the environmental sensitivity of organic systems, and to generate data on the potential of organic farming in different contexts for farmers, advisers and policy-makers. Some of the issues relating to the design of appropriate research programmes have been discussed in Chapter 3. Significant public funding for research into organic farming is available in the German-speaking and Scandinavian countries, the Netherlands and Britain (Besson, 1990; Lehmbecker and Thode, 1990). The EC has also provided funds for research into organic farming as part of recent framework agricultural research programmes.

Current organic farming practices have been developed primarily by existing organic farmers against the background of scientific knowledge. This explains some of the emphasis given to participatory ('farmer-first') and multi-disciplinary (farming systems) research approaches in organic and sustainable agriculture (Chapter 3). Schemes to support conversion to organic farming also need to reflect the role of established organic producers in this context, as they can be an important source of information for new farmers and also for advisers. Most organic farming organizations actively promote meetings of regional groups to facilitate such sharing of experience.

Extension

In most countries, the availability of information on organic farming to farmers is still very limited. The main sources of information for organic farmers include specialist publications (books and magazines) and other organic farmers. Traditional extension agencies have been less important for organic producers, mainly because these agencies advocated conventional technology and have not been regarded as reliable sources of information on organic farming (Padel, 1994). However, the communication of existing information is perhaps the most important factor which would reassure farmers that organic farming is a viable option, reducing the role of lack of confidence as a barrier to conversion, and which would ensure that the potential for negative impacts of conversion is minimized (Chapter 17). Germany, Denmark, Sweden, Finland, Norway, Switzerland, Austria and the Netherlands have full-time state-funded advisers specializing in organic farming, or contribute to the costs of private sector advisers, working both within existing advisory structures and within the organic producer organizations (Besson, 1990). In other countries such as Britain, the organic sector relies almost entirely on the input of a very limited number of private sector consultants, with only occasional assistance from state-owned extension agencies.

Few studies exist which evaluate the effectiveness of specialist organic farming extension services and which would allow an assessment of their potential role in conversion aid programmes. In Sweden, Brorsson (1989) and Le Clercq (1989) highlighted the lack of information on organic farming in research and extension circles. Following the introduction of support for organic farming extension services and financial aid for conversion to organic farming, Svensson (1991) found that 67% of the farmers surveyed thought that the advisory service was satisfactory, while one third made use of specialist consultants. Bader (1990) analysed specialist organic farming extension services in Germany and found that, despite governmental support, financial difficulties and limited resources are widespread.

A further area of difficulty is the limited information available to, and the lack of experience of, the advisers themselves. Specialist training for advisers in organic farming is not yet available and, under pressure to justify their services in a user-pays environment, existing advisory bodies cannot supply training for newcomers without additional funding. A sudden increase in demand, and possibly willingness to pay, for advice, does not mean that the ability to provide that advice can materialize instantaneously. Particularly when a programme to support conversion is introduced, the lack of trained and experienced advisers at that moment can lead to a situation where converting farms will experience more difficulties than where adequate advice is available. While specialist extension services for organic farming can certainly improve the situation, resource limitations severely restrict their ability to make information on organic farming available to the wider farming community. In countries where no state support is available, the need for additional, experienced advisers will become very critical if conversion to organic farming is to be encouraged by financial incentives.

Education and training

There are an increasing number of training initiatives in organic farming, particularly at agricultural college level in the northern European countries (Besson, 1990; Lehmbecker and Thode, 1990). Specialist, full-time, three-year vocational training courses are now available in Germany, Britain, Denmark and the Netherlands. Many of these initiatives come from the colleges themselves, relying on existing funding mechanisms, the European Social Fund (ESF) and private tuition fees. Professorships in organic/ecological agriculture have been established at universities and polytechnics in the Netherlands (Wageningen), Austria (Vienna), Finland and Germany (Kassel, Kiel, Bonn, Giessen, Osnabrück), but organic specializations at degree level are not widely available. Masters degrees in ecological and sustainable agriculture are available at some universities in the Netherlands and Britain, but these rely on ESF funding or tuition fees. The agri-environment programme (EC, 1992b) provides specifically for training in 'good organic farming practice', but few European Union states seem likely to take up this option.

Certification and market development

The ability of organic farmers to remain competitive requires continual market development, including the development of opportunities to move from niche to mainstream markets, a process which is critically dependent on the availability of supply. Recent changes to European Community marketing support programmes for agricultural produce have made certain organic food marketing projects eligible for financial assistance (EC, 1985, 1990). As a consequence, a limited number of projects have been developed in Britain, Denmark and some German states. Despite this, there is still considerable scope for further investment to take advantage of the increasing supplies of organic products resulting from the various financial aid schemes.

Market development also relies on consumer confidence in the authenticity of the organic product, and this can only be secured with effective certification programmes and/or legislation. Many countries in Europe, North America and Australasia have given support to organic farming through the development of certification schemes (Chapter 2). However, the basis of any state-sponsored certification programme is critical, in that emphasis should be placed on defining the production system, rather than any particular quality of the end product, such as freedom from residues, which cannot be guaranteed. Nearly all the state-sponsored initiatives rely heavily on the input of private sector organic organizations and their individual members, often without recompense, so that most of the extra costs have to be borne by producers. Some European countries, such as the Netherlands and Britain, provide financial contributions towards the costs of operating certification schemes, and in Denmark all certification is carried out by the state authorities. In most cases, however, there is no support and policy-makers might consider whether resources could be made available to private sector certification organizations to assist them in their activities.

Conclusions

The potential environmental and surplus reduction benefits provide the main justification for increased public expenditure to support organic farming, particularly where the market is not able to compensate farmers fully for the financial penalties associated with voluntarily internalizing their external costs. Organic farming, with its detailed codes of good environmental practice, is possibly unique in its potential to combine the influence of the market with policy incentives, but for this situation to continue, policy measures need to be implemented with care. The experience in the German-speaking and Scandinavian countries has shown that, given the right conditions, rapid expansion of the organic sector can be achieved. Within the European Union, however, the CAP reform means that conversion to organic farming now entails additional costs on most farm types due to the loss of eligibility for subsidy payments, and therefore a potential additional barrier to the expansion of organic farming has arisen. These additional costs may not be entirely offset by any financial assistance for organic farming available as part of the agri-environment programme.

Financial support policies for organic farming need to achieve a balance between encouraging the expansion of the organic sector and promoting excessive growth leading to market disruption, adverse consequences on existing organic producers, and the possible disillusionment and withdrawal of participants. By recognizing the contribution which existing organic producers can make, and directing support into related areas such as extension, research and training, and market development and certification, many of the negative side-effects of financial support payments can be avoided. In this way, the full potential of organic farming's contribution to agricultural and environmental policy objectives can be achieved.

References

Anon. (1992a) Historic agreement on CAP reform. *Agra-Europe* 1492 (22 May), 1–15.

Anon. (1992b) Entwicklung des biologischen Landbaus in Österreich. *Ökologie und Landbau* 83 (3/92), 20.

Bader, U. (1990) Organisationsfragen der Beratung im ökologischen Landbau in der Bundesrepublik Deutschland. *Berichte über Landwirtschaft* 68, 616–630.

Besson, J.-M. (ed.) (1990) *Biological Farming in Europe – an Expert Consultation.* REUR Technical Series, 12. Food and Agriculture Organization, Rome.

Brorsson, K.-A. (1989) *Ekonomiska effekter av omställnings-bidrag till alternativ odling.* Alternativ Odling, 2. Research Committee for Alternative Agriculture, Swedish University of Agricultural Sciences, Uppsala.

Dubgaard, A. (1987) *Anvendelse af afgifter til regulering af pesticidforbruget.* Rapport, 35. Statens Jordbrugsøkonomiske Institut, Copenhagen.

EC (1985) Council Regulation (EEC) No. 797/85 laying down a system of aid for investments in agricultural holdings and measures to assist farmers in less favoured areas. *Official Journal of the European Communities* L93 (30 March), 1–13.

EC (1987) Council Regulation (EEC) No. 1760/87 amending Regulations 797/85, 270/79, 1360/78 and 355/77 as regards agricultural structures, the adjustment of agriculture to the new market situation and the preservation of the countryside. *Official Journal of the European Communities* L167 (26 June), 1–8.

EC (1988a) Commission Regulation (EEC) No. 4115/88 laying down detailed rules for applying the aid scheme to promote the extensification of production. *Official Journal of the European Communities* L361 (29 December), 13–17.

EC (1988b) Council Regulation (EEC) No. 1094/88 amending Regulations 797/85 and 1760/87 as regards the set-aside of arable land and the extensification of production. *Official Journal of the European Communities* L106 (27 April), 28–32.

EC (1990) Council Regulation (EEC) No. 866/90 on improving the processing and marketing conditions for agricultural products. *Official Journal of the European Communities* L91/1 (29 March), 1–6.

EC (1992a) Regulation (EEC) No. 1765/92 establishing a support scheme for producers of certain arable crops. *Official Journal of the European Communities* L181 (2 July), 12–20.

EC (1992b) Council Regulation (EEC) No. 2078/92 on agricultural production methods compatible with the requirements of the protection of the environment and the maintenance of the countryside. *Official Journal of the European Communities* L215 (30 July), 85–90.

Ervin, D.E. (1988) Cropland diversion (set-aside) in the US and the UK. *Journal of Agricultural Economics* 39, 183–196.

Greenpeace (1992) *Green Fields – Grey Future. EC Agricultural Policy at the Crossroads.* Greenpeace, Amsterdam.

Hoffmann, H. (1990) Ecological agriculture in Europe – challenge and expediency – the Saarland model. In: Besson, J.-M. (ed.) *Biological Farming in Europe.* REUR Technical Series, 12. Food and Agriculture Organization, Rome, pp. 169–184.

Jenkins, T.N. (1990) *Future Harvests – the Economics of Farming and the Environment: Proposals for Action.* Council for the Protection of Rural England, London.

Jenkins, T.N. and Midmore, P. (1992) The economic implications of selected nature conservation proposals for agriculture. Paper presented at the 30th European Association of Agricultural Economists Seminar, Château D'Oex, Switzerland, November.

Koester, U. (1989) Financial implications of the EC set-aside programme. *Journal of Agricultural Economics* 40, 240–248.

Lampkin, N.H. (1993a) The economic implications of conversion from conventional to organic farming systems. PhD thesis, Department of Agricultural Sciences, University of Wales, Aberystwyth.

Lampkin, N.H. (1993b) The impact of conversion to organic farming on eligibility for arable aid, beef special premium, suckler cow premium, sheep annual premium and hill livestock compensatory allowances. Discussion Paper Series 93/5. Centre for Organic Husbandry and Agroecology, University of Wales, Aberystwyth.

Le Clercq, L. (1989) *Omläggningsproblem vid övergång från konventionell till alternativ odling.* Växtodling, 7. Department of Crop Production Science, Swedish University of Agricultural Sciences, Uppsala.

Lehmbecker, G. and Thode, U. (1990) *Lehre und Forschung im ökologischen Landbau an den Hochschulen und freien Forschungsinstituten Europas.* Christian-Albrechts-Universität, Kiel.

Liden, C.J. and Anderson, R. (1990) Legislation and measures for the solving of environmental problems resulting from agricultural practices, their economic consequences and the impact on agrarian structures and farm rationalisation. In: Besson, J.-M. (ed.) *Biological Farming in Europe.* REUR Technical Series, 12. Food and Agriculture Organization, Rome, pp. 259–272.

Midmore, P. and Lampkin, N.H. (1989) Organic farming as an alternative to set-aside and an option for extensification. In: Dubgaard, A. and Hjortshoj-

Nielson, A. (eds) *Economic Aspects of Environmental Regulation in Agriculture.* Wissenschaftsverlag Vauk, Kiel, pp. 267–278.

NCC (1990) *Nature Conservation and Agricultural Change.* Focus on Nature Conservation, 25. Nature Conservancy Council, Peterborough.

OECD (1991) *Agricultural and Environmental Policy Integration: Principles, Progress and New Directions.* Environment Committee Ad Hoc Group on Agriculture and the Environment, Organization for Economic Co-operation and Development, Paris.

Padel, S. (1994) *Adoption of organic farming as an example of the diffusion of an innovation: a literature review.* Discussion paper 94/1. Centre for Organic Husbandry and Agroecology, University of Wales, Aberystwyth.

Pircklhuber, W and Gründlinger, K. (1993) *Der biologische Landbau in Österreich.* Monograph 35. Umweltbundesamt, Wien.

Posch, A. (1990) Support for organic farming in Austria. Personal communication (7 December) and background documentation. Bundesministerium für Land- und Forstwirtschaft, Republik Österreich, Wien.

Potter, C.A. (1986) *Environmental Protection and Agricultural Adjustment: Lessons from the American Experience, Set-aside as an Environmental and Agricultural Policy Instrument.* Working Paper 1. Department of Environmental Studies and Countryside Planning, Wye College, Ashford.

Redman, M. (ed.) (1992) *Organic Farming and the Countryside.* Report for the Countryside Commission. British Organic Farmers, Bristol.

Rupp, C. (1993) Flankierende Maßnahmen - Hoffnungsschimmer oder Trostpflaster für den ökologischen Landbau? *Bioland* 93(6), 32–36.

Russell, N.P. and Power, A.P. (1989) UK government expenditure implications of changes in agricultural output under the Common Agricultural Policy. *Journal of Agricultural Economics* 40, 32–39.

Schulze Pals, L. (1993) *Regional und betrieblich differenzierte Untersuchung über Akzeptanz, Motivation und wirtschaftliche Folgen einer Umstellung auf ökologischen Landbau im Rahmen des Extensivierungsprogrammes.* Arbeitsbericht 2/93. Institut für Betriebswirtschaft, Bundesforschungsanstalt für Landwirtschaft, Braunschweig-Völkenrode.

Svensson, I. (1991) *Statligt stöd till alternativ odling 1989.* Alternativ Odling, 7. Research Committee for Alternative Agriculture, Swedish University of Agricultural Sciences, Uppsala.

Vaeisaenen, J.H. (1990) Ecological agriculture in Finland. In: Besson, J.-M. (ed.) *Biological Farming in Europe.* REUR Technical Series, 12. Food and Agriculture Organization, Rome, pp. 225–227.

Volkgenannt, U. (1991) Förderprogramme zur Umstellung auf ökologischen Landbau. *Ökologie und Landbau* 79, 6–7.

Weinschenck, G. and Dabbert, S. (1989) Decrease of the intensity of natural resource use as a way to reduce surplus production. In: Dubgaard, A. and Nielsen, H.A. (eds) *Economic Aspects of Environmental Regulations in Agriculture.* Wissenschaftsverlag Vauk, Kiel, pp. 25–34.

WWF (1991) *Reforming the EC Common Agricultural Policy.* Discussion paper. World Wide Fund for Nature, Godalming.

Index

Printed and bound by CPI Group (UK) Ltd, Croydon, CR0 4YY
16/09/2026
14963431-0006